Introduction to
ENVIRONMENTAL ENGINEERING

P. Aarne Vesilind

Duke University

PWS Publishing Company

I(T)P An International Thomson Publishing Company

Boston • Albany • Bonn • Cincinnati • Detroit • London • Madrid • Melbourne • Mexico City
New York • Pacific Grove • Paris • San Francisco • Singapore • Tokyo • Toronto • Washington

PWS PUBLISHING COMPANY
20 Park Plaza, Boston, MA 02116-4324

Library of Congress Cataloging-in-Publication Data

Vesilind, P. Aarne.
  Environmental Engineering / P. Aarne Vesilind.
    p. cm.
  Includes bibliographical references and index.
  ISBN 0-534-95573-8
  1. Environmental engineering. 2. Environmental ethics. I. Title
TD145.V4297 1996   96-38433
628—dc20   CIP

For more information, contact:
PWS Publishing Company
20 Park Plaza
Boston, MA 02116

International Thomson Publishing Europe
Berkshire House I68-I73
High Holborn
London WC1V 7AA
England

Thomas Nelson Australia
102 Dodds Street
South Melbourne, 3205
Victoria, Australia

Nelson Canada
1120 Birchmont Road
Scarborough, Ontario
Canada M1K 5G4

I(T)P™
International Thomson Publishing
The trademark ITP is used under license.

Sponsoring Editor: *Jonathan Plant*
Marketing Manager: *Nathan Wilbur*
Production Manager: *Elise S. Kaiser*
Manufacturing Buyer: *Andrew Christensen*
Compositor/Illustrator: *Bi-Comp, Incorporated*
Production: *Spectrum Publisher Services, Inc.*
Interior Designer: *Monique A. Calello*
Cover Designer: *Elise S. Kaiser*
Text Printer/Binder: *Quebecor Fairfield*
Cover Printer: *Coral Graphic Services, Inc.*
Cover Art: *Copyright © Steven Hunt; used with permission of the artist.*

 This book is printed on recycled, acid-free paper.

Printed and bound in the United States of America.
96 97 98 99 00 — 10 9 8 7 6 5 4 3 2 1

International Thomson Editores
Campos Eliseos 385, Piso 7
Col. Polanco
11560 Mexico D.F., Mexico

International Thomson Publishing GmbH
Königswinterer Strasse 418
53227 Bonn, Germany

International Thomson Publishing Asia
221 Henderson Road
#05-10 Henderson Building
Singapore 0315

International Thomson Publishing Japan
Hirakawacho Kyowa Building, 31
2-2-1 Hirakawacho
Chiyoda-ku, Tokyo 102
Japan

*This book is dedicated, with gratitude, to
the late Edward E. Lewis—
publisher, curmudgeon, and friend.*

CONTENTS

PART 1 FUNDAMENTALS

Chapter 1 ENGINEERING DECISIONS 3

1.1 Decisions Based on Technical Analyses 4
1.2 Decisions Based on Cost-Effectiveness Analyses 5
1.3 Decisions Based on a Benefit/Cost Analysis 9
1.4 Decisions Based on a Risk Analysis 12
1.5 Decisions Based on an Environmental Impact Analysis 20
1.6 Decisions Based on an Ethical Analysis 28
1.7 Continuity in Engineering Decisions 38
Problems 38   *Endnotes* 44

Chapter 2 ENGINEERING CALCULATIONS 46

2.1 Engineering Dimensions and Units 47
2.2 Approximations in Engineering Calculations 53
2.3 Information Analysis 56
Abbreviations 65   *Problems* 65   *Endnotes* 69

Chapter 3 MATERIALS BALANCES & SEPARATIONS 70

3.1 Materials Balances with a Single Material 71
3.2 Materials Balances with Multiple Materials 83
3.3 Materials Balances with Reactors 100
Abbreviations 104   *Problems* 104   *Endnotes* 112

Chapter 4 REACTIONS **113**

    4.1 Zero-Order Reactions 115
    4.2 First-Order Reactions 117
    4.3 Second-Order and Noninteger-Order Reactions 121
    4.4 Half-Life 122
    4.5 Consecutive Reactions 122
    Abbreviations 123 *Problems* 123 *Endnotes* 125

Chapter 5 REACTORS **126**

    5.1 Mixing Model 127
    5.2 Reactor Models 139
    Abbreviations 147 *Problems* 148

Chapter 6 ENERGY FLOWS & BALANCES **151**

    6.1 Units of Measure 152
    6.2 Energy Balances and Conversion 153
    6.3 Energy Sources and Availability 158
    Abbreviations 165 *Problems* 165 *Endnotes* 170

Chapter 7 ECOSYSTEMS **171**

    7.1 Energy and Materials Flows in Ecosystems 172
    7.2 Human Influence on Ecosystems 180
    Abbreviations 193 *Problems* 194 *Endnotes* 198

PART 2 APPLICATIONS

Chapter 8 WATER QUALITY **203**

    8.1 Measures of Water Quality 204
    8.2 Assessing Water Quality 225
    8.3 Water Quality Standards 226
    Abbreviations 228 *Problems* 229 *Endnote* 233

Chapter 9 WATER SUPPLY & TREATMENT **234**

    9.1 The Hydrologic Cycle and Water Availability 235
    9.2 Water Treatment 246
    9.3 Distribution of Water 257
    Abbreviations 259 *Problems* 259

Chapter 10 WASTEWATER TREATMENT **263**

10.1 Wastewater Transport 264
10.2 Primary Treatment 266
10.3 Secondary Treatment 271
10.4 Tertiary Treatment 295
10.5 Sludge Treatment and Disposal 298
10.6 Selection of Treatment Strategies 313
Abbreviations 314 *Problems* 314 *Endnote* 321

Chapter 11 AIR QUALITY **322**

11.1 Meteorology and Air Movement 323
11.2 Major Air Pollutants 328
11.3 Sources and Effects of Air Pollution 336
11.4 Air Quality Standards 351
Abbreviations 354 *Problems* 354 *Endnotes* 358

Chapter 12 AIR QUALITY CONTROL **359**

12.1 Treatment of Emissions 360
12.2 Dispersion of Air Pollutants 369
12.3 Control of Moving Sources 375
Abbreviations 379 *Problems* 379 *Endnotes* 384

Chapter 13 SOLID WASTE **385**

13.1 Collection of Refuse 386
13.2 Reuse, Recycling, and Recovery of Materials from Refuse 392
13.3 Ultimate Disposal of Refuse 402
13.4 Integrated Solid Waste Management 408
Abbreviations 410 *Problems* 410 *Endnotes* 415

Chapter 14 HAZARDOUS WASTE **416**

14.1 Defining Hazardous Waste 417
14.2 Hazardous Waste Management 421
14.3 Radioactive Waste Management 427
14.4 Pollution Prevention 432
14.5 Hazardous Waste Management and Future Generations 436
Abbreviations 438 *Problems* 438 *Endnotes* 441

Chapter 15 NOISE POLLUTION **442**

15.1 Sound 443
15.2 Measurement of Sound 448
15.3 Effect of Noise on Human Health 449

15.4 Noise Abatement 452
15.5 Noise Control 453
    Abbreviations 454   *Problems* 455   *Endnotes* 457

EPILOG **458**

INDEX **463**

PREFACE

Introduction to Environmental Engineering has two unifying themes. First, I introduce environmental science and engineering using the concepts of materials balances, reactions, and reactors, binding together topics on water supply, wastewater treatment, air pollution control, and solid and hazardous waste management, integrating these topics into a single approach to applied environmental sciences.

A second feature of this book is the discussions on ethics throughout the text. The first chapter introduces engineering decision making, concluding with a brief discussion of ethics and a suggestion that ethical decision making in engineering is just as important as technical decision making. Throughout the remainder of this book ethical decision making is incorporated into the discussions and is highlighted in the assigned problems. In many of the problems, students are required not only to solve the technical part of these problems, but also to consider the ethical ramifications of solving the technical problems.

These two unifying features of the book—materials balances and environmental ethics—respond to two significant and emerging developments in applied environmental sciences:

1. Environmental problems must be solved using a holistic approach, not a fragmented, single-pollution or single-medium approach.

2. Ethics plays an increasingly important part in the professional lives of engineers, and perhaps most critically in the decisions made by environmental engineers.

This book is divided into two parts: Fundamentals and Applications. The first part introduces the concepts of materials balances and reactions occurring in reactors. In the second part, these principles are applied to environmental engineering and science.

In the first chapter, the various tools engineers use in making decisions are introduced, including technology, benefit/cost, risk, and ethics. The discussion on ethics is brief, serving merely as an introduction to some of the value-laden problems faced by engineers but sufficient to provide some background for the discussions throughout the text.

Dimensions and units are introduced in the second chapter, followed by a discussion of some basic concepts such as density, concentration, flow rate, and residence time. The need for approximations in engineering calculations follows.

This leads directly to the introduction of materials balances, a theme used throughout the book. The applicability of materials balances to all types of environmental problems is demonstrated, including ecosystem dynamics, wastewater treatment, and air pollution control.

The discussions on reactions is similar to what would be covered in an introductory physical chemistry or biochemical engineering course, and this is followed by ideal reactor theory, similar to material found in a chemical engineering unit operations course, but at a level readily understood by freshmen engineering students.

Principles of energy are introduced, showing how energy conversion takes place and why losses occur but without resorting to principles of thermodynamics. The mass balance approach is again applied to energy flow. Next is the recognition that some of the most fascinating reactions occur in ecosystems, and ecosystems are again described using the mass balance approach and the reaction kinetics introduced earlier. This concludes the first part of the book, the Fundamentals.

The second part of the book—Applications—begins with the quest for clean water. When is water "clean enough," and how do we measure water quality characteristics? This discussion is followed by introduction to water supply and treatment. Both groundwater and surface water supplies are discussed, including the operation of a typical water treatment plant. Beginning with wastewater transport, the logic and methods of wastewater treatment are covered in the next chapter, concluding with a discussion of sludge treatment and disposal.

Because meteorology determines the motion of air pollutants, this topic introduces the section on air pollution. The types and sources of air pollutants of concern are discussed, concluding with the evolution of air quality standards, treatment of emission, and the dispersion of pollutants.

The chapter on the collection and disposal of municipal solid waste concentrates on the recovery and recycling of refuse. Emphasis is placed on concepts of pollution prevention and life cycle analysis, two ideas that will be very important in the future. The book concludes with discussions of hazardous waste management and noise pollution.

The material on ethics is at a basic level so that it is readily understandable by any engineering instructor or student. No formal preparation in ethics is required. The technical material is at a level that a freshman engineering student or a B.S. environmental science student can readily digest. Calculus is used in the text, and I assume that the students have had at least one course in differential and integral calculus. A college chemistry course is useful but not mandatory. Fluid mechanics is not used in the textbook, and hence the material is readily applicable to introductory courses in environmental engineering or environmental engineering courses for science students.

The core material for the book (balances, reactions, reactors) that make up Part 1 of the book must be taught sequentially. My experience in teaching this material has shown that it can be covered in 4 to 6 weeks. I recommend, however, that the instructor take lateral excursions into areas of environmental ethics during this time as well as embellish the lectures with personal "war stories" to maintain student interest. The material in Part 2 may be used in any sequence deemed best without losing its value or meaning.

Introducing the complementary material throughout the course is an effective teaching technique, grounded in modern learning theory. New information is always retained by associating it with old and prior knowledge, thus the material must be presented sequentially.

Effective learning is active learning; students must be part of the learning process and not passive receptors of spoon-fed information. The cognitive theory of learning further holds that the maturation of knowledge, the transfer from short- to long-term memory, takes time and repetition, and the presentation of the same ideas in different guises. The materials balance approach represents an opportunity for such maturation. The student learns environmental engineering from the widest possible context.

Learning theory also suggests that there are two types of memory— epistatic and semantic. Epistatic memory relates to personal experiences, to emotions, feelings, and events in one's life. Semantic memory is our cultural dictionary, the knowledge we all share as a culture, the common understanding of words and ideas.

Effective teaching activates both types of memory and increases the students' cognitive powers. Laboratory exercises, field trips, and discussions of current events all rely on epistatic memory for learning. Epistatic memory can also be transferred from instructor to student (and often in reverse!) by relating one's professional experiences. Students appreciate and learn from such excursions to real life and retain the knowledge long after the semantic material is forgotten.

Accordingly, this textbook uses many case studies, both real and contrived, centering on technical as well as ethical problems, and presents an opportunity for the instructor to relate his or her experiences as enrichment throughout the semester. By using the environmental engineering material as complementary to the core structure, the instructor is free to use his or her imagination and personal initiative in the development of the course. Each instructor will therefore most likely present a unique course, based in great part on his or her own experiences. The core material ties it all together as an organized and systematic approach to applied environmental sciences.

The ethical materials introduced throughout the text similarly requires that both the student and the instructor pause and discuss the technical problem in a different light, thus reinforcing the material learned. Because ethics is such a personal issue, the discussion of technical matters from that perspective tends to internalize the subject and creates in effect an epistatic learning experience such as would be achieved by a field trip or a narrative of a real-world experience. Focusing on ethics therefore will help the students learn the technical material.

Acknowledgments

Much of this book was written while I was on sabbatical leave at the Center for the Study of Applied and Professional Ethics at Dartmouth College, with a joint arrangement with the Thayer School of Engineering. Deni Elliott, director of the ethics center, and Charles Hutchinson, dean of engineering, were both congenial hosts during my stay. Partial funding was obtained from the National Science Foundation, and grateful appreciation is expressed to Rachelle Hollander, who continues to believe that engineering and ethics can indeed complement each other.

I would also like to thank the reviewers who have helped in the development of this book:

David W. Hendricks
Colorado State University

Deborah Imel Nelson
University of Oklahoma

Howard Liljestrand
University of Texas—Austin

Irvine W. Wei
Northeastern University

Mark R. Matsumoto
University of California, Riverside

David Yonge
Washington State University

James R. Mihelcic
Michigan Technological University

Finally, I want to thank the many freshman environmental engineering students at Duke University who over the years suffered through having to use this book in manuscript form, rightfully complaining about too many errors and the absence of an index. Helpful advice was also obtained from Drew Endy, Chris Endy, and Jessica Beil; and Lauren Bartlett wrote the Instructor's Manual. To all of you—a hearty *"thank you!"*

P. Aarne Vesilind

part
1

FUNDAMENTALS

1

ENGINEERING

DECISIONS

Any engineering project, large or small, encompasses within its implementation a series of decisions made by engineers. Sometimes these decisions turn out to be poorly made. A far greater number of decisions, however, made hundreds of times a day by hundreds of thousands of engineers, are correct and improve the lot of human civilization, protect the global environment, and enhance the integrity of the profession. Since so few engineering decisions turn out poorly, engineering decision making is a little known or discussed process. Yet, when a decision turns out to be wrong, the results are often catastrophic.

This chapter is a review of how environmental engineers make decisions, beginning with a short description of technical decisions, followed by a discussion of cost-effectiveness, possibly the second most commonly employed tool in environmental engineering decision making, and the second most quantifiable. Next, the use of the benefit/cost analysis is described, followed by a discussion of decisions based on risk analysis. Moving even further toward the more subjective forms of decision making, the environmental impact analysis as an engineering tool is reviewed next. This chapter concludes with an introduction to ethics and ethical decision making as applied to environmental engineering.

1.1 ◆ Decisions Based on Technical Analyses

In engineering, there seldom is "one best way" to design anything. If there ever *was* a best way, then engineering would become stagnant, innovation would cease, and technical paralysis would set in. Just as we recognize that there is no single perfect work of art such as a painting, there also is no perfect water treatment plant. If there *were* a perfect painting or plant, all future water treatment plants would look alike and all paintings would look alike.

The undergraduate engineering student is taught during the early years of an engineering education that each homework assignment and test question has a single "right" answer, and that all other answers are "wrong." But in engineering practice, many technical decisions may be right, in that a problem may have several equally correct technical solutions. For example, a sewer may be constructed of concrete, cast iron, steel, aluminum, vitrified clay, glass, bamboo, or many other materials. With correct engineering design procedures, such a sewer would carry the design flow and thus would be technically correct.

One characteristic of technical decisions is that they can be checked by other engineers. Before a design drawing leaves an engineering office, it is checked and rechecked to make sure that the technical decisions are correct; that is, the structure/machine/process will function as desired if built according to the specifications. Technical decisions thus are clearly quantifiable and can be evaluated and checked by other competent professional engineers.

e • x • a • m • p • l • e 1.1

Problem   A small town has 1950 residents who want to establish a municipally owned and operated solid-waste (garbage) collection program. They can purchase one of three possible trucks that have the following characteristics:

Truck A   24 cubic yard capacity
Truck B   20 cubic yard capacity
Truck C   16 cubic yard capacity

If the truck is to collect the refuse for one-fifth of the town during a five-day work week, then every residence will be collected during the week and the truck will have to make only one trip per day to the landfill. Which truck or trucks will have sufficient capacity?

Solution Assume a solid-waste production rate of 3.5 lb/capita/day, so the total tonnage of solid waste to be collected is 1950 people × 3.5 lb/capita/day × 7 days/week = 47,775 lb/week. Of that, one-fifth or 9555 lb will have to be collected every day. Assume that the truck is able to compact the refuse to 500 lb/yd³.

$$9555 \text{ lb/day} \times 1/500 \text{ yd}^3/\text{lb} = 19.1 \text{ yd}^3/\text{day}$$

Both the 20- and 25-cubic-yard trucks have sufficient capacity, while the 16-cubic-yard truck does not.

◆

The pressures of modern practice, however, dictate that the engineering decision must be not only effective (it will do the job) but also economical (it will do the job at minimum cost). For example, in Example 1.1, while both Truck A and Truck B have sufficient capacity to haul the garbage, the cost of operation might be very different. While technical calculations can answer technical questions, questions of cost require a different form of engineering decision making; the *cost-effectiveness analysis.*

1.2 ◆ Decisions Based on Cost-Effectiveness Analyses

Engineers often find themselves working for an employer or client who requires various alternatives for solving an engineering problem to be analyzed on the basis of cost. For example, a municipal engineer is considering purchasing refuse-collection vehicles and finds that he or she can buy either expensive trucks that achieve great compaction of the refuse, thereby making efficient trips to the landfill, or inexpensive trucks that require more trips to the landfill. How does the engineer know which is less expensive for the community? Obviously, the lowest total cost alternative would be the most rational decision.

Cost-effectiveness analysis is complicated by the fact that money changes in value with time. If a dollar today is invested in an interest-bearing account, this dollar will produce say $1.05 a year from now, if the annual interest is 5%. Thus a dollar a year from now is not the same dollar that exists today, and the two dollars cannot be added. Adding such dollars is like adding apples and oranges, since money changes in value with time, and a dollar today is different from a dollar next year. If a community sells bonds and uses future tax revenue to pay the interest and principal on these bonds, it does not make any sense to add the amount it pays one year to the amount it pays the next year, since these are different dollars. Therefore, it does not make any sense to add the annual operating cost of a facility or a piece of equipment over the life of the equipment, since once again the dollars are different. For example, if it costs the community $4000 a year to operate and maintain a trash-collecting truck during one year, and it costs $5000 the next year, it makes no sense to say that the average cost is $4500. Each year's dollars are different and thus we cannot add $4000 and $5000 to calculate the average.

This represents a problem for communities in estimating the costs of public facilities or public services. The technique used to get around this difficulty is to compare the costs of different alternatives based on either the *annual cost* or the *present worth* of the project. In the annual cost calculation, all costs represent the money the community needs annually to operate the facility and retire the debt. The operating costs are estimated from year to year, and the capital costs are calculated as the annual funds needed to retire the debt within the expected life of the project.

In the case of present worth calculations, the capital costs are the funds needed to construct the facility, and the operating costs are calculated as if the money to pay for them is available today, and is put in the bank to be used for the operation, again over the expected life of the facility. A project with a higher operating cost would require a larger initial investment to have sufficient funds to pay for the cost of operation.

Either annual cost or present worth is in most cases acceptable as a method of comparison among alternate courses of action. The conversion of capital cost to annual cost, as well as the calculation of the present worth of operating cost, is most readily performed using tables (or preprogrammed hand calculators). Table 1-1 is a sample of the type of interest tables used in such calculations.

Annual costs are converted to present worth by calculating what the money spent annually would be worth at the present time. At 10% interest, $1.00 spent in one year is worth only $0.9094 today, because investing $0.9094 today at 10% would yield $1.00 in one year. If another dollar is invested at the beginning of the second year, the total investment *today* would be worth $1.7355, as shown in the "Present Worth" column of the sample interest table. Put another way, if a constant sum A is invested for N time periods at an I rate of compound interest, the present worth of this money is ($C_P \times A$), where C_P = present worth factor.

Capital costs are converted to annual costs by recognizing that the total outlay of capital would have earned interest at a given rate if the money had been invested at the prevailing interest rate. A $1.00 investment now, at 10% interest, would have earned $0.10 interest in one year and be worth $1.10. If this $1.00 is to be paid back over two years, the payment each year, as shown in the "Capital Recovery" column, is $0.57619. The amount of money necessary each year to pay back a loan of A dollars over N time periods at a compound interest rate of I is $C_F \times A$, where C_F = capital recovery factor.

Both types of conversion are illustrated in Example 1.2.

TABLE 1–1 Compound Interest Factors

I = 6% INTEREST

Number of Years (N)	Capital Recovery Factor (C_F)	Present Worth Factor (C_P)
1	1.0600	0.9434
2	0.54544	1.8333
3	0.37411	2.6729
4	0.28860	3.4650
5	0.23740	4.2123
6	0.20337	4.9172
7	0.17914	5.5823
8	0.16104	6.2097
9	0.14702	6.8016
10	0.13587	7.3600
11	0.12679	7.8867
12	0.11928	8.3837
13	0.11296	8.8525
14	0.10759	9.2948
15	0.10296	9.711
16	0.09895	10.105
17	0.09545	10.477
18	0.09236	10.827
19	0.08962	11.158
20	0.08719	11.469

I = 8% INTEREST FACTORS

Number of Years (N)	Capital Recovery Factor (C_F)	Present Worth Factor (C_P)
1	1.0800	0.9259
2	0.56077	1.7832
3	0.38803	2.5770
4	0.30192	3.3121
5	0.25046	3.9926
6	0.21632	4.6228
7	0.19207	5.2063
8	0.17402	5.7466
9	0.16008	6.2468
10	0.14903	6.7100
11	0.14008	7.1389
12	0.13270	7.5360
13	0.12642	7.9037
14	0.12130	8.2442
15	0.11683	8.5594
16	0.11298	8.8513
17	0.10963	9.1216
18	0.10670	9.3718
19	0.10413	9.6035
20	0.10185	9.8181

I = 10% INTEREST

Number of Years (N)	Capital Recovery Factor (C_F)	Present Worth Factor (C_P)
1	1.1000	0.9094
2	0.57619	1.7355
3	0.40212	2.4868
4	0.31547	3.1698
5	0.26380	3.7907
6	0.22961	4.3552
7	0.20541	4.8683
8	0.18745	5.3349
9	0.17364	5.7589
10	0.16275	6.1445
11	0.15396	6.4950
12	0.14676	6.8136
13	0.14078	7.1033
14	0.13575	7.3666
15	0.13147	7.6060
16	0.12782	7.8236
17	0.12466	8.0215
18	0.12193	8.2013
19	0.11955	8.3649
20	0.11746	8.5135

e • x • a • m • p • l • e 1.2

Problem   A municipality is trying to decide on the purchase of a refuse-collection vehicle. Two types are being considered, A and B, and the capital and operating costs are as shown. Each truck is expected to have a useful life of 10 years.

	Truck A	Truck B
Initial (capital) cost	$80,000	$120,000
Maintenance cost, per year	6,000	2,000
Fuel and oil, per year	8,000	4,000

Which truck should the municipality purchase based on these costs alone? Calculate the costs both on an annual and on a present worth basis, assuming an interest rate of 8%.

Solution   Calculation of annual cost for Truck A:

From the 8% interest table (Table 1–1)
> Interest for 10 years
> $80,000 × 0.14903 = $11,922
> Maintenance   6000
> Fuel   8000
> Total $25,922/year

Calculation of annual cost for Truck B:

From the 8% interest table (Table 1–1)
> Interest for 10 years
> $120,000 × 0.14903 = $17,884
> Maintenance   2000
> Fuel   4000
> Total $23,884/year

On an annual cost basis, Truck B is the rational choice because its annual cost to the community is lower than the annual cost of Truck A.
Calculation of present worth for Truck A:

Capital cost	$80,000
Present worth of maintenance and fuel, $14,000 at 8% interest for 10 years from the interest table, $14,000 × 6.7100	93,940
Total	$173,940

Calculation of present worth for Truck B:

Capital cost	$120,000
Present worth of maintenance and fuel, $6000 at 8% interest for 10 years from the interest table, $6000 × 6.7100	40,260
Total	$160,260

Based on present worth, Truck B is the rational choice because if the community were to borrow the money to operate the truck for 10 years, it would have to borrow less money than if it wanted to purchase Truck A.

◆

But suppose several alternative courses of action also have different benefits to the client or to the employer. Suppose in the earlier example that one alternative open to the community would be to go from twice-per-week collection of refuse to once-per-week collection. Now the level of service is also variable, and the foregoing analysis is no longer applicable. It is necessary to incorporate benefits in the cost-effectiveness analysis in order to be sure the most effective use is made of scarce resources.

1.3 ◆ Decisions Based on a Benefit/Cost Analysis

In the 1940s, the Bureau of Reclamation and the U.S. Army Corps of Engineers battled for public dollars in their drive to dam all the free-flowing rivers in the United States. To convince the Congress of the need for major water storage projects, a technique called *benefit/cost analysis* was developed. At face value, this is both useful and uncomplicated. If a project is contemplated, an estimate of the benefits derived is compared in ratio form to the cost incurred. If this ratio is more than 1.0, the project is clearly worthwhile, and the projects with the highest benefit/cost ratios should be constructed first because these will provide the greatest returns on the investment. By submitting their projects to such an analysis, the Bureau and the Corps could argue for increased expenditure of public funds and could rank the proposed projects in order of priority.

As is the case with cost-effectiveness analyses, the calculations in benefit/ cost analyses are in dollars, with each benefit and each cost expressed in monetary terms. For example, the benefits of a canal would be calculated as monetary savings in transportation costs.

But some benefits and costs such as clean air, flowers, whitewater canoeing, foul odors, polluted groundwater, and littered streets cannot be expressed easily in monetary terms. Yet these benefits and costs are very real and should somehow be included in benefit/cost analyses.

One solution is simply to force monetary values on these benefits. In estimating the benefits for artificial lakes, for example, recreational benefits are calculated by predicting what people would be willing to pay to use such a facility. There are, of course, many difficulties with this. The value of a dollar varies substantially from person to person, and some persons benefit a great deal more from a public project than others, and yet all may share in the cost. Because of the problems involved in estimating such benefits, recreational benefits can be bloated in order to increase the benefit/cost ratio. It is thus possible to justify almost any project since the benefits can be adjusted as needed.

In the following example, monetary values are placed on subjective benefits and costs to illustrate how such an analysis is conducted. The reader should recognize that the benefit/cost analysis is a simple arithmetic calculation and that even though the final value can be calculated to many decimal places, it is only as valuable as the weakest estimate used in the calculation.

e • x • a • m • p • l • e **1.3**

Problem   For a number of years, a small community has owned and operated a refuse-collection service, consisting of one truck used to collect refuse once a week. This truck is wearing out and must be replaced. It also appears that only one truck is no longer adequate and a second truck may have to be purchased. If the second truck is not purchased, the citizens will be asked to burn wastepaper in their backyards so as to reduce the amount of refuse collected. There are two options:

a.   The old truck is sold and two newer models purchased. This will allow for collection twice per week instead of only once.

b.   The old truck will be sold and only one new truck bought, but the citizens of the community will be encouraged to burn the combustible fraction of the refuse in their backyards, thus reducing the quantity of refuse collected. Decide on the alternative of choice using a benefit/cost analysis.

Solution   The benefits and costs of both alternatives are listed, using annual cost numbers (from Example 1.2). For those costs that cannot be easily expressed in monetary terms, reasonable estimates are suggested. Next, the dollars are added.

ALTERNATIVE 1:
Benefits
Collection of all refuse[1]	$250,000
Total benefits	$250,000

Costs
Two new vehicles (including operating cost)	$ 47,768
Increased noise & litter	10,000
Labor cost	200,000
Total costs	$257,768

$$\text{Benefit/Cost} = \frac{250,000}{257,768} = 0.97$$

ALTERNATIVE 2:
Benefits
Collection of 60% of refuse	$150,000
Total benefits	$150,000

Costs
    Dirtier air   0
    New vehicle (including operating cost)   23,884
    Labor costs   120,000
    Total costs $143,884

$$\text{Benefit/Cost} = \frac{150,000}{143,884} = 1.04$$

The benefit/cost of the first alternative is 0.97, while the benefit/cost ratio for the second alternative is 1.04. It would seem, therefore, that the second alternative should be selected.

◆

Benefit/cost analyses can and often are subverted by a technique known as "sunk cost." Suppose a government agency decides to construct a public facility and estimates that the constructions cost will be $100 million. It argues that since the benefits (however they might be calculated) are $120 million, the project is worth constructing because the benefit/cost ratio is greater than 1.0 ($120/$100). It then receives appropriations from Congress to complete the project.

Somewhere in the middle of the project, and having already spent the original $100 million, the agency discovers that they underestimated the construction cost. It turns out that the project will actually cost $180 million and not the $100 million originally estimated. This would of course lower the benefit/cost to $120/$180 = 0.67, or less than one, and the project would not be economically feasible. But the agency has already spent $100 million on the construction. This is a "sunk cost," or money that will be lost forever if the project is not completed. Hence, they argue, the *true* cost of the project is the increment between what the estimated cost is and what has already been spent, or $180 − $100 = $80. The benefit/cost is then calculated as 120/80 = 1.5, substantially greater than one, indicating that the project should be completed. It is absolutely astounding how many times this scam works, and nobody seems to ask the agencies why their *next* estimates should be believed if all of the previous ones were wildly undervalued.

Another significant problem with benefit/cost estimates is that only some benefits can be quantified, such as the costs of the vehicles in the Example 1.3. Many other benefits are highly subjective. In the previous example there is no cost attached to dirty air. If such a cost is included, the calculation might result in a different conclusion. The problem of course is in qualifying the cost of dirty air.

Benefit/cost calculations are also plagued by the problem of shared costs and benefits. The concept of unequal sharing of costs and benefits is perhaps best illustrated by an example of the "common" as it was used in a medieval village.[2] The houses in the village surrounded the common, and everyone grazed their cows on the common. The common area was shared land, but the cows belonged to the individual citizens. It would not have taken too long for one of the citizens of the village to figure out that the benefits of having a cow were personal, but the costs of keeping it on the common ground was shared by all. In true selfish fashion, it made sense for the farmer to graze several cows and thereby increase his wealth. But this action would have triggered a similar response by his neighbors. Why should *they* stay with one cow when they saw their friend getting wealthy? So they all would buy more cows, and all of these would graze on the common. Eventually, however, the number of cows would overwhelm the capacity of the common and all of the cows would have to be killed.

The sharing of the cost of dirty air is similar to the tragedy of the commons. Each one of us, in using clean air, uses it for personal benefit, but the cost of polluting it is shared by everyone. Is it possible for humans to rethink the way they live and to agree voluntarily to limit their pollutional activities? This is unlikely, at least in the near term, and as a result government has to step in and limit each one of us to only one cow if our "green" is to survive.

1.4 ◆ Decisions Based on a Risk Analysis

Often the benefits of a proposed project are not such simple items as recreational values, but the more serious concern of human health. When life and health enter such benefit/cost calculations, the analyses are generally referred to as *risk*/benefit/cost analyses to indicate that people are at risk. These calculations have in the past few years become more widely known as simply *risk analyses*.

Risk analysis is further divided into *risk assessment* and *risk management*. The former involves a study and analysis of the potential effect of certain hazards to human health. Using statistical information, risk assessment is intended to be a tool for making informed decisions. Risk management, however, is the process of reducing risks that are deemed unacceptable.

In our private lives, we are continually doing both. Smoking cigarettes is a risk to our health, and it is possible to calculate the potential effect of smoking. Quitting smoking is a method of risk management, since the effect is to reduce the risk of dying of certain diseases.

The risk of dying of something is 100%. The medical profession has yet to save anyone from death. The questions then become *when* death will occur, and what the *cause* of death is.

There are three ways of calculating risk of death due to some cause. First, risk can be defined as the ratio of the number of deaths in a given population exposed to a pollutant, divided by the number of deaths in a population not exposed to the pollutant. That is,

$$\text{Risk} = D_1/D_0$$

where D_1 = number of deaths in a given population exposed to a specific pollutant, per unit time

D_0 = number of deaths in a similar-size population not exposed to the pollutant, per unit time

e • x • a • m • p • l • e 1.4

Problem   Kentville, a small community of 10,000 people, is located next to a krypton mine and there is concern that emissions from the krypton smelter have resulted in adverse effects. Specifically, kryptonosis seems to have killed 10 of the Kentville's inhabitants last year. A neighboring community, Lanesburg, has 20,000 inhabitants and is far enough from the smelter not to be affected by the emissions, and in Lanesburg only 2 persons last year died of kryptonosis. What is the risk of dying of kryptonosis in Kentville?

Solution   If risk is defined as above, then

$$\text{Risk of dying of kryptonosis} = \frac{\dfrac{10}{10,000}}{\dfrac{2}{20,000}} = 10$$

That is, a person is ten times more likely to die of kryptonosis in Kentville than in a noncontaminated locality.

◆

Note, however, that even though statistically there is a far greater chance of dying of kryptonosis in Kentville than anywhere else, and that Kentville just happens to have a krypton smelter, *this does not prove that the smelter is responsible*. All we have is statistical evidence.

A second method of calculating risk is to determine the number of deaths due to various causes per population and compare these ratios. That is,

$$\text{Relative risk of dying of cause } A = \frac{D_A}{P}$$

where D_A = number of deaths due to a cause A in a unit time
 P = population

e • x • a • m • p • l • e 1.5

Problem   The number of deaths in Kentville and their causes last year were

Heart attack	5
Accidents	4
Kryptonosis	10
Other	6

What is the risk of dying of kryptonosis relative to other causes?

Solution   The risk of dying of a heart attack in Kentville is 5/10,000, while the risk of dying of kryptonosis is 10/10,000. That is, the risk of dying of kryptonosis is twice as large as the chance of dying of a heart attack. This may be different in Lanesburg, of course, and the two risks can be compared.

◆

Finally, risk can be calculated as the number of deaths due to a certain cause divided by the total number of deaths, or

$$\text{Risk dying of cause } A = \frac{D_A}{D_{\text{total}}}$$

where D_{total} = total number of deaths in the population in a unit time

e • x • a • m • p • l • e   1.6

Problem   What is the risk of dying of kryptonosis in Kentville relative to deaths due to other causes?

Solution   The total number of deaths from Example 1.5 is 25. Hence

$$\text{Risk of dying of kryptonosis} = 10/25 = 0.4$$

That is, of all the ways to die, the inhabitants of Kentville have a 40% chance of dying of kryptonosis.

◆

Some risks we choose to accept while other risks are imposed on us from outside. We choose, for example, to drink alcohol, drive cars, or fly in airplanes. Each of these has a calculated risk because people die every year as a result of alcohol abuse, traffic accidents, or airplane crashes. Most of us subconsciously weigh these risks and decide to take our chances. Typically, people seem to be able to accept such risks if the chances of death are in the order of 0.01, that is, about 1% of the deaths are attributed to these causes.

Some risks are imposed from without, however, and these we can do little about. For example, the life expectancy of people living in a dirty urban atmosphere is considerably shorter than that of people living identical lives but breathing clean air. This is a risk we can do little about (except to move), and it is this type of risk that most people resent the most. In fact, studies have shown that the acceptability on an *involuntary* risk is on the order of 1000 times less than our acceptability of a *voluntary* risk. Such human attitudes can explain why people who smoke cigarettes still get upset about air quality, or why people will drive while intoxicated to a public hearing protesting the siting of an airport, fearing the crash of airplanes.

Some federal and state agencies use a modified risk analysis where the benefit is a life saved. For example, if a certain new type of highway guardrail is to be installed, it might be possible that its use would reduce expected highway fatalities by some number. If a value is placed on each life, the total benefit can be calculated as the number of lives saved times the value of a life. Setting such a number is both an engineering and a public policy decision, answerable ideally to public opinion.

e · x · a · m · p · l · e 1.7

Problem   A 95% reduction of kryptonite emissions from a smelter will cost $10 million. Toxicologists estimate that such a pollution-control scheme will reduce the deaths due to krypton poisoning from ten per year to four per year. Should the money be spent?

Solution   Assume that each life is worth $1.2 million, based on lifetime earnings. Six lives saved would be worth $7.2 million. Since this is less than the cost of the control, then based on a risk analysis, it is not cost effective to install the pollution equipment.

◆

But what about the assumption of a human life being worth $1.2 million? Is this really true? If in the foregoing example a human life is assumed to be worth say $5 million, then the pollution control is warranted. But the $10 million spent on pollution control could then not be spent on such other worthy projects as education, health, or transportation. More on this later.

Risk assessment is also useful as a tool in environmental management, where the imposition of controls would be used as a means of responding to various risks. Unfortunately, such risk calculations are fraught with great uncertainty. For example, the National Academy of Sciences report on saccharin concludes that over the next 70 years the expected number of cases of human bladder cancer in the United States resulting from daily exposure to 120 mg of saccharin might range from 0.22 to 1,144,000 cases.

That is quite an impressive range, even for toxicologists. The problem, of course, is that we have to extrapolate data over many orders of magnitude, and often the data are not for humans but for other species, thus requiring a species conversion. Yet governmental agencies are increasingly placed in positions of having to make decisions based on such inadequate data.

1.4.1 Environmental Risk Analysis Procedure

Environmental risk analysis takes place in discrete steps.

1. Define the source and type of pollutant of concern.
2. Identify the pathways and rates of exposure. How can it get to humans so it can cause health problems?
3. Identify the receptors of concern. Who are the people at risk?

4. Determine the potential health impact of the pollutant on the receptor. That is, define the dose–response relationship.

5. Decide what impact is acceptable. What effect is considered so low as to be acceptable to the public?

6. Based on the allowable effect, calculate the acceptable level at the receptor, and then calculate the maximum allowable emission.

7. If the emission or discharge is at present (or planned to be) higher than the maximum allowable, determine what technology is necessary in order to attain the maximum allowable emission or discharge.

Defining the source and type of pollutant is often more difficult than it might seem. Suppose a hazardous-waste treatment facility is to be constructed near a populated area. What types of pollutants should be considered? If the facility is to mix and blend various hazardous wastes in the course of reducing their toxicity, which products of such processes should be evaluated? In other cases, the identification of both the pollutant of concern and the source is a simple matter, such as the production of chloroform during the addition of chlorine to drinking water.

Defining the pathway may be fairly straightforward, as in the case of water chlorination. In other situations, such as the effect of atmospheric lead, the pollutant can enter the body in a number of ways, including through food, skin, and water.

Defining the receptor can cause difficulty since all humans are not of standard size and health. The EPA has attempted to simplify such analyses by suggesting that all adult human beings weigh 70 kg and live for 70 years, and that they drink 2 L of water and breathe 20 m³ air each day. These values are used for comparing different risks.

Defining the effect is one of the most difficult steps in risk analysis since this presumes a certain response of a human body to various pollutants. First, it has become commonplace to consider two types of effects: (1) cancer and (2) all other adverse effects causing death.

The dose–response curve for toxic noncancerous substances is assumed to be linear with a threshold. As shown in curve *A* of Figure 1–1, a low dose of a non-cancer-causing toxin would not cause measurable harm, but any increase higher than the threshold would have a detrimental effect. Ingesting small quantities of inorganic mercury, for example, seems to be acceptable because we cannot demonstrate that this has any detrimental effect on human health.

Some toxins, such as zinc for example, are necessary nutrients in our metabolic system and are required for good health. The absence of such chemicals from our diet can be detrimental, whereas high doses can be toxic. Dose–response curves for such chemicals are illustrated by curve *B* in Figure 1–1.

Most vexing is that we do not yet know the shape of the dose–response curve for chemicals that cause cancer. Some authorities suggest that the curve is linear, starting at zero effect at zero concentration, and that the harmful effect

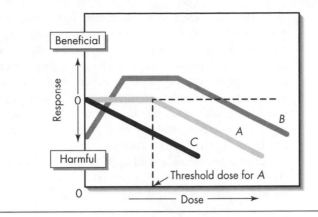

FIGURE 1-1   Three dose–response curves.

increases linearly as shown by curve *C* in Figure 1-1. If this is true, every finite dose of a carcinogen can cause a finite increase in the incidence of cancer. An alternative view is that the body is able to resist small doses of carcinogens and that a threshold exists below which there is no effect.

The EPA has chosen the more conservative route and developed what it calls the *potency factor* for carcinogens. The potency factor is defined as the risk of getting cancer (not necessarily dying from it) produced by a lifetime average daily dose of 1 mg of the pollutant/kg body weight/day. The dose–response relationship is therefore

[Lifetime risk] = [Average daily dose] × [Potency factor]

The units of average daily dose are (mg pollutant/kg body weight/day) and the units for the potency factor are therefore (mg pollutant/kg body weight/day)$^{-1}$. The lifetime risk is unitless, as a fraction. The dose is assumed to be a chronic dose over a 70-year life span.

EPA has calculated the potency factors for many common chemicals and published these through an *Integrated Risk Information Systems (IRIS)*. Potency factors are listed for both oral dose and inhalation.

Deciding what is acceptable risk is probably the most contentious parameter in the above calculations. Is a risk of one in a million acceptable? Who decides if this level is acceptable? Certainly if we ask the person who would be harmed if such a risk is acceptable, he or she would most definitely say no.

The entire concept of "acceptable risk" presupposes a value system used by environmental engineers and scientists in their professional work. Unfortunately, this value system often does not match the value system held by the public at large. Engineers tend to seek the greatest total good in all actions. For example, it is acceptable to place a value on a random human life in order to achieve some net good, and this allows the EPA to calculate the risk associated with pollutants and to place an "acceptable" risk at a probability of one in a million. The cost is very small compared to the good attained, and thus (the argument goes) it is in the public's interest to accept the decision.

But there are strong arguments to be made that most people do not view their own welfare or even the welfare of others in the same light. Most people have the greatest interest in themselves, while recognizing that it is also in their self-interest to behave morally toward other people. Any analysis where costs are unequally assigned (for example, that one person in a million gets hurt) many people consider unfair. Many people also believe that it is unethical to place value on human life and refuse even to discuss the one in a million death for a given environmental contaminant. People are generally loathe to take advantage of that single individual for the benefit of many. It is wrong to talk about one person in a million who may be harmed, since for that one person the harm is unfair. And, perhaps most important, that one person may be yourself or someone you love.

In a public hearing the engineer may announce that the net detrimental effect of the emission from a proposed sludge incinerator is to increase cancer deaths by only one in a million and consider this risk to be quite acceptable given the benefit of the public funds saved. But many members of the public, not appreciating this reasoning, will see this one death as grossly unethical.

Such a disagreement has often been attributed by engineers to "technical illiteracy" on the part of those who disagree with them. But this seems to be a mistake. It is not technical illiteracy, but a different ethical viewpoint that is at issue.

It is not at all clear how the public would make societal decisions such as the siting of landfills, clear-cutting forests, use of hazardous materials in consumer products, or selecting sludge-disposal practices, if these decisions had to be made by the general public. If pressed, they might actually resort to a similar form of analysis, but only after exhausting all other alternatives. The point is that they are not in the position to make such decisions, and thus are free to criticize the engineer, who often does not understand the public's view and may get blindsided by what may appear to be irrational behavior.

Calculating the acceptable levels of pollution is the next step in the risk-analysis process, and this is a simple arithmetic calculation since the value decisions have already been made. Finally, it is necessary to *design treatment strategies* in order to meet this acceptable level of pollution.

The next example illustrates how this process works.

e • x • a • m • p • l • e   1.8

Problem   The EPA lists chromium VI as a carcinogen with an inhalation route potency factor of 41 $(mg/kg\text{-}day)^{-1}$. A sludge incinerator, with no air-pollution-control equipment, is expected to emit chromium VI at a rate such that the airborne concentration at the plant boundary immediately downwind of the incinerator is 0.001 $\mu g/m^3$. Will it be necessary to treat the emissions so as to reduce the chromium VI to stay within the risk level of one additional cancer per 10^6 persons?

Solution   The source is defined, and the pathway is the inhalation of chromium VI. The receptor is EPA's "standard person" who weighs 70 kg, breathes 20 m³ of air per day, and lives for 70 years immediately downwind of the incinerator and never leaves to grab a breath of fresh air. (As an aside, consider the irrationality of this assumption. But how else could this be done?) The dose response is assumed to be linear, and the chronic daily intake is calculated by relating the risk to the potency factor as

$$[\text{Risk}] = [\text{Chronic daily intake}] \times [\text{Potency factor}]$$

$$[1 \times 10^{-6}] = [\text{CDI}] \times [41 \times (\text{mg/kg·day})^{-1}]$$

$$\text{CDI} = 0.024 \times 10^{-6}\ \text{mg/kg-day}$$

$$[\text{CDI}] = [\text{Volume of air per day inhaled}] \\ \times [\text{Concentration of chromium VI}]/ \\ [\text{kg body weight}]$$

$$0.024 \times 10^{-6}\ \text{mg/kg-day} = [(20\ \text{m}^3/\text{day})(C\ \mu\text{g/m}^3) \times (10^{-3}\ \text{mg/}\mu\text{g})]/70\ \text{kg}$$

and solving for C

$$C = 0.085 \times 10^{-6}\ \text{mg/m}^3 \quad \text{or} \quad 0.085 \times 10^{-3}\ \mu\text{g/m}^3$$

The system does not meet this level of chromium VI and emission controls are necessary.

$\blacklozenge$

1.4.2   Environmental Risk Management

If it is the responsibility of government to protect the lives of its citizens against foreign invasion or criminal assault, then it is equally the government's responsibility to protect the health and lives of its citizens from other potential dangers such as falling bridges and toxic air pollutants. Government has a limited budget, however, and we expect that this money is distributed so as to achieve the greatest benefits to health and safety. If two chemicals are placing people at risk, then it is rational that funds and effort be expended to eliminate the chemical that results in the greatest risk.

But is this what we really want? Suppose, for example, that it is cost effective to spend more money and resources to make coal mines safer than it is to conduct heroic rescue missions if accidents occur. It might be more "risk effective" to put the money we have into safety, eliminate all rescue squads, and simply accept the few accidents that will still inevitably occur. But since there would no longer be rescue teams, the trapped miners would then be left on their own. The net effect would be, however, that overall fewer coal miners' lives would be lost.

Even though this conclusion would be risk effective, we would find it unacceptable. Human life is considered sacred. This does not mean that infinite resources have to be directed at saving lives, but rather that one of the sacred rituals of our society is the attempt to save people in acute or critical need, such as crash victims, trapped coal miners, and the like. Thus purely rational calculations such as the coal miners example above might not lead us to conclusions that we find acceptable.[3]

In all such risk analyses, the benefits are usually to humans only, and they are *short-term benefits*. Likewise, the costs determined in the cost-effectiveness analysis are real budgetary costs, money that comes directly out of the pocket of the agency. Costs related to environmental degradation and *long-term costs,* which are very difficult to quantify, are not included in these calculations. The fact that long-term and environmental costs can nowhere be readily considered in these analyses, coupled with the blatant abuse of benefit/cost analysis by governmental agencies, makes it necessary to bring into action another decision-making tool—*environmental impact analysis.*

1.5 ◆ Decisions Based on an Environmental Impact Analysis

On January 1, 1970, President Nixon signed into law the National Environmental Policy Act (NEPA), which was intended to ". . . encourage productive and enjoyable harmony between man and his environment." As with other imaginative and groundbreaking legislation, the law contained many provisions that were difficult to implement. Nevertheless, it provided the model for environmental legislation soon adopted by most of the Western world.

NEPA set up the Council on Environmental Quality (CEQ), which was to be a watchdog on federal activities as they influence the environment, and the CEQ reported directly to the President. The vehicle by which the CEQ would monitor significant federal activity having impact on the environment was to be a report called the *Environmental Impact Statement* (EIS). This lightly regarded provision in NEPA, tucked away in Section 102, stipulates that the EIS is to be an inventory, analysis, and evaluation of the effect of a planned project on environmental quality. The EIS is to be written first in draft form by the federal agency in question for each significant project, and then this draft is to be submitted for public comment. Finally, the report is rewritten taking into account public sentiment and comments from other governmental agencies. When complete, the EIS is to be submitted to the CEQ, which then makes a recommendation to the President as to the wisdom of undertaking that project.

The impact of Section 102 of NEPA on federal agencies was traumatic since they were not geared up in manpower or in training to write such statements, nor were they psychologically able to accept this new (what they viewed as a) restriction on their activities. Thus the first few years of the EIS was tumultuous, with many environmental impact statements being judged for adequacy in courts of law.

Conflict of course arises when the cost-effective alternative, or the one with the highest benefit/cost (B/C) ratio, also results in the greatest adverse environmental impact (EI). Decisions have to be made, and quite often the B/C wins out over the EI. It is nevertheless significant that since 1970, the effect of the project on the environment must be considered, whereas before 1970 these concerns were never even acknowledged, much less included in the decision-making process.

In practice, governmental agencies tend to conduct internal EI studies and propose only those projects that have both a high B/C ratio and a low adverse environmental impact. Most EI statements are thus written as a justification for an alternative that has already been selected by the agency.

Recent reorganization in the White House has resulted in the abolishment of the Council on Environmental Quality (CEQ) and the establishment of a White House office on Environmental Policy. This office performs the functions of the CEQ, as well as establishes environmental policy at the highest level. More important, the abolishment of the CEQ and the creation of the new office does not change the need for reviewing draft environmental impact statements, and the NEPA requirements still apply.

Although the old CEQ developed some fairly complete guidelines for environmental impact statements, the form of the EIS is still variable and considerable judgment and qualitative information (some say prejudice) go into every EIS. Each agency seems to have developed its own methodology, within the constraints of the CEQ guidelines, and it is difficult to argue that any one format is superior to another. Since there is no standard EIS, the following discussion is a description of several alternatives within a general framework.

A typical EIS should be in three parts: inventory, assessment, and evaluation.

1.5.1 Inventory

The first duty in the writing of any EIS is the gathering of data, such as hydrological, meteorological, or biological information. A listing of the species of plants and animals in the area of concern, for example, is included in the inventory. There are no decisions made at this stage, since *everything* properly belongs in the inventory.

1.5.2 Assessment

The second stage is the analysis part commonly called the assessment. In this mechanical part of the EIS the data gathered in the inventory are fed to the assessment mechanism and the numbers are crunched accordingly. Numerous assessment methodologies are available, only a few of which are described below.

Quantified Checklist.   This method is possibly the simplest quantitative way of comparing alternatives and involves first the listing of those areas of the environment that might be affected by the proposed project, and second the estimation of (a) the *importance* of the impact, (b) the *magnitude* of the impact, (c) the *nature* of the impact (whether negative or positive). Commonly the importance is given numbers such as 0 to 5, where 0 means no importance whatever, while 5 implies extreme importance. A similar scale is used for magnitude, while the nature is expressed as simply −1 for negative (adverse) and +1 for positive (beneficial) impact. The environmental impact (EI) is then calculated as

$$EI = \sum_{i=1}^{n} (I_i \times M_i \times N_i)$$

where I_i = importance of ith impact
M_i = magnitude of ith impact
N_i = nature of ith impact, so that $N = +1$ if beneficial and $N = -1$ if detrimental
n = total number of areas of concern

The following example illustrates the use of the quantitative checklist.

e • x • a • m • p • l • e **1.9**

Problem   Continuing Example 1.3, the community has two alternatives, increase the refuse-collection frequency from one to two times per week, or allow for the burning of rubbish on premises. Analyze these two alternatives using a quantified checklist.

Solution   First the areas of EI are listed. In the interest of brevity, only five areas are shown as follows, although a thorough assessment would include many other concerns. Following this, values for importance and magnitude are assigned (0 to 5) and the nature of the impact (+/−) is indicated. The three columns are then multiplied.

ALTERNATIVE 1: Increasing collection frequency

Area of concern	Importance (I)	Magnitude (M)	Nature (N)	(I × M × N)
Air pollution (trucks)	4	2	−1	−8
Noise	3	3	−1	−9
Litter in streets	2	2	−1	−4
Odor	2	3	−1	−6
Traffic congestion	3	3	−1	−9
Groundwater pollution	4	0	−1	−0

(Note: No new refuse will be landfilled.)

EI = −36

ALTERNATIVE 2: Burning on premises

Area of concern	Importance (I)	Magnitude (M)	Nature (N)	(I × M × N)
Air pollution (burning)	4	4	−1	−16
Noise	0	0	−1	0
Litter	2	1	+1	+2
(Note: Present system of collection causes litter.)				
Odor	2	4	−1	−8
Traffic congestion	0	0	−1	0
Groundwater pollution	4	1	+1	+4
(Note: Less refuse will be landfilled.)				

$$EI = -18$$

On the basis of this analysis, burning the refuse would result in the lower adverse impact.

◆

Interaction Matrix. For simple projects, the quantified checklist is an adequate assessment technique, but it gets progressively unwieldy for larger projects, such as the construction of a dam, for example, where many smaller actions combine to produce the overall final product. The effect of each of these smaller actions should be judged separately with respect to its impact. Such an interaction between the individual actions and areas of concern gives rise to the interaction matrix, where once again the importance and the magnitude of the interaction is judged (such as the 0–5 scale previously used). There seems to be no agreement on what calculation should be made to produce the final numerical quantity. In some cases, the interaction value is multiplied by the magnitude, and the products summed as before, while other procedures are simply to add all the numbers on the table. In the next example, the products are summed into a grand sum shown on the lower right corner of the matrix.

e • x • a • m • p • l • e 1.10

Problem Continuing Example 1.3, use the interaction matrix assessment technique to decide on the alternatives presented.

Solution Note again that these are incomplete lists used only for illustrative purposes. The results indicate that once again it makes more sense to burn the paper.

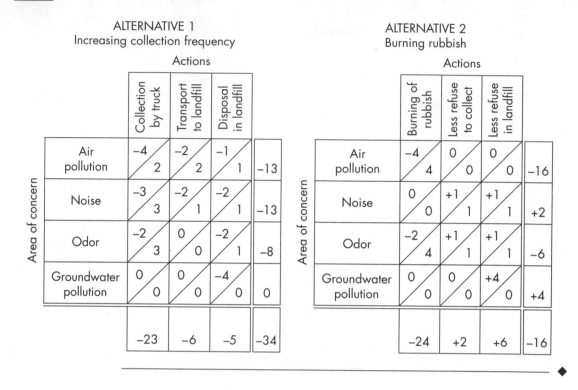

Before moving on to the next technique, it should again be emphasized that the method illustrated in the previous example can have many variations and modifications, none of which are "right" or "wrong," but which will depend on the type of analysis conducted, for whom the report is prepared, and what is being analyzed. Individual initiative is often a most valuable component in the development of a useful EIS.

Common Parameter Weighed Checklist.   This is another technique for environmental impact assessment. It differs from the quantified checklist technique only in that instead of using arbitrary numbers for importance and magnitude, the importance term is called the effect (E) and is calculated from actual environmental data (or predicted quantitative values) and the magnitude is expressed as weighing factors, W.

The basic objective of this technique is to reduce all data to a common parameter, thus allowing the values to be added. The data (actual or predicted) are translated to the effect term by means of a function that describes the relationship between the variation of the measurable value and the effect of that variation. This function is commonly drawn for each interaction on a case-by-case basis. Three typical functions are illustrated in Figure 1–2. These curves show that as the value of the measured quantity increases, the effect on the environment (E) also increases, but that this relationship can take several forms. The value of E ranges from 0 to plus or minus 1.0, with the positive sign implying beneficial impact and the negative detrimental.

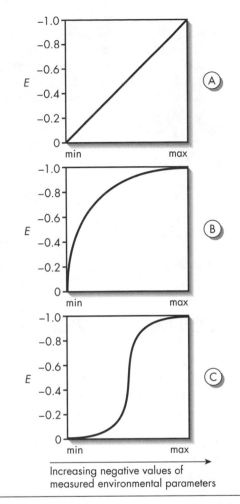

FIGURE 1-2   Three types of functions relating environmental characteristics to effects.

Consider, for example, the presence of a toxic waste on the health and survival of a certain aquatic organism. The concentration of the toxin in the stream is the measured quantity, and the health of the aquatic organism is the effect. The effect (detrimental) increases as the concentration increases. A very low concentration has no detrimental effect, whereas a very high concentration can be disastrous. But what type of function (curve) makes the most sense for this interaction? The straight-line function (Figure 1-2A) implies that as the concentration of the toxin increases from zero, the detrimental effects are immediately felt. This is seldom true. At very low concentrations, most toxins do not show a linear relationship with effect, and thus this function does not appear to be useful. Figure 1-2B, the next curve, is also clearly incorrect. But Figure 1-2C seems much more reasonable, since it implies that the effect of the toxin is very small at lower concentrations, but when it reaches a threshold level, it becomes very toxic quickly. As the level increases above the toxic threshold, there can be no further damage since the organisms are all dead, and the effect levels off at 1.0.

Once the effect (E) terms are estimated for each characteristic, they are multiplied by weighing factors (W), which are distributed among the various effects according to the importance of the effect. Typically, the weighing terms add up to 100, but this is not important as long as an equal number of weighing terms are distributed to each alternative analyzed.

The final impact is then calculated by adding up the products of the effect terms (E) and weighing factors (W). Thus for each alternative considered,

$$\text{Environmental impact} = \sum_{i=1}^{n} (E_i \times W_i)$$

where n = total number of environmental areas of concern considered

Remember that E can be negative (detrimental) or positive (beneficial).

e • x • a • m • p • l • e **1.11**

Problem   Continuing Example 1.3, using only litter, odor, and airborne particulates as an example of air quality, calculate the EI of rubbish burning using the common parameter weighed checklist. (This is only a very small part of the total impact assessment that would be necessary when using this technique.)

Solution   Assume that the three curves shown in Figure 1-3 are the proper functions relating the environmental characteristics of litter, odor, and airborne particulates, respectively. Assume that it has been estimated that the burning of rubbish will result in a litter level of 2 on a scale of 0 to 4, an odor number of 3, and an airborne particulate increase to 180 $\mu g/m^3$. Entering Figure 1-3A, B, and C at 2, 3, and 180, respectively, the effects (E) are read off as -0.5, -0.8, and -0.9. Next assign the weighing factors. Out of a total of 10, suppose you decide to assign 2, 3, and 5, respectively, implying that the most important effect is the air quality, and the least important is the litter. Then the EI is

$$\text{EI} = (-0.5 \times 2) + (-0.8 \times 3) + (-0.9 \times 5) = -7.9$$

A similar calculation would be performed for other alternatives and the EIs compared.

Obviously, this technique is wide open to individual modifications and interpretations, and the foregoing example should not be considered in any way a standard method. It does, however, provide a numerical value for an EI, a number that can be compared to other alternatives. This process of comparison and evaluation represents the third part of an EIS.

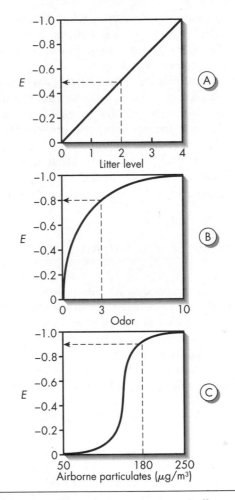

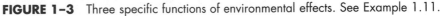

FIGURE 1-3 Three specific functions of environmental effects. See Example 1.11.

1.5.3 Evaluation

The comparison of the results of the assessment procedure and the development of the final conclusions are all covered under evaluation. The previous two steps, inventory and assessment, are simple and straightforward procedures compared to the final step, which requires judgment and common sense. During this step in the writing of the EIS, the conclusions are drawn and presented. Often the reader of the EIS sees only the conclusions and never bothers to review all the assumptions that went into the assessment calculations, and so it is important to include in the evaluation the flavor of these calculations and to emphasize the level of uncertainty in the assessment step.

But even when the EIS is as complete as possible, and the data have been gathered and evaluated as carefully as possible, conclusions concerning the use of the analysis are open to severe differences. For example, the EIS written for the Alaskan oil pipeline, when all volumes are placed into a single pile, represents 14 *feet* of work. Further, at the end of all that effort, good people on both sides drew diametrically opposite conclusions on the effect of the pipeline. The trouble was, they were arguing over the *wrong thing*.[4] They may have thought that they were arguing about how many caribou would be affected by the pipeline, while their disagreement was actually about how deeply they *cared* that the caribou were affected by the pipeline. For a person who does not care one wit about the caribou, the impact is zero, while those who are concerned about the herds and the long-range effects on the sensitive tundra ecology care very much. What then is the solution? How can engineering decisions be made in the face of conflicting *values*? Such decisions require another type of engineering decision making—an *ethical analysis*.

1.6 ◆ Decisions Based on an Ethical Analysis[5]

Before embarking on a discussion of ethical analysis, it is necessary to define quite clearly what is meant by ethics. The popular opinion is that an ethical person is a "good" person, a person with high standards. Likewise a moral person is thought to have certain conventional views on sex. These are both common misconceptions.

Morals are the values people choose to guide the way they ought to treat each other. One such moral value may be telling the truth, and some people will choose to be truthful. Such persons are thought of as moral persons with regard to truth since they would be acting according to their moral convictions. If a person does not value truthfulness, however, then telling the truth is irrelevant, and such a person does not have a moral value with regard to truthfulness. It is also possible to hold a moral view that one always ought to lie in all circumstances, and in this case the person would be thought of as a moral person because he or she would be acting according to a moral conviction.

Most rational people will agree that it is much better to live in a society where people do not lie, cheat, or steal. Societies where people lie, cheat, and steal certainly exist, but the lives of people there are poorer, and given the choice, most people would not want to behave in such a manner and would choose to live in societies where everyone shares the moral values that govern human interactive behavior for mutual benefit.

While it is fairly straightforward to agree that it is not acceptable to lie, cheat, or steal, and most people will not do so, it is a much more difficult matter to decide what to do when conflicts arise between such values. For example, suppose it is necessary to lie in order to keep a promise. How are we to decide what to do when such values conflict?

This question is not that different from questions of economics. For example, how are we to decide which project to undertake with limited resources? As discussed earlier, we use a benefit/cost analysis. How then can we make a decision when moral values conflict? We use an ethical analysis.

Ethics provide a systematized framework for making decisions where values conflict. The selection of the nature and function of that decision-making machinery depends on one's own moral values. Both the cost-effectiveness analysis and the benefit/cost analysis are methods for making decisions based (mostly) on money. Risk analysis calculates the potential damage to health, and environmental impact analysis provides a means for decision making based on long-term effects on resources. Ethics is similarly a framework for decision making, but the parameters of interest are not dollars or environmental data, but values. It then follows that since ethics is a system for decision making, an ethical person is one who makes decisions based on some ethical system. *Any system*! For example, if one chooses to observe a system of ethics that maximized personal pleasure (hedonism), it would be correct (ethical) for him or her to make all decisions so that personal pleasure is maximized. One would, in that case, push old ladies off benches so one could sit down or cheat on tests because this decreases one's required homework time and maximizes party time. Provided one adopts hedonism as the accepted mode of behavior (ethic), one would in these cases be acting ethically.

There are, of course, many other systems of ethics that result in actions most civilized people consider more acceptable forms of social behavior. The most important aspect of any ethical code or system one adopts is that one should then be prepared to defend it as a system that *everyone* should employ. If such a defense is weak or faulty, then this ethical system is considered inadequate, with the implication that a rational person would then abandon the system and seek one that can be defended as being a system that ought to be adopted by everyone.

Most ethical thinking over the past 2500 years has been a search for the appropriate ethical theory to guide our behavior in human–human relationships, and some of the most influential theories in Western ethical thinking, theories that are most defensible, are based on consequences or on acts.

In the former, moral dilemmas are resolved on the basis of the consequences. If good is to be maximized, then the alternative that creates the greatest good is correct (moral). The most influential consequentialist ethical theory is the *utilitarianism* of Jeremy Bentham (1784–1832) and John Stuart Mill (1806–1873), in which the pain and pleasure of all actions is calculated and the worth of all actions is judged on the basis of the total happiness achieved, where happiness is defined as the highest pleasure/pain ratio. (Sound familiar?) This "utilitarian calculus" allows for the calculation of happiness for all alternatives being considered, and to act ethically then would be to choose that alternative that produces the highest level of pleasure and the lowest level of pain. Since the happiness of all human beings involved is summed, such calculations may dictate a decision where the moderate happiness of many results in the extreme unhappiness of a few. Benefit/cost analysis is utilitarian in its origins since money is presumed to equal happiness.

The supporters of consequentialist theories argue that these are the proper principles for human conduct since they promote human welfare; that to act simply on the basis of some set of rules without reference to the consequences of these actions is irrational. Agreeing with Aristotle, utilitarians argue that happiness is always chosen for its own sake, and thus must be the basic good we all seek, and that since utilitarian calculus provides for that calculation, it is the proper tool for decision making where values are involved.

The second group of ethical theories is based on the notion that human conduct should be governed by the morality of acts, and that certain rules like "do not lie" should always be followed. These theories, often called *deontological* theories, emphasize the goodness of the act and not its consequence. Supporters of these theories hold that acts must be judged as good or bad, right or wrong, *in themselves*, irrespective of the consequences of these acts. An early system of deontological rules is the Ten Commandments, in that these rules were meant to be followed *regardless of consequences*.

Possibly the best-known deontological system is that of Immanuel Kant (1724–1804), who suggested the idea of the *categorical imperative*: the concept that one develops a set of rules for making value-laden decisions such that one would wish that all people obeyed such rules. Once these rules are established, one must always follow them, and only then can that person be acting ethically, since it is the act that matters. A cornerstone of Kantian ethics is the principle of *universalizability*, a simple test for the rationality of a moral principle. In short, this principle holds that if an act is acceptable for one person, it *must be equally acceptable for others*. For example, if one considers lying acceptable behavior for him- or herself, everyone should be allowed to lie, and everyone is in fact expected to lie. Similarly, if one person decides that cheating on an exam is acceptable, then he or she agrees, by the principle of universalizability, that it is perfectly acceptable for everyone else to cheat also. In both cases, to live in a world where everyone is expected to lie or cheat would be a sorry situation, and our lives would be a great deal poorer as a result. It thus makes no sense to hold that lying or cheating is acceptable since these behaviors cannot be universalized.

Supporters of rule-based theories argue that consequentialist theories permit and often encourage the suffering of a few for the benefit of many, and that this is clearly an injustice. For example, they assert that consequentialist theories would welcome the sacrifice of an innocent person if this death would prevent the death of others. In contrast, ethicists accepting rule-based principles argue that if killing is wrong, then the mere act of allowing one innocent person to die is wrong and immoral.

The utilitarians counter by arguing that often a "good act" results in net harm. A trivial example would be the question from your roommate/spouse/ friend: "How do you like my new hairdo/shirt/tie/etc.?" Even if you honestly think it is atrocious, the "good act" would be to tell the truth since one is never supposed to lie. Would a white lie not result in the greater good, ask the utilitarians? The deontologists respond that it is wrong to lie even though it might hurt short-term feelings, since telling the truth may create a trust that would hold fast in times of true need.

There are, of course, many more systems of ethics that could be discussed and that have relevance to the environmental engineering profession, but it should be clear that traditional ethical thinking represents a valuable source of insight in one's search for a personal and professional lifestyle.

All classical ethical systems are intended to provide guidance as to how human beings ought to treat each other. In short, the *moral community*, or those individuals with whom we would need to interact ethically, includes only humans and the only *moral agents* within the moral community are rational human beings. Moral agency requires *reciprocity* in that each person agrees to treat one another in a mutually acceptable manner.

But we obviously are not the only inhabitants on earth, and is it not also important how we treat nonhuman animals? or plants? or places? Should the moral community be extended to include other animals? plants? inanimate objects such as rocks, mountains, and even places? If so, should we also extend the moral community to our progeny, who are destined to live in the environment we will to them?

Such questions are being debated and argued in a continuing search for what has become known as *environmental ethics*, a framework intended to allow us to make decisions *within* our environment, decisions that will concern not only ourselves but the rest of the world as well.

One approach to the formulation of an ethic that incorporates the environment is to consider environmental values as *instrumental values*, values that can be measured in dollars and/or in terms of the support nature provides for our survival (e.g., production of oxygen by green plants). The instrumental-value-of-nature view holds that the environment is useful and valuable to people, just like other desirable commodities such as freedom, health, and opportunity.

This *anthropocentric* view of environmental ethics, the idea that nature is here only for the benefit of people, is of course an old one. Aristotle argues that "Plants exist to give food to animals, and animals to give food to men. . . . Since nature makes nothing purposeless or in vain, all animals must have been made by nature for the sake of men."

Kant incorporates nature into his ethical theories by suggesting that our duties to animals are "indirect" duties, which are really duties to our own humanity. His view is quite clear: "So far as animals are concerned, we have no direct duties. Animals are not self-conscious and are there merely as a means to an end. That end is man."[6]

Thus by this reasoning, the value of nonhuman animals can be calculated as their value to people. We would not want to kill off all the Plains buffalo, for example, because they are beautiful and interesting creatures, and we enjoy looking at them. To exterminate the buffalo would mean that we are causing harm to other humans. William F. Baxter, for example, says that our concern for "damage to penguins, or to sugar pines, or geological marvels is . . . simply irrelevant. . . . Penguins are important [only] because people enjoy seeing them walk about the rocks."[7]

We can also agree that it is necessary to live in a healthy environment to be able to enjoy the pleasures of life, and therefore we include a concern for nonhuman animals on our list of moral goods. Likewise, other aspects of the environment also have instrumental value. One could argue that to contaminate water, or pollute air, or destroy natural beauty is taking something that does not belong only to a single person. Such pollution is stealing from others, plain and simple.

Not only would we not want to kill useful animals, but we would not want to exterminate species since there is the possibility that they will somehow be useful in the future. An obscure plant or microbe might be essential in the future for medical research, and we should not deprive others of that benefit. And it would be unethical to destroy the natural environment because so many people enjoy hiking in the woods or canoeing down rivers, and we should preserve these for *our* benefit.

While the instrumental value of nature approach to environmental ethics has merit, it also has a number of problems. First, this argument would not prevent us from killing or torturing individual animals as long as this does not cause harm to other people. Such a mandate is not compatible with our feelings about animals. We would condemn a person who causes unnecessary harm to any animal that could feel pain, and many of us do what we can to prevent this.

Second, this notion creates a deep chasm between humans and the rest of nature, a chasm with which most people are very uneasy. The anthropocentric approach suggests that people are the masters of the world and can use its resources solely for human benefit without any consideration for the rights of other species or individual animals. Such thinking led to the nineteenth-century "rape of nature" in the United States, the effects of which are still with us. Henry David Thoreau (1817–1862), John Muir (1838–1914), and many others tried to put into words what many people felt about such destruction and recognized that it is our alienation from nature, the notion that nature only has instrumental value, that will eventually lead to the destruction of nature. Clearly the valuation of nature only on an instrumental basis is an inadequate approach to explain our attitudes toward nature.

Given these problems with the concept of instrumental value of nature as a basis for our attitudes toward the environment, we need some alternative basis for including nonhuman animals, plants, and even things within our sphere of ethical concern. The basic thrust in this development is the attempt to attribute *intrinsic* value to nature, and to incorporate nonhuman nature within our moral community.

This is known as *extensionist* ethical thinking since it attempts to extend the moral community to include other creatures. Such a concept is perhaps as revolutionary as the recognition only a century ago that slaves are also humans and must be included in the moral community. Aristotle, for example, did not apply ethics to slaves since they were not, in his opinion, intellectual equals. We now recognize that this was a hollow argument, and today slavery is considered morally repugnant. It is possible that in the not too distant future, the moral community will include the remainder of nature as well and we will include nature in our ethical decision-making machinery.

The extensionist environmental ethic was initially publicized not by a philosopher, but by a forester. Aldo Leopold (1887–1948) defined the environmental ethic, or as he called it, a land ethic, as an ethic that ". . . simply enlarges the boundaries of the community to include soils, waters, plants and animals, or collectively the land."[8] He recognized that both our religious and secular training has created a conflict between humans and the rest of nature, where nature has to be subdued and conquered, and nature is something powerful and dangerous against which we have to continually fight, but that a rational view of nature would lead us to an environmental ethic that ". . . changes the role of *Homo sapiens* from conqueror of the land community to plain member and citizen of it."

Leopold was, in fact, questioning the age-old belief that humans are special, that somehow we are not a part of nature, but pitted against nature in a constant combat for survival, and that we have a God-given role of dominating nature, as specified in Genesis. Much as later philosophers (and people in general) began to see slavery as an untenable institution and recognizing that slaves belong within our moral community, succeeding generations may recognize that the rest of nature is equally important in the sense of having rights.

The question of admitting nonhumans to the moral community is a contentious one and centers on whether or not nonhuman creatures have rights. If it can be argued that they have rights, then there is reason to include them in the moral community.

The question of rights for nonhumans has drawn resounding "NOs" from many philosophers, and these arguments are often based on reciprocity. For example, Richard Watson points out that "To say an entity has rights makes sense only if that entity can fulfill reciprocal duties, i.e. can act as a moral agent." He goes on to argue that moral agency requires certain characteristics such as self-consciousness, capability of acting, free will, and an understanding of moral principles. Since most animals do not fulfill any of these requirements, they cannot be moral agents, and therefore cannot be members of the moral community.[9] H. J. McClosky insists that "where there is no possibility of [morally autonomous] action, potentially or actually . . . and where the being is not a member of a kind which is normally capable of [such] action, we withhold talk of rights."[10] It therefore is not reasonable to extend our moral community to include anyone other than humans because of the requirement of reciprocity.

But is reciprocity a proper criterion for admission to the moral community? Do we not already include within our moral community human beings who cannot reciprocate—infants, the senile, the comatose, our ancestors, and even future people? Maybe we are making Aristotle's mistake again by our exclusionary practices. Perhaps being human is not a necessary condition for inclusion in the moral community, and other beings have rights similar to the rights that humans have. These rights may not be something that we necessarily give them, but the rights they possess by virtue of being.

The concept of natural rights, those rights inherent in all humans, was proposed in the seventeenth century by John Locke[11] (1632–1704) and Thomas Hobbes[12] (1588–1679), who held that the right of life, liberty, and property should be the right of all, regardless of social status.[13] These rights are natural rights in that we humans cannot give them to other humans, or we would be giving rights to ourselves, which makes no sense. Thus all humans are "endowed with unalienable rights" that do not emanate from any giver.

If this is true, then there is nothing to prevent nonhuman animals from having "unalienable rights" simply by virtue of their being, just as humans do. They have rights to exist and to live and to prosper in their own environment, and not to have humans deny them these rights unnecessarily or wantonly. If we agree that humans have rights to life, liberty, and absence of pain, then it seems only reasonable that animals, who can feel similar sensations, should have similar rights.

With these rights comes moral agency, independent of the requirement of reciprocity. The entire construct of reciprocity is of course an anthropocentric concept that serves well in keeping others out of our private club. If we abandon this requirement, it is possible to admit more than humans into the moral community.

But if we crack open the door, what are we going to let in? What can legitimately be included in our moral community, or, to put it more crassly, where should we draw the line?

If the moral community is to be enlarged, many people agree that it should be on the basis of sentience, or the ability to feel pain. This argument suggests that all sentient animals have rights that demand our concern.

Some of the classical ethical theories have recognized that animal suffering is an evil. Jeremy Bentham, for example, argues that animal welfare should somehow be taken into the utilitarian benefit\cost calculation since "The question is not, Can they *reason*? nor Can they *talk*?, but Can they *suffer*?" [14] Australian philosopher Peter Singer believes that an animal is of value simply because it values its own life, and sentience is what is important, not the ability to reason. He believes that "the only defensible boundary of concern is for the interests of others." [15]

To include animal suffering within our circle of concern however, opens up a Pandora's box of problems. While we might be able to argue with some vigor that suffering is an evil, and that we do not wish to inflict evil on any living being that suffers, we do not know for sure what animals (or plants) feel pain and therefore are unsure about who should be included. We can presume with fair certainty that higher animals can feel pain because their reactions to pain resemble ours. A dog howls, and a cat screams, and they try to stop the source of the pain. Anyone who has put a worm on a hook can attest to the fact that the worm probably feels pain. But how about creatures that cannot show us in unambiguous ways that they are feeling pain? Does a butterfly feel pain when a pin is put through its body? An even more difficult problem is the plant kingdom. There are those people who insist that plants feel pain, and that we are just too insensitive to recognize it.

If we use the utilitarian approach, we have to calculate the *amount* of pain suffered by animals and humans. If, for example, a human needs an animal's fur to keep warm, is it acceptable to cause suffering in the animal to prevent suffering in the human? It is clearly impossible to include such variables in the utilitarian calculus.

If we do *not* focus only on the pain and pleasure suffered by animals, then it is necessary to either recognize that the rights of the animals to avoid pain are equal to those of humans, or to somehow list and rank the animals in order to specify which rights animals have under what circumstances. In the first instance, trapping animals and torturing prisoners would have equal moral significance. In the second, it would be necessary to decide that the life of a chickenhawk is less important than the life of a chicken, and an infinite number of other comparisons.

Finally, if this is the extent of our environmental ethics, we are not able to argue for the preservation of places and natural environments, except as how they might affect the welfare of sentient creatures. Damming up the Grand Canyon would be quite acceptable if we adopted the sentient animal criterion as our sole environmental ethic.

It seems, therefore, that it is not possible to draw the line at sentience, and the next logical step is to simply incorporate all of life within the folds of the moral community. This is not as outrageous as it seems, the idea having been developed by Albert Schweitzer, who called his ethic a *reverence for life*. He concluded that limiting ethics to only human interactions is a mistake, and that a person is ethical "only when life, as such, is sacred to him, that of plants and animals as that of his fellow men."[16] Schweitzer believed that an ethical person would not maliciously harm anything that grows, but would exist in harmony with nature. He recognized of course that in order to eat, humans must kill other organisms, but he held that this should be done with compassion and a sense of sacredness toward all of life. To Schweitzer, human beings were simply a part of the natural system.

Charles Darwin is probably most responsible for the acceptance of the notion of humans not being different in kind from the rest of nature. If indeed humans evolved from less complex creatures, we are different only in degree and not kind from the rest of life, and are simply another part of a long chain of evolution. As Australian philosopher Janna Thompson points out, "Evolutionary theory, properly understood, does not place us at the pinnacle of the development of life on earth. Our species is one product of evolution among many others."[17]

Similarly, American philosopher Paul Taylor[18] holds that all living things have an intrinsic good in themselves and therefore are candidates to be included in the moral community. He suggests that all living organisms have inherent worth, and as soon as we can admit that we humans are not superior, we will recognize that all life has a right to moral protection. What he labels the *biocentric* outlook depends on the recognition of common membership of all living things in the earth's community, that each organism is a center of life, and that all organisms are interconnected. For Taylor, humans are no more or less important than other organisms.

This approach to environmental ethics has a lot of appeal, and quite a few proponents. Unfortunately, it fails to convince on several accounts. First, there is no way to determine where the line between living and nonliving really should be drawn. Viruses present the greatest problem here, and if Taylor's ideas are to be accepted, the polio virus might also be included in the moral community.

Second, the problem of how to weigh the value of nonhuman animal life relative to the life of humans is unresolved. Should the lives of all creatures be equal, and thus a human life be equal to that of any other creature? If so, the squashing of a cockroach would be of equal moral significance to the murder of a human being. If this is implausible, then there must again be some scale of values, and each living creature must have a slot in the hierarchy of values as placed on them by humans.

If such a hierarchy is to be constructed, how would the value of the life of various organisms be determined? Are microorganisms of equal value to polar bears? Is lettuce of the same order of importance as a gazelle?

Such ranking will also introduce impossible difficulties in determining what is and is not deserving of moral protection. "You, the amoeba, you're in. You, the paramecium, you're out. Sorry." just doesn't compute. Drawing the line for inclusion in the moral community at "all of life" therefore seems to be indefensible.

One means of removing the objection of knowing where to draw the line is simply to extend the line further and include everything within the circle of moral concern, so that the system within which individuals exist is the central issue for concern. Aldo Leopold is often credited with the initial idea for such a *ecocentric* environmental ethic. He suggested that ecosystems should be preserved because without the ecosystem, nothing can survive. The ethic then extends to ecosystems, and ". . . a thing is right when it tends to preserve the integrity, stability, and beauty of the biotic community. It is wrong when it tends otherwise."

Environmental philosophers Val and Richard Routley[19] and Holmes Rolston III[20] recognize that within the ecocentric environmental ethics some creatures, such as our family for example, take precedence over others, such as strangers, and that human beings take precedence over nonhuman animals. They view this as a system of concentric rings, with the most important moral entities in the middle and the rings extending outward, incorporating others within the moral community, but at decreasing levels of moral protection. The question of how the various creatures and places on the earth are to be graded in terms of their moral worth is not resolved, and indeed such grading is up to the people doing the valuing. This process is of course once again human centered, and ecocentric environmental ethics turns out to be an anthropocentric environmental ethic with fuzzy boundaries.

North Carolina State University philosopher Tom Regan presents a similar concept as a *preservation principle*, a principle of "non-destruction, non-interference, and generally, non-meddling."[21] A school of thought embracing this idea is the so-called *deep ecology* movement. Its most notable proponent is the Norwegian philosopher Arne Naess, who suggests that in nature humans are no more important than other creatures or the rest of the world.[22] Deep ecology centers on the idea that humans are part of the total cosmos and are made of the same raw materials as everything else; therefore, humans should live so that they respect all of nature and recognize the damage that *Homo sapiens* already have done to the planet. Deep ecologists call for a gradual reduction of the human population as well as changes in lifestyle so that fewer resources are used.

Deep ecology (as opposed to the anthropocentric "shallow ecology," in which nature is valued instrumentally) eliminates the problem of where to draw the line between those that are in and out of our moral community since everyone and everything is included. But this again presents us with the necessity of valuing all of nature equally, a clearly counterintuitive notion. We would all agree, for example, that humans have greater moral value than rocks because we are the ones placing the value. As soon as we start to propose some hierarchial characteristics, we are back to the original problem of judging everything by human standards.

A third approach to environmental ethics is to recognize that we are, at least at the present time, unable to explain rationally our attitudes toward the environment and that these attitudes are nevertheless deeply felt, not unlike the feeling of spirituality. Why do we not then simply admit that these attitudes are grounded in spirituality? This suggestion may not be as outrageous as it might sound at first (but certainly is outrageous in an engineering textbook).

Although we are deeply imbedded in the Western culture, we recognize that other cultures have existed and still do exist and that their approaches to nature may be instructive. Many older religions, including the Native American religions, are animistic, recognizing the existence of spirits within nature. These spirits do not take human form, as in the Greek, Roman, or Judean religions. They simply are within the tree, the brook, or the sky. It is possible to commune with these spirits, to talk to them, and to feel close to them.

Is it too far-fetched to hope that future people will live in harmony with the world because we will experience, in Wendell Berry's words, a "secular pilgrimage"?[23] John Stewart Collis has an optimistic view of our future. He writes:

> Both polytheism and monotheism have done their work. The images are broken, the idols are all overthrown. This is now regarded as a very irreligious age. But perhaps it only means that the mind is moving from one state to another. The next stage is not a belief in many gods. It is not a belief in one god. It is not a belief at all—not a conception in the intellect. It is an extension of consciousness so that we may *feel* God.[24]

It is not at all obvious that we should have protective, caring attitudes for some (nonhuman) being or thing only if these attitudes can be reciprocated. Perhaps we are hung up on the idea of acceptance into the moral community, and the logjam can be broken by thinking of it as the inclusion of all things in a *community of concern*. In this community, reciprocity is not necessary, and what matters is the love and caring for others, simply because of their presence in the community of concern. The amount of love and care is proportional to the ability to give it and demands nothing in return.

One aspect of the profession of environmental engineering, oft unstated, as if we are embarrassed about it, is that the environmental engineer is engaged in a truly worthwhile mission. The environmental engineer is the epitome of the *solution* as opposed to the *problem*, and we should feel good about that. Our client, in the broadest sense, is the environment itself, and our objective is to preserve and protect our global home, for the sake of our progeny as well as mother earth herself.

1.7 ◆ Continuity in Engineering Decisions

The methods of decision making available to engineers stretch from the most objective (technical) to the most subjective (ethical). The inherent method of decision making is the same in all cases. The problem is first analyzed—taken apart and viewed from many perspectives. When all the numbers are in and the variables are evaluated, the information is synthesized into a solution. Then this solution is looked at in its entirety to see if it "makes sense." This process is especially true in ethical decisions, where there seldom are numbers for comparison. Yet the problem can be analyzed, or "unpacked," to view it in small chunks. When these ideas are synthesized into a larger conclusion, it is necessary to determine if the conclusion "makes sense," or perhaps most important, "feels good."

As engineering decision making stretches from the technical to the ethical, decisions become increasingly less quantitative and more subject to the personal tastes, prejudices, and concerns of the decision maker. Is it reasonable to suggest that at some point these decisions cease being true engineering decisions?

Not a few prominent engineers have argued eloquently that the only true engineering decisions are technical decisions, and other concerns should be left to some undefined decision maker for whom the engineer works and who presumably has the training and background for making these decisions of which the engineer is not perhaps capable and certainly not responsible. Such a view would of course free the engineer of all judgment (other than technical) and make him or her a virtual smart robot, working at the behest of the client or employer. By this argument, the social consequences of how his or her actions affect society at large is of little interest as long as the client/employer is well served.

Fortunately most engineers will not accept such a cop-out. They recognize that engineering, perhaps more than the other professions, *can make a difference*. Projects involving environmental change or manipulation will invariably need the services of the professional engineer. He or she is thus morally obligated, as perhaps the one indispensable cog in the wheel of progress, to seek the best solutions not only technically, but also economically and ethically.

◆ Problems

1-1   Chapel Hill, North Carolina, has decided to build a 15-mile raw-water pipeline to allow it to purchase water from Hillsborough, North Carolina, during times of drought. The engineer has recommended that a 16-inch-diameter pipeline be built. The basis for his decision is as follows:

Diameter (in.)	Cost of pipe in the ground ($/ft)	Capital cost of pumping station	Expected annual power cost
8	5	$150,000	$10,000
10	8	145,000	8000
12	12	140,000	7000
16	14	120,000	6000

Interest rate is 8%, expected life is 20 years.

a.   Compare the four alternatives by calculating the annual cost to Chapel Hill.

b.   If the engineer is paid on the basis of 6% of the capital cost of the total job, which alternative might he recommend if he based all decisions on hedonistic ethics?

c.   Who might the engineer consider to be his "clients" in this situation? Defend each choice with a few sentences.

1-2   The local nuclear power plant has decided that the best place to store its high-level nuclear waste (small in volume but highly radioactive) would be in an unoccupied field on your college campus. They propose to build storage sheds, properly shielded so that the level of radioactivity at the site boundary is equal to background radiation, and use these sheds for the next 20 years. The power plant serves a population of 2,000,000 and it is at present costing the power company $1,200,000 per year to dispose of the spent fuel in Washington State. The new facility will cost $800,000 to construct and $150,000 per year to operate. The power industry pays 8% interest on its borrowed money. The power company is willing to pay the college $200,000 per year rent.

a.   Is this a good deal for the power company? (Will they save money by using this form of long-term storage?)

b.   If you are the college president, decide if this scheme is acceptable to you (and the college) by using a nonquantitative

1.   cost-effectiveness analysis

2.   benefit/cost analysis

3.   risk analysis

4.   environmental impact analysis

5.   ethical analysis

Assume reasonable values and conditions.

1-3   The following cost information was calculated for a proposed birdhouse.

	Costs	
	Capital	Operating
Lumber	$4.00	
Nails	.50	
Additional birdseed required		$1.50/week
	Benefits	
Joy of watching birds		$5.00/week

Expected life of the birdhouse is 2 years. Assume 6% interest. Calculate the benefit/cost ratio. Should you build the birdhouse?

1-4   One definition of pollution is "unreasonable interference with another beneficial use." On the basis of this definition, defend the use of a stream as a conduit for wastewater disposal using any of the decision-making tools presented in this chapter. Then argue against this definition.

1-5   You have been given the responsibility of designing a large trunk sewer. The sewer alignment follows a creek widely used as a recreational facility. A popular place for picnics and nature walks has been constructed by volunteers and the local community eventually wants to make it part of the state park system. The trunk sewer will badly disrupt the creek, destroy its ecosystem, and make it unattractive for recreation. What thoughts would you have on this assignment? Write a journal entry as if you were keeping a personal journal (diary).

1-6   Would you intentionally run over a box turtle trying to cross a road? Why or why not? Present an argument for convincing others that your view is correct.

1-7   You have decided to start a ranch for raising Tasmanian Devils (you may wish to look up something about these unusual creatures). Your ranch will be located in a residential neighborhood, and you have discovered that the ordinances have nothing to say about raising the Devils, so you believe your ranch will be legal. You plan to sell the Devils to the local landfill operator who will use them as scavengers. Discuss your decision from the standpoint of

a.   benefit/cost to yourself

b.   risk to yourself and others

c.   environmental impact

1-8   The water authority for a small community argues that a new dam and lake are needed and presents an initial construction cost estimate of $1.5 million. The authority calculates a benefit (recreation, tourism, water supply) of $2 million. The community agrees to the project and floats bonds to pay for the construction. As construction proceeds, however, and the original $1.5 million has been spent, it becomes evident that the dam will actually cost $3.0 million, and the extra cost is due to change orders and unforeseen problems, not a responsibility of the contractor. The half-finished dam is of no use to the community, and $1.5 million additional funds are needed to finish the project. Use benefit/cost analysis to argue whether the community should continue with this project.

1-9   Name one Department of Defense project that has not used the technique described in Problem 1-8 to argue for its completion.

1-10   Often engineering decisions affect people of the next generation, and even several generations to come. A major bridge, such as the Brooklyn Bridge, just celebrated its 100th anniversary, and it looks like it will stand for another century. This bridge is a mere piker compared to some of the Roman engineering feats, of course.

Engineering decisions such as nuclear waste disposal also affect future generations, but in a negative way. We are willing this problem to our kids, their kids, and untold generations. But should this be our concern? What, after all, have future generations ever done for us?

Philosophers have long wrestled with the problem of concern for future generations, without much success. Listed as follows are some arguments often found in literature about why we should *not* worry about future people. Think about these arguments, and construct opposing arguments if you disagree. Be prepared to discuss these in class.

We do not have to bother worrying about future generations because:

1. These people do not even exist, so they have no claim to any rights such as moral consideration. We cannot allow moral claims by people who do not exist and who may never exist.

2. We have no idea what the future generations will be like or what their problems will be, so it makes no sense to plan for them. To conserve (not use) nonreplenishable resources in case future people may need them makes no sense whatsoever.

3. If there will be future people, we have no idea how many, so how can we plan for them?

4. Future generations will be better off than each previous generation, given the advance of technology, so we have no moral obligation to them.

5. If we discount the future, we find that reserves of resources are of little use to future generations. For example, if a barrel of oil will cost $100 fifty years from now, how much would that barrel cost today? Assume an interest rate of 5%. Clearly from your answer, it makes no sense to conserve a barrel of oil today that is worth only the pennies you calculated.

1–11   A particularly bombastic book that paints the environmental movement as a bunch of nincompoops is *Disaster Lobby*.[25] Chapter 6 begins with this sentence: "Since 1802, the Corps of Engineers of the United States Army has been busy helping to carve a dry, safe, habitable environment out of the savagely hostile wilderness that once was America." Consider the language used in that one sentence and write a one-page analysis of the values held by the author.

1–12   Chlorides enter the Great Lakes from human activity, but the most important source is from the salting of roads in the winter to make them safe for driving. High chloride concentrations (salty water) are damaging to the aquatic ecosystem. By the year 2000 three of the Great Lakes will have chloride levels of >20 mg/L, effectively making them uninhabitable by many of the fish species native to the Great Lakes, thus changing dramatically the lake ecosystems.

Suppose you are the chief environmental engineer in charge of the joint Canadian–USA Great Lakes Water Quality Commission, and you have to set water-quality limits for chlorides, recognizing that the limits you set will be considerably less than 20 mg/L and that changes in human lifestyle will be needed for the chloride concentration to remain below these standards. What type of decisions would you be making? (There is no "correct" answer to this question.)

1–13   One of the alternatives to fossil-fueled power plants is to dam up rivers and use the water power to produce electricity. Suppose you are working for a power company and the decision has to be made to construct one of three new facilities: a large dam on a scenic river, a coal-fired power plant, or a nuclear power plant. Suppose the three plants would all produce about the same amount of electricity, but the nuclear plant is the most expensive, followed by the dam, and finally the coal-fired plant.

What factors would go into your decision making, and what types of decisions would the management of the *power company* probably make to decide which plant to construct? Why?

1–14   A wastewater treatment plant for a city discharges its treated plant effluent into a stream, but the treatment is inadequate. You are in charge of the city's public works, and you hire a consulting engineer to assess the problem and to offer solutions. She estimates that expanding the capacity of the treatment plant to achieve the required effluent quality will be an expensive proposition. She figures that the city can meet the downstream water-quality standard by building a large holding basin for the plant effluent (discharge) and holding back the treated wastewater during dry weather (low river flow) and discharging only during high stream flows (rainy weather). The amount of organic pollution being discharged would remain the same, of course, but now the stream standards would be met, the river water quality would be acceptable for aquatic life, and the city would be off the hook.

Your consulting engineer did some calculations before making her recommendation, of course. The cost of increasing the plant capacity is $1.5 million and the cost of the holding basin is $1.8 million. The annual operating cost of the treatment plant expansion is $400,000 and the operating cost of the holding pond is $100,000 per year. The city can borrow money at 6% interest. The expected life of the treatment plant expansion is 10 years and the expected life of the holding pond is 20 years.

a.   On the basis of economics alone, is the engineer right?

b.   Would you recommend this solution to the city council? What type of decisions would you make to determine your final recommendation?

1–15   A farmer who has a well for his irrigation water hires you as a consultant and asks you if he can withdraw more water than he is presently withdrawing. You respond honestly that yes he can, but that the groundwater reserves will be depleted and that his neighbor's wells may go dry. He asks you if it would be possible for anyone to know that he is withdrawing more water than the maximum rate, and you again tell him honestly that it is unlikely that anyone will know. He then tells you that he plans to double the withdrawal rate, run as long as he can, and when the water gives out, abandon the farm, buy a house trailer and move to Florida. You appear to be the only person who knows of the farmer's plans.

Driving home, you reach a decision on what you will do. What kind of decision did you make, and how did you make it?

1–16[26]   Consider the lifetime health risk of eating 100 g of apples contaminated with 1 ppb (part per billion) heptachlor (a pesticide and probable human carcinogen). The potency factor for heptachlor has been estimated by the EPA as 3.4 $(mg/kg/day)^{-1}$.

a. On the basis of 70-kg adults who eat one apple per day for 70 years, roughly estimate the annual number of additional (heptachlor-related) cancers one might expect in a population of 100,000.

b. Aside from the obvious consumption, lifetime, and body weight assumptions, briefly discuss at least two other assumptions that underlie the estimate calculated in part (a).

1–17 Engineer Diane works for a large international consulting firm that has been retained by a federal agency to assist in the construction of a natural gas pipeline in Arizona. Her job is to lay out the centerline of the pipeline according to the plans developed in Washington.

After a few weeks on the job, she is approached by the leaders of a local Navajo village and told that it seems that the gas line will traverse an ancient sacred Navajo burial ground. She looks on the map and explains to the Navajo leaders that the initial land survey did not identify any such burial grounds.

"Yes, although the burial ground has not been used recently, our people believe that in ancient times this was a burial ground, even though we cannot prove it. What matters is not that we can show by archeological digs that this was indeed a burial ground, but rather that the people believe that it was. We therefore would like to change the alignment of the pipeline to avoid the mountain."

"I can't do that by myself. I have to get approval from Washington. And whatever is done would cost a great deal of money. I suggest that you not pursue this further," replies Diane.

"We already have talked to the people in Washington, and as you say, they insist that in the absence of archeological proof, they cannot accept the presence of the burial ground. Yet to our people the land is sacred. We would like you to try one more time to divert the pipeline."

"But the pipeline will be buried. Once construction is complete, the vegetation will be restored, and you'll never know the pipeline is there," suggests Diane.

"Oh, yes. We will know it is there. And our ancestors will know it is there."

The next day Diane is on the telephone with Tom, her boss in Washington and she tells him about the Navajo visit.

"Ignore them," advises Tom.

"I can't ignore them. They truly *feel* violated by the pipeline on their sacred land," replies Diane.

"If you are going to be so sensitive to every whim and wish of every pressure group, maybe you shouldn't be on this job," suggests Tom.

What responsibility do engineers have for public attitudes? Should they take into account the feelings of the public, or should they depend only on hard data and quantitative information? Should Diane simply tell the Navajo leaders that she has called Wellington and that they are sorry, but the alignment cannot be changed? If she believes that the Navajo people have been wronged, what courses of action does she have? How far should she stick her neck out?

Respond by writing a two-page paper analyzing the ethics of this case.

1-18   The selection of what scientists choose to study is the subject of a masterful essay by Leo Tolstoy,[27] "The Superstitions of Science," first published in 1898. He recounts how

> . . . a simple and sensible working man holds in the old-fashioned and sensible way that if people who study during their whole lives, and, in return for the food and support he gives them, think for him, then these thinkers are probably occupied with what is necessary for people, and he expects from science a solution to those questions on which his well-being and the well-being of all people depends. He expects that science will teach him how to live, how to act towards members of his family, towards his neighbors, towards foreigners; how to battle with his passions, in what he should or should not believe, and much more. And what does our science tell him concerning all these questions?
>
> It majestically informs him how many million miles the sun is from the earth, how many millions of ethereal vibrations in a second constitute light, how many vibrations in the air make sound; it tells him of the chemical constitution of the Milky Way, of the new element helium, of microorganisms and their waste tissue, of the points in the hand in which electricity is concentrated, of X-rays, and the like. But, protests the working man, I need to know today, in this generation, the answers to how to live.
>
> Stupid and uneducated fellow, science replies, he does not understand that science serves not utility, but science. Science studies what presents itself for study, and cannot select subjects for study. Science studies everything. This is the character of science.
>
> And men of science are really convinced that this quality of occupying itself with trifles, and neglecting what is more real and important, is a quality not of themselves, but of science; but the simple, sensible person begins to suspect that this quality belongs not to science, but to people who are inclined to occupy themselves with trifles, and to attribute to these trifles a high importance.[27]

In the formulation and planning of your career as an *engineer*, not a scientist, how would you answer Tolstoy? Respond with a two-page paper.

◆ Endnotes

1. This could be estimated as the cost of each household taking the refuse to the landfill. The money saved by not having to make the weekly trip, multiplied by the total number of trips, would be a benefit.
2. The story is credited to Garrett Hardin, "Tragedy of the Commons," *Science* v 162 December 1968.
3. D. MacLean (Ed.), *Values at Risk,* Rowman and Littlefield, Totowa, NJ (1986).
4. J. M. Petulla, *American Environmentalism,* Texas A & M Univ. Press, College Station, TX (1980).

5. Much of this discussion is based on A. S. Gunn and P. A. Vesilind, *Environmental Ethics for Engineers,* Lewis Publishers, Chelsea, MI (1986).

6. Immanuel Kant, "Duties Toward Animals and Spirits," in *Lecture on Ethics,* p. 240, as quoted by Mary Midgley, "Duties Concerning Islands," in R. Elliot and A. Gare (Eds.), *Environmental Philosophy,* Pennsylvania State University Press, State College, (1983).

7. William F. Baxter, *People or Penguins: The Case for Optimal Pollution,* Columbia University Press, New York (1974), p. 5.

8. A. Leopold, *A Sand County Almanac,* Oxford University Press, New York (1966).

9. Richard A. Watson, "Self-Consciousness and the Rights of Nonhuman Animals and Nature," *Environmental Ethics* 1:2:99–129 (1979).

10. H. J. McClosky, *Ecological Ethics and Politics,* Rowman and Littlefield, Totowa, NJ (1983), p. 29.

11. John Locke, *Two Treatises on Government,* 2nd ed., Peter Laslett (Ed.), Cambridge, England (1967).

12. Thomas Hobbes, *Leviathan,* Henry Morley (Ed.), London (1885).

13. In the new United States of America, the revolution proclaimed life, liberty, and *happiness*, a modification to get around the sticky problem of slaves as both men and property.

14. Jeremy Bentham, *An Introduction to the Principles of Morals and Legislation,* (1789). Reprinted by Hafner, New York (1948).

15. Peter Singer, *Practical Ethics,* Cambridge University Press, Cambridge, UK (1979), p. 50.

16. Albert Schweitzer, *Out of My Life and Thought: An Autobiography,* H. Holt, New York (1949).

17. Janna Thompson, "Preservation of Wilderness and the Good Life," in R. Elliot and A. Gare (Eds.), *Environmental Philosophy,* Pennsylvania State University Press, State College, (1983) p. 97.

18. Paul W. Taylor, *Respect for Nature: A Theory of Environmental Ethics,* Princeton University Press, Princeton, NJ (1986).

19. Val and Richard Routley, "Against the Inevitability of Human Chauvinism," Kenneth Goodpaster and Kenneth Sayre (Eds.), in *Ethics and the Problems of the 21st Century,* University of Notre Dame Press, Notre Dame, IN (1987).

20. Holmes Rolston III, *Environmental Ethics,* Temple University Press, Philadelphia, PA (1988).

21. Tom Reagan, "The Nature and Possibility of an Environmental Ethic," *Environmental Ethics* 3:1:31–32 (1981).

22. Arne Naess, "Basic Principles of Deep Ecology," in Bill Devall and George Sessions (Eds.), *Deep Ecology,* Pergrine Press, Salt Lake City, UT (1985).

23. Wendell Berry, "A Secular Pilgrimage," in Ian Barbour (Ed.), *Western Man and Environmental Ethics,* Addison-Wesley, Reading, MA (1973).

24. John Stewart Collis, *The Triumph of the Tree,* William Sloane Associates, New York (1954).

25. M. J. Grayson and T. R. Shepard, Jr., *Disaster Lobby,* Follett Publishing Co., Chicago, IL (1973).

26. Problem credited to William Ball, The Johns Hopkins University, Baltimore, MD.

27. Leo Tolstoy, "The Superstitions of Science," *The Arena* 20, 1898, reprinted in J. G. Burke (Ed.), *The New Technology and Human Values*, Wadsworth Publishing Co., Belmont, CA (1966).

2

ENGINEERING

CALCULATIONS

The ability to solve problems using engineering calculations represents the very essence of engineering. Although certainly not all engineering problems can be solved using numerical calculations (as ought to be clear from Chapter 1), such calculations are absolutely necessary for the development of technical solutions. Engineering calculations make it possible to describe the physical world in terms of units and dimensions that are understood by all those with whom communication takes place.

The first section of this chapter is devoted to a review of units and dimensions used in engineering. The second part describes some basic principles of performing some "back-of-the-envelope" calculations in the face of incomplete or unobtainable information, and the last section is about how to manage information.

2.1 ◆ Engineering Dimensions and Units

A *fundamental dimension* is a unique quantity that describes a basic characteristic, such as, force (F), mass (M), length (L), and time (T). *Derived dimensions* are calculated by an arithmetic manipulation of one or more fundamental dimensions. For example, velocity has the dimensions of length/time (L/T) and volume is (L^3).

Dimensions are descriptive but not numerical; they cannot describe *how much*, they simply describe *what*. *Units*, and the *values* of those units, are necessary to describe something quantitatively. For example, the length (L) dimension may be described in units as meters, yards, or fathoms. Adding the value, we have a complete description, such as 3 meters, 12.6 yards, or 600 fathoms.

Three systems of units are in common use: the SI system, the American engineering system, and the cgs system.

Developed in 1960 in an international agreement, the SI (for *Système International d'Unités*) is based on meter for length, second for time, kilogram for mass, and degrees Kelvin for temperature. Force is expressed in Newtons.

The tremendous advantage of the SI system over the older English (and now American) system is that it works on a decimal basis, with prefixes decreasing or increasing the units by powers of 10. Although the SI units are now used throughout the world, most contemporary American engineers still use the old system of feet, pounds (mass), and seconds, with force expressed as pounds (force).

To ensure facility and familiarity, both systems are used in this book.

2.1.1   Density

The *density* of a substance is defined as its mass divided by a unit volume, or

$$\rho = \frac{M}{V}$$

where ρ = density
     M = mass
     V = volume

In the SI system the base unit for density is kg/m³, whereas in the American engineering system density is commonly expressed as lb_M/ft^2 [lb_M = pounds (mass)].

Water in the SI system has a density of 1×10^3 kg/m³, which is equal to 1 g/cm³. In the American engineering system, water has a density of 62.4 lb_M/ft^3.

2.1.2   Concentration

The derived dimension *concentration* is usually expressed *gravimetrically* as the mass of a material A in a unit volume consisting of material A and some other materials B. The concentration of A in a mixture of A and B is

$$C_A = \frac{M_A}{V_A + V_B} \tag{2.1}$$

where C_A = concentration of A
M_A = mass of material A
V_A = volume of material A
V_B = volume of material B

In the SI system, the basic unit for concentration is kg/m³, although the most widely used concentration term in environmental engineering is milligrams per liter, written as mg/L.

e • x • a • m • p • l • e   **2.1**

Problem   Plastic beads with a volume of 0.04 m³ and a mass of 0.48 kg are placed in a container and 100 liters of water are poured into the container. What is the concentration of plastic beads in mg/L?

Solution   Using equation 2.1,

$$C_A = \frac{M_A}{V_A + V_B}$$

where A represents the beads and B represents the water.

$$C_A = \frac{0.48}{0.04 + (100 \times 10^{-3})}$$

$$C_A = 3.43 \text{ kg/m}^3 = 3.43 \frac{10^6 \text{ mg/kg}}{10^3 \text{ L/m}^3} = 3430 \text{ mg/L}$$

◆

Note that in the previous example the volume of water is added to the volume of the beads. If the plastic beads with a volume of 0.04 m³ are placed in a 100-

liter container and the container is filled to the brim with water, the total volume is

$$V_A + V_B = 100\,\text{L}$$

and the concentration of beads is

$$C_A = 4800\,\text{mg/L}.$$

The concentration of beads is higher since the total volume is lower.

Another measure of concentration is *parts per million* (ppm). This is numerically equivalent to mg/L if the fluid in question is water since one milliliter (mL) of water weighs one gram, that is, the density is 1.0 g/cm³.

Note that mg/L = ppm if ρ = 1.0 g/cm³ by the following conversion:

$$\frac{1\,\text{mg}}{\text{L}} = \frac{0.001\,\text{g}}{1000\,\text{mL}} = \frac{0.001\,\text{g}}{1000\,\text{cm}^3} = \frac{0.001\,\text{g}}{1000\,\text{g}} = \frac{1\,\text{g}}{1,000,000\,\text{g}}$$

or one gram in a million grams, or one ppm.

Some material concentrations are most conveniently expressed as percentages, usually in terms of mass.

$$\Phi_A = \frac{M_A}{M_A + M_B} \times 100$$

where Φ_A = percent of material A
   M_A = mass of material A
   M_B = mass of material B

Φ_A can, of course, also be expressed as a ratio of volumes.

e • x • a • m • p • l • e 2.2

Problem   A wastewater sludge has a solids concentration of 10,000 ppm. Express this in percent solids (mass basis), assuming that the density of the solids is 1 g/cm³.

Solution

$$10,000 = \frac{1 \times 10^4}{1 \times 10^6} = \frac{1}{100}$$

◆

The last example illustrates a useful relationship:

10,000 mg/L = 10,000 ppm (if density = 1) = 1% (by weight)

In air-pollution control, concentrations are generally expressed gravimetrically as mass of pollutant per volume of air at standard temperature and pressure. For example, the national air quality standard for sulfur dioxide is 0.05 μg/m³ (one microgram = 10^{-6} gram). Occasionally, air quality is expressed in ppm, and in this case, the calculations are in terms of volume/volume, or one ppm = 1 volume of a pollutant per 1×10^6 volumes of air. Conversion from mass/volume ($\mu g/m^3$) to volume/volume (ppm) requires knowledge of the molecular weight of the gas. At standard condition, 0°C and 1 atmosphere of pressure, one mole of a gas occupies a volume of 22.4 L. One mole is the amount of gas in grams numerically equal to its molecular weight. The conversion is therefore

$$\mu g/m^3 = \frac{1\ m^3\ \text{pollutant}}{10^6\ m^3\ \text{air}} \times \frac{\text{molecular weight (g/mol)}}{22.4 \times 10^{-3}\ m^3/\text{mol}} \times 10^6$$

or simplifying,

$$\mu g/m^3 = (\text{ppm} \times \text{molecular weight})/22.4 \quad \text{at 0°C and 1 atmosphere}$$

If the gas is at 25°C at one atmosphere, as is common in air quality standards calculations, the conversion is

$$\mu g/m^3 = (\text{ppm} \times \text{molecular weight})/24.45 \quad \text{at 25°C and 1 atmosphere}$$

2.1.3   Flow Rate

In engineering processes the *flow rate* can be either *gravimetric (mass) flow rate* or *volummetric (volume) flow rate*. The former is in kg/s, or lb_M/s, while the latter is expressed as m^3/s or ft^3/s. The mass and volummetric flow rates are not independent quantities, since the mass (M) of material passing a point in a flow line during a unit time is related to the volume (V) of that material as

$$[\text{mass}] = [\text{density}] \times [\text{volume}]$$

Thus a volummetric flow rate (Q_V) can be converted to a mass flow rate (Q_M) by multiplying by the density of the material.

$$Q_M = Q_V \rho \tag{2.2}$$

where Q_M = mass flow rate
$\quad\quad Q_V$ = volume flow rate
$\quad\quad \rho$ = density

The symbol Q is almost universally used to denote flow rate.

The relationship between mass flow, of some component A, concentration of A, and the total volume flow (A plus B) is

$$Q_{M_A} = C_A \times Q_{V_{(A+B)}} \tag{2.3}$$

Note that this is not the same as equation 2.2, which is applicable to only one material or one component in a flow stream. Equation 2.3 relates to two *different* materials or components in a flow. For example, a mass flow rate of plastic balls moving along and suspended in a stream is expressed as kg of these balls per second passing some point, which is equal to the concentration (kg balls/m³ total volume, balls plus water) times the stream flow (m³/s of balls plus water).

---

e • x • a • m • p • l • e 2.3

Problem A wastewater treatment plant discharges a flow of 1.5 m³/s (water plus solids) at a solids concentration of 20 mg/L (20 mg solids per liter of flow, solids plus water). How much solids is the plant discharging each day?

Solution Using equation 2.2,

[mass flow] = [concentration] × [volume flow]

$$Q_{M_A} = C_A \times Q_{V_{(A+B)}}$$

$$= \left[20\,\frac{mg}{L} \times \frac{1 \times 10^{-6}\,kg}{mg} \right] \times \left[1.5\,\frac{m^3}{s} \times \frac{10^3\,L}{m^3} \times 86{,}400\,\frac{s}{day} \right]$$

$$= 2592\ kg/day$$

♦

---

e • x • a • m • p • l • e 2.4

Problem A wastewater treatment plant discharges a flow 34.2 mgd (million gallons per day) at a solids concentration of 0.002% solids (by weight). How many pounds per day of solids does it discharge?

Solution Using equation 2.2

[mass flow] = [concentration] × [volume flow]

$$Q_{M_A} = C_A \times Q_{V_{(A+B)}}$$

Note first that 0.002% = 20 mg/L (assuming again that $\rho = 1.0$), and the stated volume flow rate includes solids plus water.

$$Q_{M_A} = \left[20\,\frac{mg}{L} \times 3.79\,\frac{L}{gal} \times 2.2 \times 10^{-6}\,\frac{lb}{mg} \right] \times \left[34.2 \times 10^6\,\frac{gal}{day} \right]$$

$$= 5700\,\frac{lb}{day}$$

♦

Note that:

$$3.79\,\frac{L}{gal} \times 2.2 \times 10^{-6}\,\frac{lb_M}{mg} \times 10^6\,\frac{gal}{million\ gal} = 8.34 \left[\frac{L}{million\ gal} \right] \left[\frac{lb_M}{mg} \right]$$

The factor, 8.34, is useful in conversions where the flow rate is given in million gallons per day and the concentrations in mg/L, and the discharge is in pounds per day, so that

$$\begin{bmatrix} \text{mass flow} \\ \text{rate in} \\ \text{pounds per day} \end{bmatrix} = \begin{bmatrix} \text{volume flow} \\ \text{rate in million} \\ \text{gallons per day} \end{bmatrix} \times \begin{bmatrix} \text{concentration} \\ \text{in mg per L} \end{bmatrix} \times 8.34 \quad (2.4)$$

e • x • a • m • p • l • e 2.5

Problem   A drinking-water treatment plant adds fluorine at a concentration of 1 mg/L and the average daily water demand is 18 million gallons. How much fluorine must the community purchase?

Solution   Using equation 2.4

$$18\,\frac{\text{million gallons}}{\text{day}} \times 1\,\frac{\text{mg}}{\text{L}} \times 8.34 \left[\frac{\text{L}}{\text{million gallons}}\right]\left[\frac{\text{lb}}{\text{mg}}\right] = 150\,\frac{\text{lb}}{\text{day}}$$

2.1.4   Residence Time

One of the most important concepts in treatment processes is *residence time* (sometimes called *detention time* or even *retention time*), defined as the time an average particle of the fluid spends in the container through which the fluid flows. An alternate definition is the time it takes to fill the container.

Mathematically, if the volume of a container such as a large holding tank, is V (m³), and the flow rate into the tank is Q (m³/sec), then the residence time is

$$\bar{t} = V/Q \qquad (2.5)$$

in seconds. The average residence time can be increased by reducing the flow rate Q or increasing the volume V, and decreased by doing the opposite.

e • x • a • m • p • l • e 2.6

Problem   A lagoon has a volume of 1500 m³ and the flow into the lagoon is 3 m³/hour. What is the residence time in this lagoon?

Solution

$$\bar{t} = (1500)/3 = 500 \text{ hours}$$

2.2 ◆ Approximations in Engineering Calculations

Engineers are often called on to provide information not in its exact form, but as approximations. For example, an engineer may be asked by a client such as a city manager what it might cost to build a new wastewater treatment plant for the community. The manager is not asking for an exact figure, but a "ball park" estimate. Obviously, the engineer cannot in a few minutes conduct a thorough cost estimate. She would recognize the highly variable nature of land costs, construction costs, required treatment efficiency, and the like. Yet the manager wants a preliminary estimate—a *number*—and quickly!

In the face of such problems the engineer has to draw on whatever information might be available. For example, she might know that the population of the community to be served is approximately 100,000. Next she estimates, based on experience, that the domestic wastewater flow might be about 100 gallons per person per day, thus requiring a plant of about 10 million gallons per day (mgd) capacity. With room for expansion, industrial effluents, storm inflow, and infiltration of groundwater into the sewers, she may estimate that a 15-mgd capacity may be adequate.

Next she evaluates the potential treatment necessary. Knowing that the available watercourses for discharging the effluent are all small streams that may dry up during droughts, a high degree of treatment is required. She figures that nutrient removal will be needed. Such treatment plants, she is aware, cost about $3 million per million gallons of influent to construct. She calculates that the plant would cost about $45 million. Giving herself a cushion, she could respond by saying, "about 50 million dollars."

This is exactly the type of information the manager seeks. He has no use for anything more accurate since he might be trying to decide whether to ask for a bond issue of around $100 million or $200 million. There is time enough for more exact calculations.

2.2.1 Procedure for Calculations with Approximations

Problems not requiring exact solutions can be solved by

1. Carefully defining the problem.
2. Introducing simplifying assumptions.
3. Calculating an answer.
4. Checking the answer, both systematically and realistically.

Defining the Problem.   The engineer in the previous case is asked for an estimate, not an exact figure. She recognizes that the use of this figure would be for preliminary planning purposes, and thus valid approximations are adequate. She also recognizes that the manager wants a *dollar* figure answer, thus establishing the units.

Simplifying.   This step is perhaps the most exciting and challenging of the entire process, because intuition and judgment play an important role. For example, the engineer has to first estimate the population served, then consider the average flow. What does she ignore? Obviously a great deal, such as daily transient flows, variability in living standards, seasonal variations, and so on. A thorough estimate of potential wastewater flows requires a major study. She has to simplify her problem and choose to consider only an estimate of the population and an average per capita discharge.

Calculating.   In the case of this problem, the calculations are straight-forward.

Checking.   More important is the process of checking. There are two kinds of checks: systematic and realistic. In systematic checking the units are first checked to see if they make sense. For example,

$$[\text{persons}] \times \left[\frac{\text{gallons}}{\text{persons}}\right] = [\text{gallons}]$$

makes sense, whereas

$$\frac{\text{persons}}{[\text{gallons/person}]} = \frac{[\text{persons}]^2}{\text{gallons}}$$

is nonsense. If the units check out, the numbers can be recalculated to check for mistakes.

Finally, a reality check is necessary. Possibly no practicing engineer will explicitly recognize that he or she performs reality checks day in and day out, but such checks are central to good engineering. Consider, for example, if the engineer had made a mistake and thought (erroneously) that a wastewater treatment plant of the type needed by the community costs $3000 per million gallons of influent. Her calculations would have checked, but her answer would have been $50,000 instead of $50 million. Such an answer should have immediately been considered ludicrous, and a search for the error would have been initiated. Reality checks, when routinely performed, will save considerable pain and embarrassment.

2.2.2   Use of Significant Figures

Finally, note that significant figures in the answer reflect the accuracy of the data and the assumptions. Consider how silly it would sound to say that the treatment plant would cost *about* $5,028,467.19. Many problems require an-swers to only one significant figure, or even to an order of magnitude. Nonsig-nificant figures tend to accumulate in the course of calculations, like mud on a boot, and must be wiped off at the end.[1]

Suppose you are asked to estimate the linear feet of fence posts needed for a pasture and are told that there will be 87 posts with an average height of 46.3 inches. You multiply and get 335.675 ft. Now it is time to scrape the mud off since the most accurate of your numbers only has three significant figures while your answer has six significant figures. So you report 336 ft. Or more likely you say 340 ft, recognizing that it is better to err slightly on the high side than to run out of fence posts.

Significant figures are those that transfer information based on the value of that digit. Zeros that hold place are not significant since they can be eliminated without loss of information. For example, in the number 0.0036 the two zeros are only holding a place and can be eliminated by writing $0.0036 = 3.6 \times 10^{-3}$.

Zeros at the end of a number are a problem, however. Suppose the newspaper reports that there were 46,270 fans attending a football game. The last digit (zero) may be significant if every person was counted, and indeed there were *exactly* 46,270 fans in the stadium. If, however, one were to estimate the number of people as 46,000 fans, then the three zeros are simply holding a place, and are not significant. To avoid confusion, it is useful to say "about" or "approximately" if that is what is meant.

e • x • a • m • p • l • e   **2.7**

Problem   A community of approximately 100,000 people has about 5 acres of landfill left that can be filled to about 30 feet deep with refuse compacted to somewhere between 600 and 800 lb_M per cubic yard. What is the remaining life of the landfill?

Solution   Clearly the answer does not require high precision. The definition of the problem requires an answer in time units.

There is no need to consider commercial or industrial wastes. Estimate only refuse generated by individual households.

All the necessary data are available except the per capita production of refuse. Suppose a family of four fills up three garbage cans per week. If each can is about 8 cubic feet, and if we assume the uncompacted garbage is at about one-fourth of the compacted density, say at 200 lb_M/yd^3, it is reasonable to calculate the per capita production as

$$8 \, \frac{ft^3}{can} \times \frac{1 \, yd^3}{27 \, ft^3} \times \frac{3 \, cans}{4 \, persons} \times 200 \, \frac{lb_M}{yd^3} \times \frac{1 \, week}{7 \, days} = 6.3 \, \frac{lb}{persons/day}$$

If there are about 100,000 persons, the city produces

$$6.3 \, \frac{lb}{persons/day} \times 100,000 \, persons \times \frac{1 \, yd^3}{700 \, lb_M} = 900 \, yd^3/day$$

The total available volume is

$$5 \, acres \times 43,560 \, \frac{ft^2}{acre} \times 30 \, ft \times \frac{1 \, yd^3}{27 \, ft^3} = 242,000 \, yd^3$$

Thus the expected life is

$$\frac{242,000 \, yd^3}{900 \, yd^3/day} = 268 \, days, \text{ or say 270 days}$$

Is this reasonable? Pretty much so. The calculations overestimated refuse production (the actual per capita production is more like 4 lb/capita/day) so that in actuality the landfill may last a year. But considering the extreme difficulties of siting additional landfills, the town is clearly already in a crisis situation.

◆

In professional engineering it is necessary to carry around in one's head a suitcase full of numbers and approximations. For example, most people would know that a meter is a few inches longer than a yard. We may not know *exactly* how much longer, but we could make a pretty fair guess. Similarly, we know what 100 yards look like (from goal line to goal line). In a similar way, an environmental engineer in practice knows instinctively what a flow of 10 million gallons per day looks like, because he or she has been working in plants that received that magnitude of flow. Such knowledge becomes second nature and is often the reason why engineers can avoid stupid and embarrassing mistakes. A "feel" for units is a part of engineering and is the reason why a change of units from million gallons per day to cubic meters per second is so difficult for American engineers. It would be an unusual American engineer who would know what a flow of 10 cubic meters per second looks like (without doing some quick mental approximate conversions!).

2.3 ◆ Information Analysis

Not only do engineers seldom know anything accurately, but they also seldom know anything for sure! Although there may never before have been a flood in August, as soon as a million dollar construction project depends on dry weather, it'll rain for 40 days and nights. Murphy's Law—"if anything can go wrong, it will"—and its corollary—"at the worst possible time"—is as central to engineering as are Newton's laws of motion. Engineers take chances; they place probabilities on events of interest and decide on prudent actions. They know that a flood might occur in August, but if it hasn't in the last 100 years, chances are fairly good it won't next year. But what if the engineer has only 5 years of data? How much risk is he or she taking? How sure can he or she be that a flood won't occur next year?

The concept illustrated in the previous example is called *probability*, and probability calculations are central to many engineering decisions. Related to probability is the analysis of incomplete data using *statistics*. A piece of data (e.g., stream flow for one day) is valuable in itself, but when combined with hundreds of other daily stream flow data points, the information becomes even more useful, but only if it can be somehow manipulated and reduced. For example, if it is decided to impound water from a stream for a water supply, the individual daily flow rates would be averaged (a statistic) and this would be one useful number in deciding just how much water would be available.

In this section the central ideas of probability and statistics are introduced first, and some useful tools engineers have available in the analysis of information are described. Neither probability nor statistics is developed from basic principles here, and a proper statistics and probability course is highly recommended. In this introduction only enough material is presented to allow for the solution of simple environmental engineering problems involving frequency distributions.

Much of engineering statistical analysis is based on the "bell-shaped curve," as shown in Figure 2-1. The assumption is that a data set (such as annual rainfall for example) can be described by a bell-shaped curve and that the location of the curve can be defined by the *mean* or average and some measure of how widely the data are distributed, or the spread of the curve. An ideal form of the function describing the bell-shaped curve is called the *Gaussian distribution* or the *normal distribution*, expressed as

$$P_{\sigma(x)} = \frac{1}{\sqrt{2\pi}\,\sigma} \exp\left[-\frac{1}{2}\left(\frac{x - \mu}{\sigma}\right) \right]^2 \qquad (2.6)$$

where μ = mean, estimated as $\bar{x}$

$\bar{x}$ = observed sample mean = $\dfrac{\Sigma\,x}{n}$

σ = standard deviation, estimated as s

s = observed standard deviation = $\left[\dfrac{\Sigma\,x_i^2}{n-1} - \dfrac{(\Sigma\,x_i)^2}{n(n-1)} \right]^{1/2}$

n = sample size

The standard deviation is a measure of the curve's spread, defined as that value of x where 68.3% of all values of x fall within plus and minus the mean, μ. A small standard deviation indicates that all the data are closely bunched—there is little variability in the data. In contrast, a large standard deviation indicates that the data are widely spread.

A further useful statistic in engineering work is called *coefficient of variation*, defined as

$$C_v = \frac{\mu}{\sigma}$$

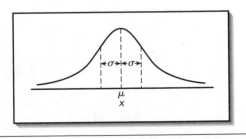

FIGURE 2-1   Bell-shaped curve suggesting normal distribution.

estimated in practice as

$$C_v = \frac{\bar{x}}{s}$$

where s = estimate of the standard deviation (equation 2.6)
    $\bar{x}$ = estimate of the mean

Engineers often plot data described by a bell-shaped curve as a *cumulative* function, where the ordinate (vertical axis) is the cumulative fraction of observations. These curves are commonly used in hydrology, soils engineering, resource recovery, and many other applications. The construction of a cumulative curve is shown in Example 2.8.

e • x • a • m • p • l • e **2.8**

Problem   One hundred kg of glass is recovered from municipal refuse and processed in a resource-recovery plant. The glass is crushed and run through a series of sieves, as shown below. Plot the cumulative distribution of particle sizes.

4-mm holes	10 kg glass remained on this sieve (90 kg went through)
3-mm holes	25 kg glass remained on this sieve
2-mm holes	35 kg glass remained on this sieve
1.0-mm holes	20 kg glass remained on this sieve
No holes	10 kg glass went all the way through to here

Solution   These data can be plotted as shown in Figure 2–2 and a rough approximation of a normal distribution is noted. The data can then be tabulated as a cumulative function.

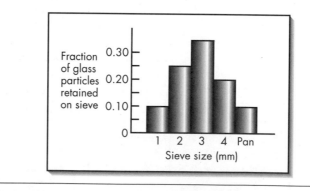

FIGURE 2–2 Histogram showing sizes of glass particles.

Sieve Size (mm)	Fraction of Glass Retained on Sieves	Fraction of Glass Smaller Than this Sieve Size
4	$\frac{10}{100} = 0.10$	$1.0 - 0.1 = 0.90$
3	$\frac{25}{100} = 0.25$	$1.0 - (0.1 + 0.25) = 0.65$
2	$\frac{35}{100} = 0.35$	$1.0 - (0.35 + 0.35) = 0.30$
1	$\frac{20}{100} = 0.20$	$1.0 - (0.70 + 0.20) = 0.10$
< 1.0	$\frac{10}{100} = 0.10$	—

These data are plotted in Figure 2–3. Note that the last point cannot be plotted since the sieve size is less than 1.0 mm, and the total fraction has to add up to 1.00. The result is a classical "S-curve."

◆

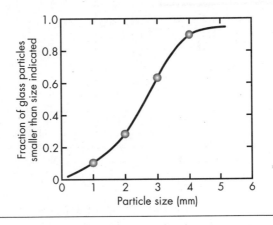

FIGURE 2–3 Cumulative glass particles size distribution. See Example 2.8.

Figure 2–3 is plotted with the *independent variable* as the abscissa (*x*-axis) and the *dependent variable* as the ordinate (*y*-axis). A variable is independent if its value is chosen, such as the size of a sieve. The value of a dependent variable is obtained by experimentation. With very few exceptions, engineering graphs are drawn using the convention of the independent variable plotted as the *x*-axis and the dependent variable as the *y*-axis.

Often it is simpler to work with data if the classical S-curve shown in Figure 2–3 can be plotted as a straight line. This often occurs on arithmetic probability paper, originally invented by Alan Hazen, a prominent sanitary engineer. The same data plotted on probability paper are shown in Figure 2–4.

A straight line on probability paper implies that the data are *normally distributed* and that statistics such as the mean and standard deviation can be read off the plot. The mean is at 0.5, or 50%, while the standard deviation is 0.335 on either side of the mean.

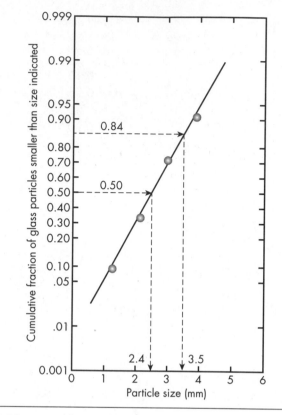

FIGURE 2-4   Glass particle size distribution on an arithmetic probability plot. See Example 2.9.

The standard deviation can also be approximated as

$$s \approx \frac{2}{5}(x_{90} - x_{10}) \tag{2.7}$$

where x_{90} and x_{10} are the abscissa values at P_{90} and P_{10}, respectively.

Such a plot makes it convenient to estimate the value of any other interval as well. For example, if it is necessary to estimate the values of x such that 95% of the values fall within these bounds (the "95% confidence interval"), it is possible to read the values at 2.5% and 97.5%. Enclosed within these bounds should be 95% of the data.

e • x • a • m • p • l • e   **2.9**

Problem   Given the plot in Figure 2-4, what are the (a) mean, (b) standard deviation, and (c) 95% interval?

Solution   The mean is read at 0.5 (cumulative fraction of glass particles smaller than the size indicated) as 2.4 mm. The standard deviation is the spread so that

68% of the observations lie on either side of the mean (or about 34% on either side). From Figure 2-4, the estimated standard deviation is calculated as 0.50 + 0.34 = 0.84, which corresponds to 3.5 mm, so s = 3.5 − 2.4 = 1.1 mm. Alternatively, using equation 2.7,

$$s \approx \frac{2}{5}(3.9 - 1.0) = 1.16$$

where 3.9 is the sieve size passing 90% of the glass and 1.0 is the size passing 10% of the glass. The 95% interval means that 95% of the data can be expected to be in this range, or between 0.025 and 0.975, or 0.2 mm to 4.8 mm.

◆

Note that the ordinate in probability paper does not have a 0% or 100%, just as the normal distribution curve has no zero of 1.0.

The data in Example 2.9 can be *normalized* by dividing each dependent data point by a constant, the total weight of the glass being sieved. Thus each data point is converted from a number with dimensions (kg) to a dimensionless fraction (kg/kg). This same trick is sometimes performed on the independent variable when time frequencies are required.

It is often necessary to calculate the probabilities that certain events will recur. Hydrologists express this in terms of the *return period*, or how often the event is expected to recur. If the annual probability of an event occurring is 5%, then the event can be expected to recur once in 20 years, or have a return period of 20 years. Stated in equation form:

Return period = (1/fractional probability)

When time-variant data are analyzed by a frequency distribution, the data must first be ranked (smallest to largest, or largest to smallest). The probabilities are then calculated, the data plotted, and the plot used to determine the desired statistics.

e • x • a • m • p • l • e   2.10[2]

Problem   A quality of a municipal wastewater treatment plant effluent (discharge) is measured daily by its biochemical oxygen demand (BOD). Because of variations in influent (flow coming into the plant) as well as the treatment plant dynamics, the effluent quality can be expected to vary. Does this variation fit the normal distribution, and if so, can this be used to calculate the mean and the standard deviation? What is the worst effluent quality to be expected once a month (30 days)? The data are:

Sample Day	Effluent BOD	Sample Day	Effluent BOD
1	56	10	35
2	63	11	65
3	57	12	35
4	33	13	31
5	21	14	21
6	17	15	21
7	25	16	28
8	49	17	41
9	21	18	36

Solution   The first step is to rank the data. The total number of data points is $n = 18$. The rankings (m) are from the lowest BOD (17 mg/L) to the highest (65 mg/L).

Rank (m)	Effluent BOD (mg/L)	m/n
1	17	0.055
2	21	0.111
3	21	0.167
4	21	0.222
5	21	0.278
6	25	0.333
7	28	0.389
8	31	0.444
9	33	0.500
10	35	0.556
11	35	0.611
12	36	0.667
13	41	0.722
14	49	0.778
15	56	0.833
16	57	0.889
17	63	0.944
18	65	1.00

The last column can be thought of as the fraction of time the BOD can be expected to be less than this value. That is, 0.055 or 5.5% of the time the BOD should be equal to or less than 17 mg/L.

These plot as a reasonably straight line on a probability plot as shown in Figure 2-5. The estimated mean is read off the plot at (m/n) = 0.5 as 35 mg/L BOD and the standard deviation as 10 mg/L. The worst effluent quality expected every month is found by calculating the fraction of time as 29/30 = 0.967 (that is, for 29 days out of 30 the BOD is less). Enter the plot at fraction of time = 0.967, and read the BOD = 67 mg/L.

◆

Note that the last data point (rank No. 18) cannot be plotted, since all of the readings are less than this value. This problem can be avoided by dividing

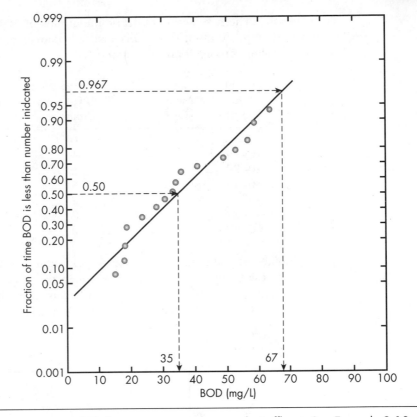

FIGURE 2–5 BOD of a wastewater treatment plant effluent. See Example 2.10.

all ranks (m) by $n + 1$ instead of n. As the number of data points increase (n gets bigger) this becomes less and less of a problem.

Another method of handling data, especially if large numbers are involved, is to group them. In this case, the mean of the group becomes the variable plotted, and the probability is calculated as

$$P = \frac{\Sigma r}{n}$$

where r = sum of the number of data points equal to or greater than the value indicated
n = total data points

The probability of an event occurring is again read from the graph.

e • x • a • m • p • l • e   **2.11**

Problem   Using the data from Example 2.10, estimate the highest expected BOD to occur once every 30 days using grouped data analysis.

Solution   The first step is to define the groups of BOD values. In this case a 10-mg/L step is chosen simply because it is convenient. The *number* of data points falling into each group is noted (r).

Effluent BOD (mg/L)	Mean of Group	Number of Days BOD Falls into Group, r	Σr	$\Sigma r/n$
0–9	5	0	0	0
10–19	15	1	1	0.06
20–29	25	6	7	0.39
30–39	35	5	12	0.67
40–49	45	2	14	0.78
50–59	55	2	16	0.89
60–69	65	2	18	1.00

The number of data points are progressively summed (Σr) and this value is divided by the total data points, $n = 18$. These now become single points on a graph of mean BOD group vs. $\Sigma r/n$, as in Figure 2–6. The probabilities can be read from the curve as before. As in Example 2.11, if we want to know the

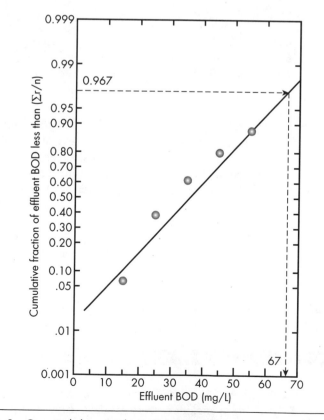

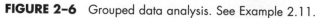

FIGURE 2–6   Grouped data analysis. See Example 2.11.

worst effluent quality in 30 days, $P_{29/30} = 0.967$, we read 67 mg/L BOD off the curve.

◆

Abbreviations

C	= concentration (mass/volume)	r	= number of data points in a group
C_V	= coefficient of variation	s	= estimate of standard deviation
M	= mass	$\bar{t}$	= residence time
m	= rank of an event	V	= volume
n	= number of events	μ	= mean
P	= probability	$\bar{x}$	= estimate of mean
Q	= flow rate	σ	= standard deviation
Q_M	= mass flow rate	ρ	= density
Q_V	= volummetric flow rate	Φ	= concentration (mass/mass)

Problems

2–1 The following table shows the frequency distribution of influent biochemical oxygen demand (BOD) in a wastewater treatment plant:

BOD (mg/L)	No. of Samples
110–129	18
130–149	31
150–169	23
170–189	20
190–209	7
210–229	1
230–249	2

Compute the following:

a. P_{10} (probability that 10% of the time the BOD will be less than this value)

b. P_{50}

c. P_{95}

2–2 The following are characteristics of annual precipitation for various cities.

	Years of Record	Mean Annual Rainfall (in./yr)	Standard Deviation (in./yr)
Cheyenne, WY	70	14.61	3.61
Pueblo, CO	52	11.51	5.29
Kansas City, MO	63	36.10	6.64

a. Which city has most variable rainfall? Why?

b. If we assume that rainfall data are normally distributed, how many years in the next 50 would we expect rainfall to be above 20 inches in each city?

c. Assuming a normal distribution, how many years in the next 50 would we expect rainfall to be below 12 inches in each city?

d. What is the probability of the annual rainfall exceeding 36.1 inches in Kansas City?

Use arithmetic probability paper for this problem.

2-3   Find the won/lost records for your college basketball team for the past 10 years. Calculate the winning percentage for each year. How often can the team be expected to have an equal number of wins and losses (play at least 500 ball)?

2-4   A normal distribution has a standard deviation of 30 and a mean of 20. Find the probability that

a. $x > 80$ b. $x < 80$

c. $x < -70$ d. $50 < x < 80$

Use arithmetic probability paper for this problem.

2-5   A lake has the following dissolved oxygen (DO) readings:

Month	DO (mg/L)
1	12
2	11
3	10
4	9
5	10
6	8
7	9
8	6
9	10
10	11
11	10
12	11

a. What fraction of time can the DO be expected to be above 10 mg/L?

b. What fraction of time will the DO be less than 8 mg/L?

c. What is the average DO?

d. What is the standard deviation?

2-6   The return periods and flood levels for Morgan Creek are shown below:

Return Period (yr)	Flood Level (m^3/sec)
5	142
10	157
20	176
40	196
60	208
100	224

Estimate the flood level for a return period of 500 years.

2-7 The number of students taking a course for the past five years is as follows:

Year	No. of Students
1	13
2	18
3	17
4	15
5	20

How many years in the next 20 years should the class be 10 students or smaller?

2-8 How much beer is consumed in the United States every day? (There are about 2.5×10^8 people in the United States. Assume everything else.)

2-9 Coal has about 2% sulfur. How many kg of sulfur would be emitted from coal-fired power plants in the United States in one year? (U.S. electricity production is about 0.28×10^{12} watts, and coal has an energy content of about 30×10^6 J/kg. Note that 1 J = 1 watt-second.)

2-10 How many automobile tires are sold annually in the United States? Make reasonable assumptions.

2-11 One gram of table salt is added to an 8-ounce glass and the glass is filled up with tap water. What is the concentration of salt in mg/L? Why might you not be so sure about your answer?

2-12 Metal concentrations in wastewater sludges are often expressed in terms of grams of metal per kg of total dry solids. A wet sludge has a solids concentration of 200,000 mg/L and 8000 mg/L of these solids is zinc.

a. What is the concentration of zinc as (g Zn/g dry solids)?

b. If this sludge is to be applied to farmland and spread on pastures used by cows, explain how you might monitor the project with regard to the potential environmental or health effects caused by the presence of the zinc.

2-13 One gram of pepper is placed in a 100-mL beaker and the beaker is filled up with water to the 100-mL mark. What is the concentration of pepper in mg/L? (What should you assume in this problem? What is the density of pepper? What is the volume of one gram of pepper?)

2-14 A wastewater treatment plant receives 10 million gallons per day of flow. This wastewater has a solids concentration of 192 mg/L. How many pounds of solids enter the plant every day?

2-15 Ten grams of plastic beads with a density of 1.2 g/cm³ are added to 500 mL of an organic solvent with a density of 0.8 g/cm³. What is the concentration of the plastic beads in mg/L? (This is a curve ball!)

2-16 A stream flowing at 60 gal/min carries a sediment load of 2000 mg/L. What is the sediment load in kg/day?

2-17 A water treatment plant produces a water that has an arsenic concentration of 0.05 mg/L.

a. If the average consumption of water is 2 gallons of water each day, how much arsenic is each person drinking?

b. If you are the water treatment plant operator and the chairman of the town council asks you during a public hearing whether or not there is arsenic in the town's drinking water, how do you respond?

c. Analyze your answer from a moral standpoint. Did you do the ''right thing'' by answering as you did? Why do you think so?

2–18 A power plant emits 120 pounds of flyash per hour up the stack. The flow rate of the hot gases in the stack is 25 cubic feet per second. What is the concentration of the flyash in $\mu g/m^3$?

2–19 If electricity costs \$0.05/kilowatt hour, how much does your university pay annually on its power bill due to the use of desk lamps? (You need to make several assumptions here specific to your university.)

2–20 If tuition were to be paid in the form of tickets to lectures, how much would each ticket to one lecture cost at your university?

2–21 A university has about 12,600 undergraduate students, and graduates 2250 students each year. Supposing everyone enrolled at the university graduates eventually, what is the approximate residence time at the university?

◆ **FIGURE 2–7** Example of sports statistics. See Problem 2–23. (Shoe by Fred McNelly. Copyright Tribune Media Services, Inc. Used with permission.)

2-22   A water treatment plant has six settling tanks that operate in parallel (the flow gets split into six equal flow streams), and each tank has a volume of 40 m^3. If the flow to the plant is 10 million gallons per day, what is the residence time in each of the settling tanks? If instead, the tanks operated in series (the entire flow goes first through one tank, then the second, and so on), what would be the residence time in each tank?

2-23   Some of the most confusing statistics are in sports (see, for example, Figure 2-7). Define the following:

a. ERA
b. RBI
c. slugging average
d. average hang time
e. on-base percentage
f. average points per game (ice hockey)
g. others?

2-24   You are working as a recruiter for your consulting engineering firm, and an applicant asks you the following question: "If I work with your firm and I find that I object, on ethical grounds, to a project the firm has been hired to do, can I request not to work on that project, without penalizing my future career with the firm?"

   Give three responses you might make to such a question, and then respond to these statements according to what you believe the student interviewee would think. Would he or she believe you? Why or why not?

◆ Endnotes

1. As observed by John Hart in a wonderful book on environmental calculations, *Consider a Spherical Cow*, William Kaufmann, Inc., Los Altos, CA (1985).
2. This example is adapted from E. Joe Middlebrooks, *Statistical Calculations,* Ann Arbor Science Publishers, Ann Arbor, MI (1976).

3

MATERIALS BALANCES & SEPARATIONS

The alchemists, perhaps the original practitioners of "science for profit," tried to develop processes for making gold out of less expensive metals. Viewed from the vantage point of modern chemistry, it is clear why they failed. Except for processes involving nuclear reactions, with which we are not concerned in this text, a pound of any materials such as lead in the beginning of any process will yield a pound of lead in the end, although perhaps in a very different form. This simple concept of the conservation of mass leads to a powerful engineering tool, the *materials balance*. In this chapter, the materials balance around a "black box" unit operation is introduced first. Then these black boxes are identified as actual unit operations that perform useful functions. Initially, these black boxes have nothing going on inside them that affects the materials flow. Then it is presumed that material quantities are produced or consumed within the box. In all cases, the flow is assumed to be at steady state, that is, not changing with time. This constraint is lifted in subsequent chapters.

3.1 ◆ Materials Balances with a Single Material

Materials flows can be most readily understood and analyzed using the concept of a black box. These boxes are schematic representations of real processes or flow junctions, and it is not necessary to specify what the process is in order to be able to develop general principles about the analysis of flows.

Figure 3–1 shows a black box into which some material is flowing. All flows into the box are called *influents* and represented by the letter X. If the flow is described as mass per unit time, X_0 is a mass per unit time of materials X flowing into the box. Similarly, X_1 is the outflow or *effluent*. If no processes are going on inside the box, processes that will either make more of the material or destroy some of it, and if the flow is assumed not to vary with time (to be at *steady state*), then it is possible to write a material balance around the box as

$$\begin{bmatrix} \text{mass per unit} \\ \text{time of } X \\ \text{IN} \end{bmatrix} = \begin{bmatrix} \text{mass per unit} \\ \text{time of } X \\ \text{OUT} \end{bmatrix}$$

or
$$[X_0] = [X_1]$$

The black box can be used to establish a volume balance and a mass balance if the density does not change in the process. Because the definition of density is mass per unit volume, the conversion from a mass balance to a volume balance is achieved by dividing each term by the density (a constant). It is generally convenient to use the volume balance for liquids and the mass balance for solids.

FIGURE 3–1   A black box with one inflow and one outflow.

3.1.1   Splitting Single Materials Flow Streams

A black box shown in Figure 3-2 receives flow from one feed source and separates this into two or more flow streams. The flow into the box is labeled X_0 and the two flows out of the box are X_1 and X_2. If again we assume that

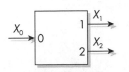

FIGURE 3–2   A separator with one inflow and two outflows.

steady-state conditions exist and that no material is being destroyed or produced, the materials balance is

$$\begin{bmatrix} \text{mass per unit} \\ \text{time of } X \\ \text{IN} \end{bmatrix} = \begin{bmatrix} \text{mass per unit} \\ \text{time of } X \\ \text{OUT} \end{bmatrix}$$

or
$$[X_0] = [X_1] + [X_2]$$

The material X can of course be separated into more than two fractions, so that the materials balance can be

$$[X_0] = [X_1 + X_2 + X_3 + \cdots + X_n]$$

where there are n exit streams or effluents.

e • x • a • m • p • l • e **3.1**

Problem   A city generates 102 tons/day of refuse, all of which goes to a transfer station. At the transfer station the refuse is split into four flow streams headed for three incinerators and one landfill. If the capacity of the incinerators is 20, 50, and 22 tons/day, how much refuse must go to the landfill?

Solution   Consider the situation schematically as shown in Figure 3–3. The four output streams are the known capacities of the three incinerators plus

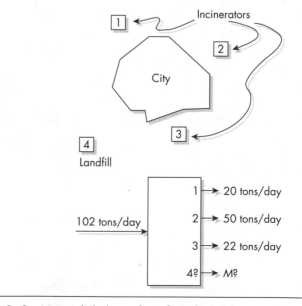

FIGURE 3–3 Materials balance for refuse disposal.

one landfill. The input stream is the solid waste delivered to the transfer station. Setting up the mass balance,

[mass per unit time of refuse IN] = [mass per unit time of refuse OUT]

$$102 = 20 + 50 + 22 + M$$

where M = mass per unit time of refuse to the landfill. Solving for the unknown, M = 10 tons/day.

◆

3.1.2   Combining Single Materials Flow Streams

A black box can also receive numerous influents and discharge one effluent, as shown in Figure 3-4. If the influents are labeled X_1, X_2, . . . , X_m, the materials balance would yield

$$[X_1 + X_2 + \cdot \cdot \cdot + X_m] = X_0$$

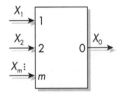

◆───

FIGURE 3-4   A blender with several inflows and one outflow.

e • x • a • m • p • l • e   **3.2**

Problem   A trunk sewer, shown in Figure 3-5, has a flow capacity of 4.0 m³/s. If the flow to the sewer is exceeded, it will not be able to transmit the sewage through the pipe and backups will occur. Currently three neighborhoods contribute to the sewer, and their maximum (peak) flows are 1.0, 0.5, and 2.7 m³/s. A builder wants to construct a development that will contribute a maximum flow of 0.7 m³/s to the trunk sewer. Would this cause the sewer to exceed its capacity?

Solution   Setting up the materials balance in terms of volume:

[volume/unit time sewage IN] = [volume/unit time sewage OUT]

$$[1.0 + 0.5 + 2.7 + 0.7] = X_0$$

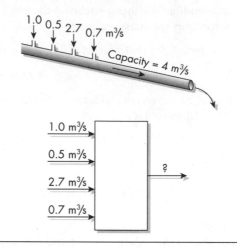

FIGURE 3–5   Capacity of a trunk sewer.

where X_0 is the flow in the trunk sewer. Solving,

$$X_0 = 4.9 \text{ m}^3/\text{s}$$

It seems that the sewer would be overloaded if the new development is allowed to discharge its sewage into the existing sewer. Even now, the system seems to be overloaded, and the only reason disaster has not struck is that not all the neighborhoods produce the maximum flow at the same time of day.

3.1.3   Complex Processes with a Single Material

The simple examples above illustrate the basic principle of materials balances. The two assumptions used to approach the analysis are that the flows are in steady state (they do not change with time) and that no material is being destroyed (consumed) or created (produced). If these possibilities are included in the full materials balance, the equation reads

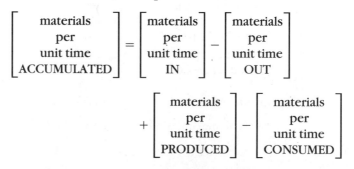

If the material in question is labeled A, the mass balance equation reads

$$
\begin{bmatrix} \text{mass of} \\ A \text{ per} \\ \text{unit time} \\ \text{ACCUMULATED} \end{bmatrix} = \begin{bmatrix} \text{mass of} \\ A \text{ per} \\ \text{unit time} \\ \text{IN} \end{bmatrix} - \begin{bmatrix} \text{mass of} \\ A \text{ per} \\ \text{unit time} \\ \text{OUT} \end{bmatrix}
$$

$$
+ \begin{bmatrix} \text{mass of} \\ A \text{ per} \\ \text{unit time} \\ \text{PRODUCED} \end{bmatrix} - \begin{bmatrix} \text{mass of} \\ A \text{ per} \\ \text{unit time} \\ \text{CONSUMED} \end{bmatrix}
$$

or in volume terms as

$$
\begin{bmatrix} \text{volume} \\ \text{of } A \text{ per} \\ \text{unit time} \\ \text{ACCUMULATED} \end{bmatrix} = \begin{bmatrix} \text{volume} \\ \text{of } A \text{ per} \\ \text{unit time} \\ \text{IN} \end{bmatrix} - \begin{bmatrix} \text{volume} \\ \text{of } A \text{ per} \\ \text{unit time} \\ \text{OUT} \end{bmatrix}
$$

$$
+ \begin{bmatrix} \text{volume} \\ \text{of } A \text{ per} \\ \text{unit time} \\ \text{PRODUCED} \end{bmatrix} - \begin{bmatrix} \text{volume} \\ \text{of } A \text{ per} \\ \text{unit time} \\ \text{CONSUMED} \end{bmatrix}
$$

provided the density does not change.

Mass or volume per unit time can be simplified to *rate*, where rate simply means the flow of mass or volume. Thus the balance equation reads, for either mass or volume,

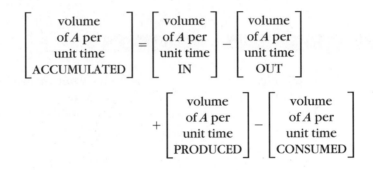

$$
\begin{bmatrix} \text{rate at} \\ \text{which} \\ \text{mass or} \\ \text{volume is} \\ \text{ACCUMULATED} \end{bmatrix} = \begin{bmatrix} \text{rate at} \\ \text{which} \\ \text{mass or} \\ \text{volume} \\ \text{IN} \end{bmatrix} - \begin{bmatrix} \text{rate at} \\ \text{which} \\ \text{mass or} \\ \text{volume} \\ \text{OUT} \end{bmatrix}
$$

(3.1)

$$
+ \begin{bmatrix} \text{rate at} \\ \text{which} \\ \text{mass or} \\ \text{volume is} \\ \text{PRODUCED} \end{bmatrix} - \begin{bmatrix} \text{rate at} \\ \text{which} \\ \text{mass or} \\ \text{volume is} \\ \text{CONSUMED} \end{bmatrix}
$$

As noted above, many systems do not change with time; the flows at one moment are exactly like the flows sometime later. This means that nothing can be accumulating in the black box, either positively (material builds up in the box) or negatively (material is flushed out). The materials balance equation, when the *steady-state assumption* holds, is

$$0 = \begin{bmatrix} \text{rate at} \\ \text{which} \\ \text{mass or} \\ \text{volume} \\ \text{IN} \end{bmatrix} - \begin{bmatrix} \text{rate at} \\ \text{which} \\ \text{mass or} \\ \text{volume} \\ \text{OUT} \end{bmatrix} + \begin{bmatrix} \text{rate at which} \\ \text{mass or} \\ \text{volume is} \\ \text{PRODUCED} \end{bmatrix} - \begin{bmatrix} \text{rate at} \\ \text{which} \\ \text{mass or} \\ \text{volume is} \\ \text{CONSUMED} \end{bmatrix} \quad (3.2)$$

In many simple systems the material in question is not being consumed or produced. If this *conservation assumption* is applied to the materials balance equation, it simplifies to

$$0 = \begin{bmatrix} \text{rate of} \\ \text{mass or volume} \\ \text{IN} \end{bmatrix} - \begin{bmatrix} \text{rate of} \\ \text{mass or volume} \\ \text{OUT} \end{bmatrix} + 0 - 0$$

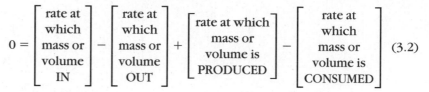

e • x • a • m • p • l • e **3.3**

Problem   A sewer carrying storm water to manhole 1 (Figure 3–6) has a constant flow of 20,947 L/min (Q_A). At manhole 1 it receives a constant lateral flow of 100 L/min (Q_B). What is the flow to manhole 2 (Q_c)?

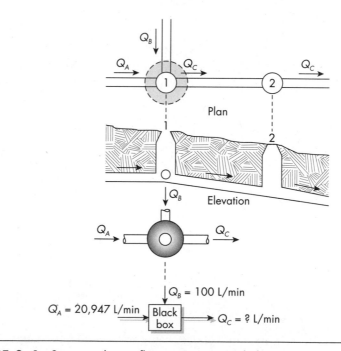

FIGURE 3–6   Sewer with two flows entering manhole 1, combining and moving toward manhole 2.

Solution   Think of manhole 1 as black box, as shown in Figure 3-6. Writing the balance equation for water;

$$
\begin{bmatrix} \text{rate of} \\ \text{water} \\ \text{ACCUMULATED} \end{bmatrix} = \begin{bmatrix} \text{rate of} \\ \text{water} \\ \text{IN} \end{bmatrix} - \begin{bmatrix} \text{rate of} \\ \text{water} \\ \text{OUT} \end{bmatrix}
$$

$$
+ \begin{bmatrix} \text{rate of} \\ \text{water} \\ \text{PRODUCED} \end{bmatrix} - \begin{bmatrix} \text{rate of} \\ \text{water} \\ \text{CONSUMED} \end{bmatrix}
$$

Since no water accumulates in the black box, the system is defined as being in steady state and the first term of this balance equation then goes to zero. Similarly, no water is produced or consumed, so the last two terms are zero. The equation then reads

$$
0 = (Q_A + Q_B) - Q_c + 0 - 0
$$

Substituting the given flows,

$$
0 = (20{,}947 + 100) - Q_c
$$

and solving,

$$
Q_c = 21{,}047 \text{ L/min}
$$

◆

Where the flow is contained in pipes or channels, it is convenient to place a black box at any junction with three or more flows. For example, the manhole in Example 3.3 is a junction receiving two flows and discharging one, for a total of three.

Sometimes the flows may not be contained in conduits or rivers. In such cases, it is useful to first visualize the system as a pipe flow network. Rain water falling to earth, for example, can either percolate into the ground or run off in a watercourse. This system can be pictured as shown in Figure 3-7 and a materials balance performed on the imaginary black box.

A system may contain any number of processes or flow junctions, all of which could be treated as black boxes. For example, Figure 3-8A shows a schematic of the hydrologic cycle. Precipitation falls to the earth, some of it runs off, and some of it percolates into the ground, where it joins a groundwater reservoir. If the water is used for irrigation, it is taken out of the groundwater reservoir through wells. The irrigation water either goes back to the atmosphere by the process of evaporation or transpiration (water released to the atmosphere from plants), commonly combined into one process—*evapotranspiration*—or percolates back into the ground.

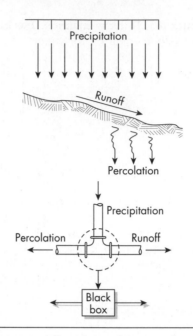

◆───

FIGURE 3-7   Precipitation, runoff, and percolation can be visualized as a black box.

e • x • a • m • p • l • e **3.4**

Problem   Suppose the rainfall is 40 inches per year, of which 50% percolates into the ground. The farmer irrigates his crops using well water. Of the extracted well water, 80% is lost by evapotranspiration, the remainder repercolating back into the ground. How much groundwater could a farmer on a 2000 acre farm extract from the ground per year without depleting the groundwater reservoir volume?

Solution   First, 40 inches per year over 2000 acres is

$$40\,\frac{in.}{yr}\left(\frac{1\ ft}{12\ in.}\right)\times 2000\ \text{acres}\left(43{,}560\,\frac{ft^2}{acre}\right)=2.90\times 10^8\,\frac{ft^3}{yr}$$

The system in question is now redrawn as Figure 3–8B, and all the available information is added to the sketch. Unknown quantities are noted. It is convenient to start the calculations by constructing a balance on the first black box (1), a simple junction with precipitation coming in and runoff and percolation going out. The volume balance is

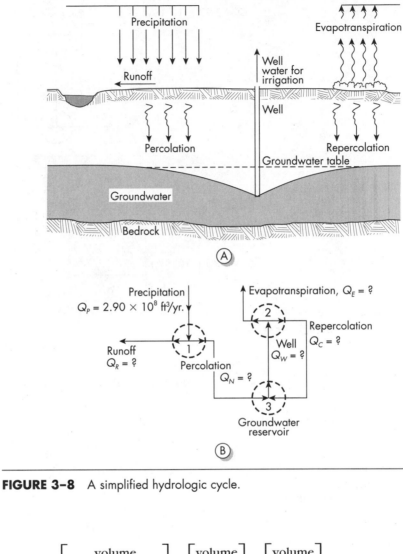

FIGURE 3-8   A simplified hydrologic cycle.

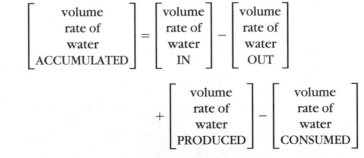

Since the system is assumed to be in steady state, the first term is zero. Likewise, the last two terms are zero,

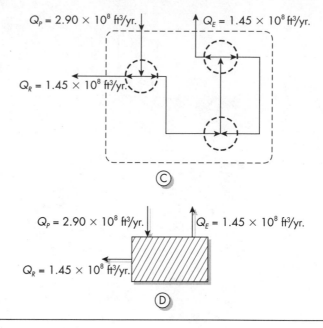

$Q_P = 2.90 \times 10^8$ ft³/yr. $Q_E = 1.45 \times 10^8$ ft³/yr.

$Q_R = 1.45 \times 10^8$ ft³/yr.

ⓒ

$Q_P = 2.90 \times 10^8$ ft³/yr. $Q_E = 1.45 \times 10^8$ ft³/yr.

$Q_R = 1.45 \times 10^8$ ft³/yr.

ⓓ

FIGURE 3–8   Continued.

$$0 = \left[2.90 \times 10^8 \frac{\text{ft}^3}{\text{yr}} \right] - [Q_R + Q_N] + 0 - 0$$

and since half of the water runs off,

$$Q_R = 0.5 \, Q_P = Q_N$$

$$0 = 2.90 \times 10^8 - 2 \, Q_R$$

Solving for Q_R,

$$Q_R = 1.45 \times 10^8 \frac{\text{ft}^3}{\text{yr}} = Q_N$$

A balance on the second black box (2) yields

$$0 = [Q_W] - [Q_E + Q_C] + 0 - 0$$

Since $Q_C = 0.2 \, (Q_W)$

$$0 = 0.8 \, Q_W - Q_E$$

Finally, a balance on the groundwater reservoir (3), if the quantity of groundwater in the reservoir does not change, is

$$0 = [Q_N + Q_C] - [Q_W] + 0 - 0$$

From the first balance, $Q_N = 1.45 \times 10^8$ ft³/yr and from the second $Q_C = 0.2 \, Q_W$, so that

$$0 = (1.45 \times 10^8) + 0.2\, Q_W - Q_W$$

or

$$Q_W = 1.81 \times 10^8 \text{ ft}^3/\text{yr}$$

This is the maximum safe yield of well water for the farmer.

As a check, consider the entire system as a black box. This is illustrated in Figure 3–8C. There is only one way water can get into this box (precipitation) and two ways out (runoff and evapotranspiration). Representing this black box as Figure 3–8D, it is possible to write

$$0 = (2.90 \times 10^8) - (1.45 \times 10^8) - (1.45 \times 10^8)$$

The balance of the overall system checks the calculations.

◆

The last check illustrates a useful and powerful tool in materials balance calculations: When the flows in a system of black boxes are balanced, it is possible to draw boundaries (dotted lines) around any one or any combination of black boxes and this combination is itself a black box. Returning to the example, if a dotted line is drawn around the entire system (Figure 3–8C), it is useful to picture it as a single black box (Figure 3–8D). If the calculations are correct, the flows for this box should also balance (and they do).

Incidentally, what would happen if the farmer decided to take more than the above amount of water out of the ground? An *unsteady* situation would occur, and in time the groundwater reservoir would run dry.

This is exactly what is occurring in the Great Plains today. The Ogallala Aquifer, a vast reservoir that has provided water for the farmers since 1930, is being drained dry. The annual overdraft is nearly equal to the flow of the Colorado River. A major engineering study recently warned that 5.1 million acres of irrigated farms will dry up by the year 2020 if current trends continue.

Excessive use of fertilizers represents another serious environmental problem. We can feed the world population only if we continue to extract huge quantities of phosphorus and other nutrients out of the ground. We know that eventually these reserves will be depleted, and that the yield from our farms will be greatly reduced.

If these events come to pass, starvation will be common in much of the world. With not enough food to go around, and too many people desperate for it, how will it be distributed? What if we simply ignore those starving people, and wait for starvation to reduce the global population so that the food we produce will be in balance with the survivors? We could argue that the amount of food available would not feed everyone anyway, and that taking away our food may cause starvation here as well, and that eventually *all* people will die of starvation. It is best, so goes the argument, to simply allow nature to take its course. Any food we give even now to needy countries just encourages them to produce more people, and this will eventually lead to tragedy. Such thinking is known as "lifeboat ethics," the name coming from the idea that a lifeboat can only hold so many people. If more than the capacity tries to get in, the boat will be swamped, and *everyone* will drown.

Do we therefore have an obligation to save our own skin and to not worry about our neighbors? This seems like a reasonable thing to do, until we consider the possibility that perhaps *we* are the ones without food, and our neighbors decide to allow *us* to starve. What should/could/would we do, if anything?

This seems like a science-fiction scenario, but unfortunately it may be all too true. Resource deprivation is predictable and assured if our global system continues in a non-steady-state mode. You are perhaps the last lucky generation. Your children may not be so fortunate.

But back to the black boxes. Thus far the black boxes have had to process only one feed material, for example, water is the only material balanced in the previous two examples. Materials balances become considerably more complicated (and useful) when several materials flow through the system.

Before introducing examples of systems with multiple materials, it is appropriate to establish some general rules that are useful in attacking all materials balance problems. The rules are:

1. Draw the system as a diagram, including all flows (inputs and outputs) as arrows.

2. Add the available information such as flow rates, concentrations, and so on. Assign symbols to unknown variables.

3. Draw a continuous dotted line around that component or components that are to be balanced. This could be a unit operation, a junction, or a combination of these. Everything inside this dotted line becomes the black box.

4. Decide what material is to be balanced. This could be a volumetric or mass flow rate.

5. Write the mass balance equation by starting with the basic equation of

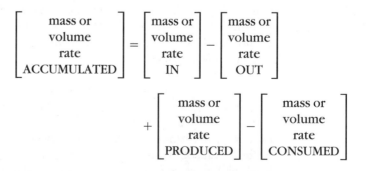

$$
\begin{bmatrix} \text{mass or} \\ \text{volume} \\ \text{rate} \\ \text{ACCUMULATED} \end{bmatrix} = \begin{bmatrix} \text{mass or} \\ \text{volume} \\ \text{rate} \\ \text{IN} \end{bmatrix} - \begin{bmatrix} \text{mass or} \\ \text{volume} \\ \text{rate} \\ \text{OUT} \end{bmatrix}
$$

$$
+ \begin{bmatrix} \text{mass or} \\ \text{volume} \\ \text{rate} \\ \text{PRODUCED} \end{bmatrix} - \begin{bmatrix} \text{mass or} \\ \text{volume} \\ \text{rate} \\ \text{CONSUMED} \end{bmatrix}
$$

6. If only one variable is unknown, solve for that variable.

7. If more than one variable is unknown, repeat the procedure using a different black box or a different material for the same black box.

Thus armed with a set of guidelines, we now return to the problems where more than one material is involved in the flows.

3.2 ◆ Materials Balances with Multiple Materials

Mass and volume balances can be developed with multiple materials flowing in a single system. In some cases, the process is one of mixing, where several inflow streams are combined to produce a single outflow stream, whereas in other cases a single inflow stream is split into several outflow streams according to some material characteristics.

3.2.1 Mixing Multiple Materials Flow Streams

Since the mass balance and volume balance equations are actually the same equation, it is not possible to develop more than one materials balance equation for a black box, unless there is more than one material involved in the flow.

Consider the following example, where silt flow in rivers (expressed as mass of solids per unit time) is analyzed.

e • x • a • m • p • l • e **3.5**

Problem   The Allegheny and Monongahela Rivers meet at Pittsburgh to form the mighty Ohio River. The Allegheny, flowing south through forests and small towns, runs at an average flow of 340 cubic feet per second (cfs) and has a low silt load, 250 mg/L. The Monongahela, however, flows north at a flow of 460 cfs through old steel towns and poor farm country, carrying a silt load of 1500 mg/L.

a. What is the average flow of the Ohio River?

b. What is the silt concentration in the Ohio River?

Solution   Follow the steps outlined.

Step 1: Figure 3-9 shows the confluence of the rivers, and the variables are identified.

Step 2: All the available information is added to the sketch, including the unknown variables.

Step 3: The confluence of the rivers is the black box, as shown by the dotted line.

Step 4: Water flow is to be balanced first.

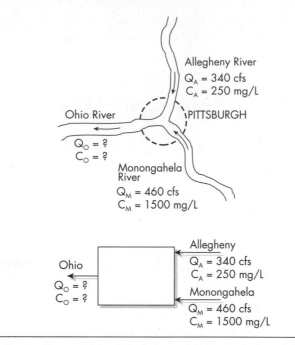

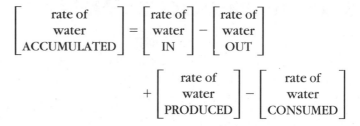

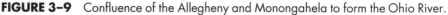

FIGURE 3–9   Confluence of the Allegheny and Monongahela to form the Ohio River.

Step 5: Write the balance equation

$$
\begin{bmatrix} \text{rate of} \\ \text{water} \\ \text{ACCUMULATED} \end{bmatrix} = \begin{bmatrix} \text{rate of} \\ \text{water} \\ \text{IN} \end{bmatrix} - \begin{bmatrix} \text{rate of} \\ \text{water} \\ \text{OUT} \end{bmatrix}
$$

$$
+ \begin{bmatrix} \text{rate of} \\ \text{water} \\ \text{PRODUCED} \end{bmatrix} - \begin{bmatrix} \text{rate of} \\ \text{water} \\ \text{CONSUMED} \end{bmatrix}
$$

Step 6: Solve for the unknown.

Because this system is assumed to be in steady state, the first term in zero. Also, because water is neither produced nor consumed, the last two terms are zero. Thus

$$0 = [\text{water IN}] - [\text{water OUT}] + 0 - 0$$

There are two rivers flowing in, and one out, so the equation reads

$$0 = [340 + 460] - [Q_O] + 0 - 0$$

where Q_O = flow in the Ohio River.
    Solving,

$$Q_O = 800 \text{ cfs}$$

The solution process must be repeated for the silt. Recall that mass flow is calculated as concentration times volume, or

$$Q_{Mass} = C \times Q_{Volume} = \frac{mg}{L} \times \frac{L}{sec} = \frac{mg}{sec}$$

Setting up the mass balance,

$$\begin{bmatrix} silt \\ ACCUMULATED \end{bmatrix} = \begin{bmatrix} silt \\ IN \end{bmatrix} - \begin{bmatrix} silt \\ OUT \end{bmatrix} + \begin{bmatrix} silt \\ PRODUCED \end{bmatrix} - \begin{bmatrix} silt \\ CONSUMED \end{bmatrix}$$

Again the first and last two terms are zero,

$$0 = [IN] - [OUT] + 0 - 0$$

$$0 = [(C_A Q_A) + (C_M Q_M)] - [C_O Q_O] + 0 - 0$$

where C = concentration of silt, and the subscripts A, M, and O are for the three rivers. Substituting:

$$0 = [(250 \times 340) + (1500 \times 460)] - [C_O \times 800]$$

Note that the flow of the Ohio River is 800 cfs as calculated from the volume balance. Solving,

$$C_O = 969 \text{ mg/L}$$

There's no need to convert the flow rate from ft³/sec to L/sec since the conversion factor would be a constant that would appear in every term of the equation and simply cancel.

In the previous example, all unknowns except one are defined, resulting in a simple calculation for the remaining unknown term. In most applications, however, it is necessary to work a little harder for the answer, as illustrated next.

e • x • a • m • p • l • e   **3.6**

Problem   Suppose the sewers shown in Figure 3–10 have $Q_B = 0$ and Q_A as unknown. By sampling the flow at the first manhole, we find that the concentra-

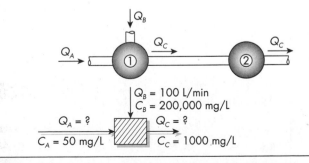

FIGURE 3–10   A manhole as a black box with solid and liquid flows.

tion of dissolved solids of the flow coming into manhole 1 (Q_A) is 50 mg/L. An additional flow, $Q_B = 100$ L/min, is added to manhole 1 and this flow contains 20% dissolved solids (recall that 1% = 10,000 mg/L). The flow through manhole 2 is sampled and found to contain 1000 mg/L dissolved solids. What is the flow rate of wastewater in the sewer (Q_A)?

Solution

Step 1:   Draw the diagram (Figure 3–10).

Step 2:   Add all information, including concentrations.

Step 3:   Manhole 1 is the black box.

Step 4:   What is to be balanced?

If the flows are balanced, there are two unknowns. Can something else be balanced? Suppose a balance is run in terms of the solids?

$$\begin{bmatrix} \text{solids} \\ \text{ACCUMULATED} \end{bmatrix} = \begin{bmatrix} \text{solids} \\ \text{IN} \end{bmatrix} - \begin{bmatrix} \text{solids} \\ \text{OUT} \end{bmatrix} + \begin{bmatrix} \text{solids} \\ \text{PRODUCED} \end{bmatrix} - \begin{bmatrix} \text{solids} \\ \text{CONSUMED} \end{bmatrix}$$

Step 5:   Balance the flows.

A steady-state assumption allows the first term to be zero, and since no solids are produced or consumed in the system, the last two terms likewise are zero and the equation reduces to

$$0 = [\text{solids flow in}] - [\text{solids flow out}] + 0 - 0$$

$$0 = [Q_A C_A + Q_B C_B] - [Q_C C_C] + 0 - 0$$

$$0 = \left[Q_A \left(50\,\frac{\text{mg}}{\text{L}} \right) + \left(100\,\frac{\text{L}}{\text{min}} \right)\left(200{,}000\,\frac{\text{mg}}{\text{L}} \right) \right] - \left[Q_C \left(1000\,\frac{\text{mg}}{\text{L}} \right) \right]$$

Note that (L/min) × (mg/L) = (mg solids/min). This results in an equation with two unknowns, so it is necessary to move on.

Step 6:   If more than one unknown variable results from the calculation, establish another balance.

A balance in terms of the volume flow rate is:

$$\begin{bmatrix} \text{volume} \\ \text{ACCUMULATED} \end{bmatrix} = \begin{bmatrix} \text{volume} \\ \text{IN} \end{bmatrix} - \begin{bmatrix} \text{volume} \\ \text{OUT} \end{bmatrix}$$

$$+ \begin{bmatrix} \text{volume} \\ \text{PRODUCED} \end{bmatrix} - \begin{bmatrix} \text{volume} \\ \text{CONSUMED} \end{bmatrix}$$

Again, the first term and the last two terms are zero, and

$$0 = [Q_A + Q_B] - [Q_C]$$

and

$$0 = Q_A + 100 - Q_C$$

We now have two equations with two unknowns. Substituting $Q_A = (Q_C - 100)$ into the first equation

$$50 (Q_C - 100) + 200,000 (100) = 1000 \, Q_C$$

and solving for $Q_C = 21,047$ L/min and $Q_A = 20,947$ L/min.

◆

Sometimes in materials balance calculations the black box is literally a box. For example, a crude but useful means of estimating the relationship between the emission of air pollutants and the air quality above a city is the *box model*. In this analysis, a box is assumed to sit above the city, and this box is as high as the *mixing depth* of pollutants and as wide and deep as the city boundaries. Wind blows through this box and mixes with the pollutants emitted, resulting in a concentration term. The "box" model can be analyzed using the same principles discussed previously.

e • x • a • m • p • l • e **3.7**[1]

Problem   Estimate the concentration of SO_2 in the urban air above the city of St. Louis if the mixing height above the city is 1210 m, the "width" of the box perpendicular to the wind is 10^5 m, the average annual wind speed is 15,400 m/hr, and the amount of sulfur dioxide discharged is 1375×10^6 pounds per year.

Solution   First express the emission rate in SI units

$$(1375 \times 10^6) \frac{lb}{yr} \times 454 \frac{g}{lb} \times 10^6 \frac{\mu g}{g} \times \frac{1}{8760} \frac{yr}{hr} = 7.126 \times 10^{13} \frac{\mu g}{hr}$$

Now construct the box above St. Louis, as shown in Figure 3–11. Calculate the volume of air moving into the box as the velocity times the area through which the flow occurs, or $Q = Av$ where v = wind velocity and A = area of the side of the box, mixing depth times width.

$$Q_{air} = (1210 \times 10^5)(15,400) = 1.86 \times 10^{12} \text{ m}^3/\text{hr}$$

A simplified box is also shown on the figure. Clearly, a simple application of the volume balance equation would show [air in] = [air out] and a mass balance would have [SO_2 in] = [SO_2 out]. The sulfur dioxide out is 1375 lb/yr and Q_{air} out is 1.86×10^{12} m³/hr. But remember that *mass* flow can also be expressed as (concentration) × (volume flow), and a mass balance in terms of SO_2 can be written as

$$\begin{bmatrix} \text{rate of } SO_2 \\ \text{ACCUMULATED} \end{bmatrix} = \begin{bmatrix} \text{rate of } SO_2 \\ \text{IN} \end{bmatrix} - \begin{bmatrix} \text{rate of } SO_2 \\ \text{OUT} \end{bmatrix}$$

$$+ \begin{bmatrix} \text{rate of } SO_2 \\ \text{PRODUCED} \end{bmatrix} - \begin{bmatrix} \text{rate of } SO_2 \\ \text{CONSUMED} \end{bmatrix}$$

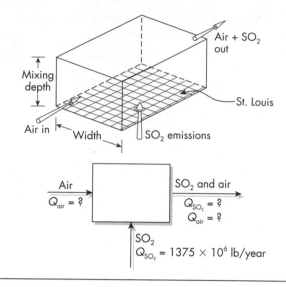

FIGURE 3-11   Air quality box model. See Example 3.7.

Assuming steady state,

$$0 = [7.126 \times 10^{13}] - [(C)(1.86 \times 10^{12})]$$

and solving

$$C = 38 \ \mu g/m^3$$

Sulfur dioxide has a long history as a problem air pollutant. In 1948, western Pennsylvania was the steel-making capital of the world. Many small towns on the Monongahela and Ohio Rivers were there because of steel making. At the turn of the century, immigrants from Central Europe poured into small towns along the Ohio and Monongahela Rivers to provide the cheap but dedicated labor. The working conditions in the mills were abominable by today's standards. Some jobs were worse than others, such as at the coke ovens, where it was known that for some jobs the life expectancy was only a few years. But all in all, the mills prospered and the towns grew into proud communities.

Donora, Pennsylvania, was one of these towns. Nestled in a bend of the Monongahela River, it boasted a steel mill, a wire mill, and a plating plant. The product was galvanized wire, and in 1948, the mills were humming.

Dirty air was accepted as part of life, the price paid for having a job. Too bad if the curtains had to be washed every month, or that collars on white shirts would soil within a few hours, or that interior walls would get so dirty that they had to be literally erased with putty. Smoke meant jobs, and any intimation of controls on production were always met with threats by the company to move the plants elsewhere.

In October of 1948, an inversion condition occurred that effectively placed a lid on the river valley and concentrated all of the pollutants in the "black box" directly above Donora. Within hours, animals began to be in distress, particularly parakeets. As time went on, and the pollutant concentrations increased, people began to feel the effects. Many sought medical help, but the atmosphere was so impenetrable that the doctors had difficulty finding their way to the homes of sick people. After three days, the meteorological lid finally lifted, and the pollution abated, leaving 27 people dead and hundreds permanently affected by the episode.

The steel companies insisted that they were not at fault, and indeed there never was any fault implied by the special inquiry into the incident. The companies were operating within the law and were not coercing any of the workers to work there or anyone to live in Donora. In the absence of legislation, the companies felt no obligation to pay for air-pollution equipment or to change processes to reduce air pollution. They felt that if only their company was required to pay for and install air-pollution control equipment they would be at a competitive disadvantage and would eventually go out of business. Obviously, national legislation was needed to prevent acute health problems like the "Donora episode," but this was slow in coming. Eventually state legislation began to appear, but it wasn't until 1972 that effective federal legislation was passed.

Following the passage of national air-pollution legislation, many United States companies moved "off-shore" to islands where pollution control is less restrictive. As long as pollution is considered a local problem, the relocation of plants off-shore makes sense, except for the jobs lost of course. Now, however, when we recognize that pollution is of global concern, the mere moving of a plant a few thousand miles may make no difference whatever as to its effect on the global atmosphere. The *earth* has become our black box.

3.2.2   Separating Multiple Materials Flow Streams

So far we have discussed black boxes where the characteristics of the materials flows are not altered by the black boxes. In this section, the black boxes receive flows made up of mixtures of two or more materials, and the intent of the black box is to change the concentration of the influents.

The objective of a materials *separator* is to split a mixed feed material into the individual components by exploiting some difference in the material properties. For example, suppose it is necessary to design a device for a solid-waste processing plant that will separate crushed glass into two types of glass: transparent or clear glass (flint) and opaque (amber or brown) glass.

First, decide what material property to exploit. This becomes the *code*, or signal, that can be used to tell the machine how to divide the individual particles in the feed stream. In the case of the glass, the code could obviously be *transparency*, and this property is used in the design of the separating system, such as the device shown in Figure 3–12, for example. In this device,

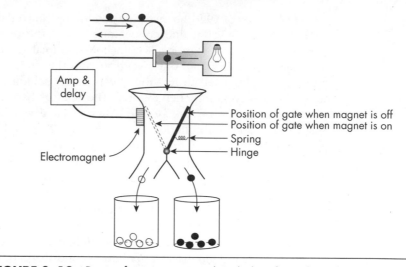

FIGURE 3-12   Device for separating colored glass from clear glass; a binary separator.

the glass pieces drop off a belt conveyor and pass through a light beam. The amount of light transmitted across the box is read by a photocell. If the light beam is interrupted by an opaque piece of glass, the photocell receives less light and an electromagnet is activated, which pulls a gate to the left. A transparent piece of glass will not interrupt the light beam and the gate will not move. The gate, activated by the signal from the photocell, is the *switch*, separating the material according to the code.

This simple device illustrates the nature of coding and switching; reading a property difference and then using that signal to achieve separation. Materials separation using the principle of coding and switching is an integral and central part of environmental engineering.

Materials separation devices are not infallible; they make mistakes. We cannot assume that the light will always correctly identify the clear glass. Mistakes will occur and an opaque piece of glass may end up in a box full of transparent pieces, or the other way around. The measure of how well the separation works is measured using two parameters: *recovery* and *purity*.

Consider a black box separator shown in Figure 3-13. The two components x and y are to be separated so that x goes to product stream 1 and y goes to product stream 2. Unfortunately, some of the y mistakenly ends up in

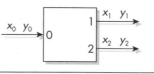

FIGURE 3-13   A binary separator.

product stream 1 and some of the x comes out at 2. The *recovery* of component x in product stream 1 is defined as

$$R_{x_1} = \frac{x_1}{x_0} \times 100$$

where x_1 = material flow in effluent stream 1
x_0 = material flow in the influent stream

Similarly, the recovery of material y flow in product stream 2 can be expressed as

$$R_{y_1} = \frac{y_2}{y_0} \times 100$$

where y_2 = material flow in effluent stream 2
y_0 = material flow in influent stream

If the recovery of both x and y is 100%, the separator is a perfect device.

Suppose we now want to maximize the recovery of x in product stream 1, so it makes sense to reduce x_2 to the smallest possible amount. We can do this by shutting off product stream 2, diverting the entire flow to product stream 1, and achieving $R_x = 100$%! Obviously, we have a problem. We achieve 100% recovery, but we aren't doing anything useful. Recovery cannot therefore be the sole criterion for judging the performance of a separator.

A second parameter used to describe separation is *purity*. The purity of x in exit stream 1 is defined as

$$P_{x_1} = \frac{x_1}{x_1 + y_1} \times 100$$

A combination of recovery and purity yields the performance criteria we need.

e • x • a • m • p • l • e **3.8**

Problem Assume that an aluminum-can separator in a local recycling plant processes 400 cans per minute. The two product streams consist of the following:

	Total in Feed	Product Stream 1	Product Stream 2
Aluminum cans, no./min	300	270	30
Steel cans, no./min	100	0	100

Calculate the recovery of aluminum cans and the purity of the product.

Solution Note first that a materials balance for aluminum yields

$$0 = [\text{IN}] - [\text{OUT}] = 300 - (270 + 30) \qquad \text{check}$$

$$R_{\text{Al cans}_1} = \frac{270}{300} \times 100 = 90\%$$

$$P_{\text{Al cans}_1} = \frac{270}{270 + 0} \times 100 = 100\%$$

◆

The same principles can be applied to a more complex environmental process such as gravitational thickening.

e • x • a • m • p • l • e 3.9

Problem   Figure 3–14 shows a gravitational thickener used in numerous water and wastewater treatment plants. This device separates suspended solids from the liquid (usually water) by taking advantage of the fact that the solids have a higher density than the water. The code for this separator is density, and the thickener is the switch, allowing the denser solids to settle to the bottom of a tank, from which they are removed. The flow into the thickener is called the influent or feed, the low-solids exit stream is the overflow, and the heavy- or concentrated-solids exit stream is the underflow. Suppose a thickener in a metal plating plant receives a feed of 40 m³/hr of precipitated metal plating waste with a suspended-solids concentration of 5000 mg/L. If the thickener is operated in a steady-state mode so that 30 m³/hr of flow exits as the overflow, and this overflow has a solids concentration of 25 mg/L, what is the underflow solids concentration and what is the recovery of the solids in the underflow?

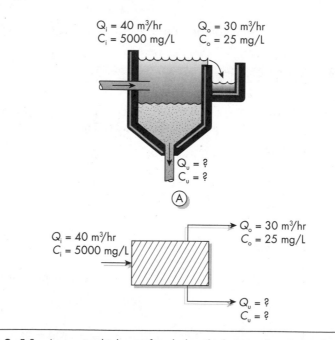

FIGURE 3–14   A gravity thickener for sludge thickening. See Example 3.9.

Solution   Once again consider the thickener as a black box and proceed stepwise to balance first the volume flow and then the solids flow. The volume balance is

$$\begin{bmatrix} \text{volume} \\ \text{ACCUMULATED} \end{bmatrix} = \begin{bmatrix} \text{volume} \\ \text{IN} \end{bmatrix} - \begin{bmatrix} \text{volume} \\ \text{OUT} \end{bmatrix}$$

$$+ \begin{bmatrix} \text{volume} \\ \text{PRODUCED} \end{bmatrix} - \begin{bmatrix} \text{volume} \\ \text{CONSUMED} \end{bmatrix}$$

$$0 = 40 - (30 + Q_u) + 0 - 0$$

$$Q_u = 10 \text{ m}^3/\text{hr}$$

For the solids, the mass balance is

$$0 = (C_i Q_i) - [(C_u Q_u) + (C_o Q_o)] + 0 - 0$$

$$0 = (5000)(40) - [C_u(10) + (25)(30)]$$

$$C_u = 19,900 \text{ mg/L}$$

The recovery of solids is

$$R_u = \frac{C_u Q_u}{C_i Q_i} \times 100$$

$$R_u = [(19,900)(10) \times 100]/[(5000)(40)] = 99.5\%$$

◆

In many engineering studies, it is inconvenient to have two parameters such as recovery and purity that describe the performance of a unit operation. For example, the performance of two competing air classifiers used for producing refuse-derived fuel may be advertised as follows:

	R(%)	P(%)
Air classifier 1	90	92
Air classifier 2	93	87

Which is the better device? For a binary separator, two parameters have been suggested for estimating the effectiveness of separation.

$$E_{\text{WS}} = \left[\frac{x_1}{x_0} \times \frac{y_2}{y_0} \right]^{1/2} \times 100 \tag{3.3}$$

where E_{WS} = Worrell–Stessel effectiveness, and

$$E_{\text{R}} = \left| \frac{x_1}{x_0} - \frac{y_1}{y_0} \right| \times 100 \tag{3.4}$$

where E_{R} = Rietema effectiveness.

Both of these allow for comparisons using a single value, and both have the benefit of ranging from 0 to 100%.

3 Materials Balances & Separations

The use of these definitions for efficiency can be illustrated using the foregoing data for the two air classifiers.

e • x • a • m • p • l • e **3.10**

Problem   Calculate the performance effectiveness for the data below using the Worrell–Stessel and Rietema formulas.

	Feed (tons/day) Organics/Inorganics	Product Stream 1 (tons/day) Organics/Inorganics	Product Stream 2 (tons/day) Organics/Inorganics
Air classifier 1	80/20	72/6	8/14
Air classifier 2	80/20	76/8	4/12

Solution   For the first air classifier:

$$E_{WS} = \left[\frac{72}{80} \times \frac{14}{20} \right]^{1/2} = 79\%$$

$$E_R = \left| \frac{72}{80} - \frac{6}{20} \right| \times 100 = 60\%$$

For the second air classifier, $E_{WS} = 75\%$ and $E_R = 55\%$. Clearly the first classifier is superior.

◆

The processes above illustrate the separation of a single material into two product streams. These so-called *binary separators* are used to split a feed into two parts according to some materials property or code. We can envision also a *polynary separator*, which divides a mixed material into three or more components. The performance of a polynary separator can also be described by recovery, purity, and effectiveness. With reference to Figure 3–15, the recovery of component x_1 in effluent stream 1 is

$$R_{x_{11}} = \frac{x_{11}}{x_{10}} \times 100$$

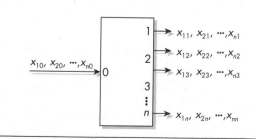

FIGURE 3–15   A polynary separator.

where x_{11} = component x_1 in effluent stream 1
x_{10} = component x_1 in the influent

The purity of product stream 1 with respect to component x_1 is

$$P_{x_{11}} = \frac{x_{11}}{x_{11} + x_{21} + \cdots + x_{n1}} \times 100$$

The Worrell–Stessel and Rietema effectivenesses are calculated as

$$E_{WS} = \left[\frac{x_{11}}{x_{10}} \cdot \frac{x_{22}}{x_{20}} \cdot \frac{x_{33}}{x_{30}} \cdots \frac{x_{nn}}{x_{n0}} \right]^{1/n} \times 100$$

$$E_{R_1} = \left| \frac{x_{11}}{x_{10}} - \frac{x_{12}}{x_{20}} - \frac{x_{13}}{x_{30}} - \cdots - \frac{x_{1n}}{x_{n0}} \right| \times 100$$

In this polynary separator, we assume that the feed has n components and that the separation process has n product streams. A more general condition m product streams for a feed with n components is shown in Figure 3–16. The equation for recovery, purity, and effectiveness could be constructed from the definitions as before.

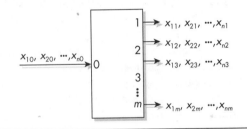

♦ ───

FIGURE 3–16 A polynary separator for n components and m product streams.

3.2.3 Complex Processes with Multiple Materials

The foregoing principles of mixing and separating using black boxes can be applied to complex systems with multiple materials that can be analyzed as a combination of several black boxes. The stepwise procedure introduced earlier can be applied to these systems as well. In Example 3.11, two black boxes are used in addition to the dewatering unit operation, a separation step.

e • x • a • m • p • l • e **3.11**

Problem   Consider a system pictured in Figure 3–17. A sludge of solids concentration $C_O = 4\%$ is to be thickened to a solids concentration $C_E = 10\%$ using a centrifuge. Unfortunately, the centrifuge produces a sludge at 20% solids from

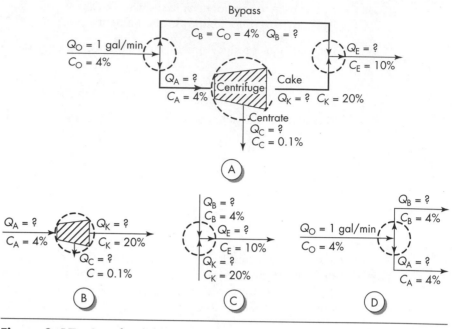

Figure 3–17   Centrifugal dewatering of sludge. See Example 3.11.

the 4% feed sludge. In other words, it works too well. We can bypass some of the feed sludge flow and blend it later with the dewatered (20%) sludge so as to produce a sludge with exactly 10% solids concentration. The question is: How much sludge to bypass? The influent flow rate (Q_0) is 1 gal/min at a solids concentration (C_0) of 4%. We can assume here, and in the following problems, that the specific gravity of the sludge solids is 1.0 g/cm³. That is, the solids have a density equal to that of water, which is usually a good assumption. The centrifuge produces a centrate (effluent stream of low solids concentration) with a solids concentration (C_C) of 0.1%; and a cake (the high solids concentrated effluent stream) with a solids concentration (C_K) of 20%. Find the required flow rates.

Solution

1. Consider first the centrifuge as a black box separator (Figure 3–17B). A volume balance yields

$$\begin{bmatrix} \text{volume} \\ \text{of sludge} \\ \text{ACCUMULATED} \end{bmatrix} = \begin{bmatrix} \text{volume} \\ \text{of sludge} \\ \text{IN} \end{bmatrix} - \begin{bmatrix} \text{volume} \\ \text{of sludge} \\ \text{OUT} \end{bmatrix}$$

$$+ \begin{bmatrix} \text{volume} \\ \text{of sludge} \\ \text{PRODUCED} \end{bmatrix} - \begin{bmatrix} \text{volume} \\ \text{of sludge} \\ \text{CONSUMED} \end{bmatrix}$$

$$0 = Q_A - [Q_C + Q_K] + 0 - 0$$

Note that the volume includes the volume of the sludge solids and the volume of the surrounding liquid. A solids balance on the centrifuge gives

$$0 = [Q_A C_A] - [Q_K C_K + Q_C C_C] + 0 - 0$$

$$0 = Q_A(4) - Q_K(20) - Q_C(0.1)$$

Obviously, there are only two equations and three unknowns. For future use, solve both of these in terms of Q_A, as $Q_C = 0.804Q_A$ and $Q_K = 0.196Q_A$

2. Next consider the second junction, in which two streams are blended (Figure 3-17C). A volume balance yields

$$0 = [Q_B + Q_K] - [Q_E] + 0 - 0$$

and a solids mass balance is

$$0 = [Q_B C_B + Q_K C_K] - [Q_E C_E] + 0 - 0$$

$$0 = Q_B(4) + Q_K(20) - Q_E(10)$$

Substituting and solving for Q_K

$$Q_K = 0.6Q_B$$

From above, it was found that $Q_K = 0.196Q_A$. Hence

$$0.6Q_B = 0.196Q_A$$

or

$$Q_B = 0.327Q_A$$

3. Now consider the first separator box (Figure 3-17D). A volume balance yields

$$0 = [Q_0] - [Q_B + Q_A] + 0 - 0$$

From above,

$$Q_B = 0.327Q_A$$

Substituting,

$$0 = 1 - 0.327Q_A - Q_A$$

or

$$Q_A = 0.753 \text{ gal/min}$$

and

$$Q_B = 0.327 \, (0.753) = 0.246 \text{ gal/min}$$

which is the answer to the problem. Further, from the centrifuge balance

$$Q_C = 0.804 \, (0.753) = 0.605 \text{ gal/min}$$

$$Q_K = 0.196 \, (0.753) = 0.147 \text{ gal/min}$$

and from the blender box the balance is

$$0 = Q_K + Q_B - Q_E$$

or

$$Q_E = Q_B + Q_K = 0.393 \text{ gal/min}$$

There is also a check available. Using the entire system, a volume balance gives

$$0 = Q_0 - Q_C - Q_E$$

$$0 = 1.0 - 0.605 - 0.393 = 0.002 \qquad \text{check}$$

◆

Most environmental engineering processes involve a series of unit operations, each intended to separate a specific material. Often the order in which these unit operations are employed can influence the cost and/or the efficiency of operation. This is shown by the following example.

e • x • a • m • p • l • e **3.12**[2]

Problem   Suppose seven tons of shredded waste plastics is classified into individual types of plastics. The four different plastics in the mix have the following quantities and densities:

Plastic	Symbol	Quantity (tons)	Density (g/cm³)
Polyvinylchloride	(PVC)	4	1.313
Polystyrene	(PS)	1	1.055
Polyethylene	(PE)	1	0.916
Polypropylene	(PP)	1	0.901

Determine the best method of separation.

Solution   Although the differences are small, it still might be possible to use density as the code, and a float/sink apparatus can be used as a switch. By judiciously choosing a proper fluid density, a piece of plastic can be made to either sink or float.

In this case, the easiest separation is the removal of the two heavy plastics (PVC and PS) from the two lighter ones (PE and PP). This can be done using a fluid with a density of 1.0 g/mL, or most likely plain water. Following this step, the two streams each must be separated further. The trickiest separation, the splitting of PE and PP, must be done with great care and is made somewhat easier by already having removed the PVC and PS. The entire process train is shown in Figure 3–18.

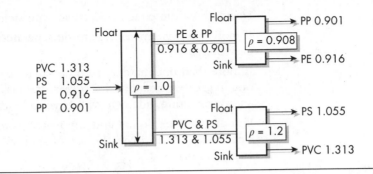

FIGURE 3–18 A process train for the separation of four types of plastics by float/sink. See Example 3.12.

But is this the best order of separation? Suppose instead of the process train in Figure 3-18 the PVC is removed first, resulting in the process shown in Figure 3-19.

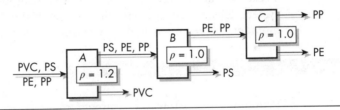

FIGURE 3–19 An alternative process train for the separation of plastics by float/sink.

In the case of the first process train, the first separator receives all the flow, or 7 tons. The following two units receive 5 and 2 tons each, for a total of 14 tons. In the alternate processing system, the first separation device must handle 7 tons, next one 3 tons (4 are removed), and the last one 2 tons, for a total of 12 tons. Thus the second process train is to be preferred in terms of the least total quantity of material handled.

In the previous example, various fluids must be used to make the separation of solids possible. If a fluid of a desired density is to be used, economics would dictate that it be separated from the plastic particles after the plastic particles have been separated from each other. This requires an extra step (perhaps screening) and invariably results in the loss of some of the fluid, which must be made up, thereby increasing the cost of the separation.

It might be instructive to suggest a set of general rules for the placement of unit operations in a process train. These rules may be listed as follows:

1. Decide what material properties are to be exploited (e.g., magnetic vs. nonmagnetic; big vs. small, etc.). This becomes the code.

2. Decide how the code is to activate the switch.

3. If more than one material is to be separated, try to separate the easy one first.

4. If more than one material is to be separated, try to separate the one in greatest quantity first. (This rule may contradict Rule 3, and engineering judgment will have to be exercised.)

5. If at all possible, do not add any materials to facilitate separation since this often involves the use of another separation step to recover the material.

3.3 ◆ Materials Balances with Reactors

To this point we have assumed not only that the system is in steady state, but also that there is no production or destruction of the material of interest. Consider next a system where the material is being destroyed or produced in a reactor, but in which the steady-state assumption is maintained. That is, the system does not change with time so that if the flows are sampled at any given moment, the results will always be the same.

e • x • a • m • p • l • e   **3.13**

Problem   In wastewater treatment, microorganisms are often used to convert dissolved organic compounds to more microorganisms, which are then removed from the flow stream by such processes as thickening (see Example 3.12). One such operation is known as the *activated sludge system* shown in Figure 3–20.

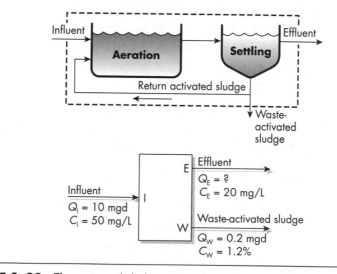

FIGURE 3–20   The activated sludge system. See Example 3.13.

Since the activated sludge system converts dissolved organics to suspended solids, it yields excess solids. These solids are the microorganisms that do the conversion from dissolved solids to biomass. Unfortunately, the process is so efficient that some of these microorganisms (solids) must be wasted. The slurry of these solids removed from the activated sludge system is called *waste activated sludge*.

Suppose an activated sludge system has an influent (feed) of 10 mgd (million gallons per day) at a suspended-solids concentration of 50 mg/L. The waste activated sludge flow rate is 0.2 mgd at a solids concentration of 1.2%. The effluent (discharge) has a solids concentration of 20 mg/L. What is the yield of waste activated sludge in pounds per day? (Restated: what is the rate of solids production in the system?) Assume steady-state condition.

Solution   With reference to Figure 3–20, the entire system is treated as the black box, and the variables are added.

The first balance is in terms of the volume, producing

$$0 = [10] - [0.2 + Q_E] + 0 - 0$$

$$Q_E = 9.8 \text{ mgd}$$

The second balance is in terms of the suspended solids,

$$\begin{bmatrix} \text{solids} \\ \text{ACCUMULATED} \end{bmatrix} = \begin{bmatrix} \text{solids} \\ \text{IN} \end{bmatrix} - \begin{bmatrix} \text{solids} \\ \text{OUT} \end{bmatrix} + \begin{bmatrix} \text{solids} \\ \text{PRODUCED} \end{bmatrix} - \begin{bmatrix} \text{solids} \\ \text{CONSUMED} \end{bmatrix}$$

$$0 = [Q_I C_I] - [Q_E C_E + Q_{WC} W] + [X] - 0$$

where X is the rate at which solids are produced. Since all the terms are in units of

$$\frac{\text{million gal}}{\text{day}} \times \frac{\text{mg}}{\text{L}}$$

they must all be converted to lb/day. Remember that mgd × mg/L × 8.34 = lb/day.

$$1 \frac{\text{mil gal}}{\text{day}} \times 3.78 \times 10^6 \frac{\text{L}}{\text{mil gal}} \times 1 \frac{\text{mg}}{\text{L}} \times 10^{-6} \frac{\text{kg}}{\text{mg}} \times 2.2 \frac{\text{lb}}{\text{kg}} = 8.34 \frac{\text{lb}}{\text{day}}$$

$$0 = [Q_I C_I] - [Q_E C_E + Q_W C_W] + [X]$$

$$0 = [(10)(50)(8.34)] - [(9.8)(20)(8.34)$$

$$+ (0.2)(12,000)(8.34)] + [X]$$

$$X = 17,438 \text{ lb/day}$$

◆

While the activated sludge system produces suspended solids, other processes in wastewater treatment such as anaerobic digestion are designed to destroy suspended solids. The anaerobic digestion system can also be analyzed using mass balances.

e • x • a • m • p • l • e **3.14**

Problem   In wastewater treatment, the digestion of sludge produces useful gas (about 50% carbon dioxide and 50% methane) and in the course of digestion, the sludge becomes less odoriferous. Raw sludges can contain pathogenic microorganisms, and digestion to some extent disinfects the sludge and may make it easier to dewater. Typically, digestion is carried out in a two-stage system shown in Figure 3–21A. The first digester, called the primary digester, is the main reactor where the anaerobic (without oxygen) microorganisms convert the high-energy organic solids (called *volatile solids*) to methane gas, carbon dioxide, and lower energy solids, which make up the *digested sludge*. The second tank, the *secondary digester*, is used as both a gas storage tank and a solids separator, where the heavier solids settle to the bottom and a *supernatant* is drawn off the sludge and reintroduced to the treatment process. In the process, some of the volatile solids entering are converted to gas that is collected in the secondary digester and used as needed.

Assuming steady-state conditions, let the feed solids (C_0) = 4%, of which 70% are volatile solids, and the feed flow rate (Q_0) = 0.1 m³/s. The gas, methane and carbon dioxide, contain no solids. The supernatant solids are 2% at 50% volatile solids and the digested sludge is at a solids concentration of 6% and 50% volatile solids. Find the flow rate of the supernatant and digested sludge.

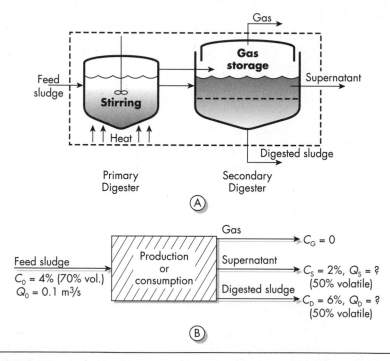

FIGURE 3–21   A sludge digestion system, simplified to a black box. See Example 3.14.

Solution First redraw the system as shown in Figure 3–21B. There are several possible balances. The following three are convenient:

- volatile solids
- total solids
- sludge volume (solids plus liquid)

1. The mass balance in terms of volatile solids reads

$$\begin{bmatrix} \text{volatile} \\ \text{solids} \\ \text{ACCUMULATED} \end{bmatrix} = \begin{bmatrix} \text{volatile} \\ \text{solids} \\ \text{IN} \end{bmatrix} - \begin{bmatrix} \text{volatile} \\ \text{solids} \\ \text{OUT} \end{bmatrix}$$

$$+ \begin{bmatrix} \text{volatile} \\ \text{solids} \\ \text{PRODUCED} \end{bmatrix} - \begin{bmatrix} \text{volatile} \\ \text{solids} \\ \text{CONSUMED} \end{bmatrix}$$

The volatile solids enter as the feed and may leave as the volatile solids in the gas, supernatant, and digested sludge. In addition, there are volatile solids consumed by the process.

Recall that in a mass balance the mass of the solids can be expressed as the product of the flow and concentrations. Since there are no volatile solids produced, and steady state is assumed,

$$0 = [Q_0 C_0] - [Q_G C_G + Q_S C_S + Q_D C_D] + 0 - [X]$$

where X = consumption of volatile solids, g/s, and the subscripts 0, G, S, and D refer to feed, gas, supernatant, and digested sludge, respectively.

$$0 = [0.1(40{,}000)(0.7)] - [0 + Q_S(20{,}000)(0.5)$$

$$+ Q_D(60{,}000)(0.5)] + 0 - [X]$$

The gas of course has no volatile solids, so $Q_G C_G = 0$. Simplified,

$$0 = 2800 - 10{,}000 Q_S - 30{,}000 Q_D - X$$

2. The balance in terms of total solids is

$$0 = [(0.1)(40{,}000)] - [0 + Q_S(20{,}000)$$
$$+ Q_D(60{,}000)] + [0] - [X]$$

3. The liquid volume balance can be approximated as

$$0 = Q_0 - Q_S - Q_D$$

since the volume of liquid in the gas is very small compared to the other effluent streams. There is no production or consumption of the liquid, so the last terms are zero, and

$$0 = 0.1 - Q_S - Q_D$$

We now have three equations and three unknowns.

1. $0 = 2800 - 10{,}000Q_S - 30{,}000Q_D - X$
2. $0 = 4000 - 20{,}000Q_S - 60{,}000Q_D - X$
3. $0 = 0.1 - Q_S - Q_D$

Solving these: $Q_D = 0.01$ m³/s, $Q_S = 0.09$ m³/s, and $X = 1600$ g volatile solids removed per second.

◆

The next assumption to be considered is the steady-state problem. This step requires some preparation in concepts such as reactions, reactors, and other process dynamics that are discussed in Chapter 4.

◆ Abbreviations

A	= area	Q_M	= mass flow rate, mass/unit time
C	= concentration, mass/volume	Q_V	= volume flow rate, volume/unit time
E_R	= Rietema efficiency		
E_{WS}	= Worrell–Stessel efficiency	R_{xn}	= recovery of component x in
P_{xn}	= purity of component x in product stream n		produce stream n
		v	= velocity
Q	= flow, as either mass or volume per unit time		

◆ Problems

3–1 A pickle-packing plant produces and discharges a waste brine solution with a salinity of 13,000 mg/L NaCl, at a rate of 100 gal/min, discharged into a stream with a flow rate above the discharge of 1.2 million gal/day and a salinity of 20 mg/L. Below the discharge point is a prime sport fishing spot and the fish are intolerant to salt concentrations above 200 mg/L.

a. What must the level of salt in the effluent be in order to reduce the level in the stream to 200 mg/L?

b. If the pickle-packing plant has to spend $8 million to reduce its salt concentration to this level, it cannot stay in business. More than 200 people will lose their jobs. All this because a few people like to fish in the stream. Write a letter to the editor expressing your views, pro or con, concerning the action of the state in requiring the pickle plant to clean up its effluent. Use your imagination.

c. Suppose now that the pickle plant is 2 miles upstream from a brackish estuary, and asks permission to pipe its salty wastewater into the estuary where the salinity is such that the waste would actually *dilute* the estuary (the waste has a lower salt concentration than the estuary water). As plant manager, write a letter addressed to the Director of the State Environmental Management Division requesting a relaxation of the salinity effluent standard for the pickle plant. Think about how you are going to frame your arguments.

3-2 Raw primary sludge at a solids concentration of 4% is mixed with waste activated sludge at a solids concentration of 0.5%. The flows are 20 and 24 gal/min, respectively. What is the resulting solids concentration? This mixture is next thickened to a solids concentration of 8% solids. What are the quantities (in gallons per minute) of the thickened sludge and thickener overflow (water) produced? Assume perfect solids capture in the thickener.

3-3 A 10-mgd wastewater treatment plant has influent and effluent concentrations of several metals as follows:

Metal	Concentration of the Metal in the	
	Influent (mg/L)	Effluent (mg/L)
Cd	0.012	0.003
Cr	0.32	0.27
Hg	0.070	0.065
Pb	2.42	1.26

The plant produces a dewatered sludge cake at a solids concentration of 22% (by weight). Plant records show that 45,000 kg of sludge (wet) are disposed of on a pasture per day. The state has restricted sludge disposal on land only to sludges that have metal concentrations less than the following:

Metal	Maximum Allowable Metal Concentration (mg metal/kg dry sludge solids)
Cd	15
Cr	1000
Hg	10
Pb	1000

Is the sludge meeting the state standards?

HINT: There are two ways of solving this problem. One approach is to assume that the flow rate (mgd of sludge) is negligible compared to the flow rate of the influent. That is, assume $Q_{influent} = Q_{effluent}$, where Q = flow rate in mgd. If you make this assumption to solve the problem, check to see if it was a valid assumption. Alternatively, you can assume that the density of the sludge solids is about that of water, so that 1 kg of sludge represents a volume of 1 L.

3-4 Two flasks contain 40% and 70% by volume formaldehyde, respectively. If 200 g out of one of the flasks and 150 g out of the second are mixed, what is the concentration (expressed as percent formaldehyde by volume) of the formaldehyde in the final mixture?

3-5 A textile plant discharges a waste containing 20% dye from vats. The color intensity is too great and the state has told the plant manager to reduce the color in the discharge. The plant chemist tells the manager that color would not be a problem if they could have no more than 8% dye in the wastewater. The plant manager decides that the least expensive way of doing this is to dilute the 20% waste stream with drinking water so as to produce an 8% dye waste. The 20% dye wastewater flow is 900 gallons per minute.

    a. How much drinking water is necessary for the dilution?

    b. What do you think of this method of pollution control?

    c. Suppose the plant manager dilutes his waste and the state regulatory personnel assume that the plant is actually removing the dye before discharge. The plant manager, because he has run a profitable operation, is promoted to corporate headquarters. One day he is asked to prepare a presentation for the corporate board (the big cheeses) on how he was able to save so much money on wastewater treatment. Develop a short play for this meeting, starting with the presentation by the former plant manager. He will, of course, try to convince the board that he did the right thing. What will be the board's reaction? Include in your script the company president, the treasurer, the legal counsel, and any other characters you want to invent.

3-6 A capillary sludge drying system operates at a feed rate of 200 kg/hr and accepts a sludge with a solids content of 45% (55% water). The dried sludge is 95% solids (5% water) and the liquid stream contains 8% solids. What is the quantity of dried sludge and what is the liquid and solids flow in the liquid stream?

3-7 A solid-waste processing plant has two air classifiers that produce a refuse-derived fuel from a mixture of organic (A) and inorganic (B) refuse. A portion of the plant schematic and the known flow rates are shown in Figure 3-22.

    a. What is the flow of A and B from Classifier I to Classifier II?

    b. What is the purity of the refuse-derived fuel, and what is the overall recovery of organics in the two air classifiers?

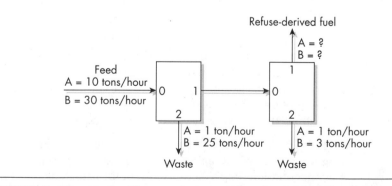

FIGURE 3-22 Air classifiers producing a refuse-derived fuel. See Problem 3-7.

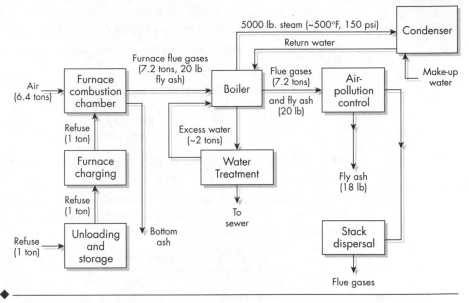

FIGURE 3-23 Materials flow in a solid-waste incinerator. See Problem 3–10.

3-8 A separator accepts waste oil at 70% oil and 30% water by weight. The top product stream is pure oil, whereas the bottom underflow contains 10% oil. If a flow of 20 gal/min is fed to the tank, how much oil is recovered, as gal/min?

3-9 The Mother Goose Jam Factory makes jam by combining black currants and sugar at a weight ratio of 45 : 55. This mixture is heated to evaporate the water until the final jam contains 1/3 water. If the black currants originally contain 80% water, how many kilograms of berries are needed to make one kilogram of jam?

3-10 The flow diagram in Figure 3-23 shows the materials flow in a heat-recovery incinerator.

  a. How much fly ash will be emitted out of the stack (flue) per ton of refuse burned?

  b. What is the concentration of the particulates (fly ash) in the stack, expressed as $\mu g/m^3$ (μg of fly ash per m^3 of flue gases emitted from the stack)? Assume 1 ton of flue gases has a volume of 500 m^3.

3-11 An electrostatic precipitator for a coal-fired power plant has an overall particulate recovery of 95% (it removes 95% of all the flue gas particulates coming to it). The company engineer decides that this is too good, that it is not necessary to be quite this efficient, and proposes that part of the flue gas be bypassed around the electrostatic precipitator so that the recovery of fly ash (particulates) from the flue gas will be only 85%.

  a. What fraction of the flue gas stream would need to be bypassed?

  b. What do you think the engineer means by "too good"? Explain her thought processes. Can any pollution-control device ever be "too good"? If so, under what circumstances? Write a one-page essay entitled "Can Pollution Control Ever Be Too Effective?" Use examples to argue your case.

3–12   Two air-classifier manufacturers report the following performance for their units:

> Manufacturer A: Recovery of organics = 80%
> Recovery of inorganics = 80%
> Manufacturer B: Recovery of organics = 60%
> Purity of the extract = 95%

Assume that the feed consists of 80% organics. Which unit is better? (*Note:* The air classifier is supposed to separate organics from inorganics.)

3–13   A package water treatment plant consists of filtration and settling, both designed to remove solids, but solids of different sizes. Suppose the solids removals of the two processes are as follows:

	Small Solids (%)	Large Solids (%)
Filter	90	98
Settling tank	15	75

If the two types of solids are each 50% by weight, would it be better to put the filter before the settling tank in the process train? Why?

3–14   A cyclone used for removing fly ash from a stack in a coal-fired power plant has the following recoveries of the various size particulates:

Size (μ)	Fraction of Total, by Weight	Recovery
0–5	0.60	5
5–10	0.18	45
10–50	0.12	82
50–100	0.10	97

What is the overall recovery of fly ash particulates?

3–15   In the control of acid rain, power plant sulfur emissions can be controlled by scrubbing and spraying the flue gases with a lime slurry, $Ca(OH)_2$. The calcium reacts with the sulfur dioxide to form calcium sulfate, $CaSO_4$, a white powder suspended in water. The volume of this slurry is quite large, so it has to be thickened. Suppose a power plant produces 27 metric tons of $CaSO_4$ per day (one metric ton = 1000 kg), and that this is suspended in 108 metric tons of water. The intention is to thicken this to a solids concentration of 40% solids.

a.   What is the solids concentration of the $CaSO_4$ slurry when it is produced in the scrubber?

b.   How much water (tons/day) will be produced as effluent in the thickening operation?

c.   Acid rain problems can be reduced by lowering SO_2 emissions into the atmosphere. This can occur by pollution control, such as lime slurry scrubbing, or by reducing the amount of electricity produced and used. What

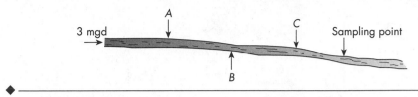

FIGURE 3–24 Stream receiving three discharges. See Problem 3–16.

responsibility do individuals have, if any, not to waste electricity? Argue both sides of the question in a two-page paper.

3–16 A stream (Figure 3-24), flowing at 3 mgd and 20 mg/L suspended solids, receives wastewater from three separate sources:

Source	Quantity (mgd)	Solids Concentration (mg/L)
A	2	200
B	6	50
C	1	200

What are the flow and suspended solids concentration downstream at the sampling point?

3–17 An industrial flow of 12 L/min has a solids concentration of 80 mg/L. A solids-removal process extracts 960 mg/min of the solids from the flow without affecting the liquid flow. What is the recovery (in percent)?

3–18 A manufacturer of beef sticks produces a wastewater flow of 2000 m³/day containing 120,000 mg/L salt. It is discharging into a river with a flow of 34,000 m³/day and a salt concentration of 50 mg/L. The regulatory agency has told the company to maintain a salt level of no greater than 250 mg/L downstream of the discharge.

a. What recovery of salt must be accomplished in the plant before the waste is discharged?

b. What could they do with the recovered salt? Suggest alternatives.

3–19 A community has a persistent air-pollution problem called an inversion (discussed in Chapter 12) that creates a mixing depth of only about 500 m above the community. The town's area is about 9 km by 12 km. A constant emission of particulates of 70 kg/m²-day enters the atmosphere.

a. If there is no wind during one day (24 hours) and if the pollutants are not removed in any way, what is the *lowest* concentration level of particles in the air above the town at the end of that period?

b. If one of the major sources of particulates is wood-burning stoves and fireplaces, should the government have the power to prohibit the use of wood stoves and fireplaces? Why or why not?

c. How much should government restrict our lifestyle because of pollution controls? Why does the government seem to be doing more of this now

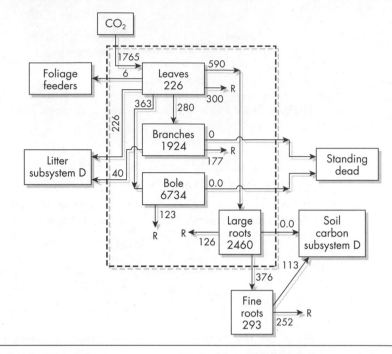

FIGURE 3-25   A forest ecosystem. See Problem 3–20. R = decay, g/m²-yr.

than just a few years ago? Where do you think it's all headed and why is it happening? Is there anything that we can do about it? Respond to these questions in a one-page paper.

3-20   Figure 3-25 shows a diagram of organic matter in a forest ecosystem. All compartmented (boxed) numbers are in grams per square meter of the forest floor, and all transfers are in grams per square meter per year. This system is not in equilibrium. Calculate the total change (g/m²-yr) in leaves, branches, bole, large roots, and fine roots (all together). R is loss by decay, g/m²-yr.

3-21   The Wrightsville Beach, North Carolina, water-treatment plant produces a distilled water of zero salinity. To reduce the cost of water treatment and supply, the city mixes this desalinated water with untreated groundwater, which has a salinity of 500 mg/L. If the desired drinking water is to have a salinity not exceeding 200 mg/L, how much of each water—distilled water and untreated groundwater—is needed to produce a finished water flow of 100 million gallons per day?

3-22   A cyclone used to control the emissions from a furniture manufacturer is expected to be 80% efficient (80% recovery). The hopper under the cyclone holds 1 ton of wood dust. If the air flow to the cyclone holds 1.2 tons per day of wood dust, how often must the hopper be emptied?

3-23   An industry has been fined heavily by the EPA because it has been polluting a stream with a substance called "gloop." It is presently discharging 200 liters per minute (L/min) of effluent at a "gloop" concentration of 100 mg/L. The stream flowing past the plant into which the effluent is discharging has an upstream flow rate of 3800 L/min and has no "gloop" in it. The EPA has told the plant that it must reduce its effluent "gloop" concentration to below 20 mg/L. The plant engineer decides that the cheapest way to meet the EPA requirement is to divert part of the stream into the plant and dilute the effluent so as to lower the "gloop" concentration to the required level.

   a. What is the present "gloop" concentration downstream from the plant discharge?

   b. How much stream water would the plant engineer need in order to achieve the required "gloop" concentration in the plant effluent?

   c. What will be the "gloop" concentration in the stream downstream from the plant if the plant engineer's scheme is put into operation?

3–24   A 1000-tons/day ore processing plant processes an ore that contains 20% mineral. The plant recovers 100 tons/day of the pure mineral. This is not a very effective operation, and some of the mineral is still in the waste ore (the tailings). These tailings are sent to a secondary processing plant that is able to recover 50% of the mineral in the tailings also at 100% purity. The recovered mineral from the secondary recovery plant is sent to the main plant, where it is combined with the other recovered mineral and then all of it is shipped to the user.

   a. How much mineral is shipped to the user?

   b. How much total tailings are to be wasted?

   c. How much mineral is wasted in the tailings?

3–25[3]   Since newsprint makes up about 17% of municipal solid waste, it appears to be an especially good candidate for recycling. Yet the price of old newspapers has not been very high. One reason for the low price of old newsprint is the economics of converting a paper mill from processing trees to processing old newsprint. This is illustrated by the situation below.

      The *Star Tribune* and the *St. Paul Pioneer Press and Dispatch* represent 75% of the newspaper circulation in Minnesota. The *Star Tribune* has a circulation of 2,973,100 and weighs about 1.08 lb per issue, while the *St. Paul Pioneer Press and Dispatch* has a circulation of 1,229,500 and weighs 0.80 lb per issue.

      Of course, not all of what the consumer receives is recyclable newsprint. Seventeen percent of the weight is non-newsprint material for glossy advertisements and magazines, and 1% is ink. Also, 5% of what is printed is overissues and never reaches the consumer, while 3% of the newsprint delivered to the press becomes newsroom scrap and is not printed.

   a. Find the mass of newspaper that reaches consumers in Minnesota each year. How much of this is recyclable newsprint, including ink but excluding glossy pages?

   b. If 65% of consumed newsprint and 100% of overissue newsprint and pressroom scrap is collected for recycling, what is the mass of recycled material sent to the pulp mills? Recall that overissue is 5% of circulation and pressroom scrap is 3% of the newsprint required for printing.

c. Newsprint must be deinked before it can be reused. Assume that the deinking process removes all the ink, but in doing so 15% of the paper is lost as waste. Calculate the mass of dry recycled pulp produced.

d. If it takes 1.5 cords of wood to create one ton of virgin pulp and there are an average of 21.3 cords per acre of timberland, how many acres of trees are saved annually by this level of recycling?

e. What would the cost of recycled newsprint have to be to justify the conversion of a 750 ton/day pulp mill to recycled fiber? Base your calculations on a virgin material cost of $90/ton, a conversion cost of $100 million, and a 10% return on investment. Assume that the cost of producing pulp from the two feedstocks is identical, so that the only economic incentive for converting to recycled newsprint is the difference between feedstock costs.

3-25[4]   According to the statements of the Draeger Works in Luebeck, in the gassing of the whole population in a city only 50% of the evaporated poison gas is effective. The atmosphere must be poisoned up to a height of 20 meters at a concentration of 45 mg/m^3. How much phosgene is needed to poison a city of 50,000 inhabitants who live in an area of four square kilometers? How much phosgene would the population inhale with the air they breathe in 10 minutes without protection against gas, if one person uses 30 liters of breathing air per minute? Compare this quantity with the quantity of the poison gas used.

◆ Endnotes

1. Based on R. E. Kohn, *A Linear Programming Model for Air Pollution Control*, MIT Press, Cambridge, MA (1978).

2. This example is adapted from B. M. Berthourex and D. F. Rudd, *Strategy for Pollution Control*, John Wiley & Sons, New York (1977).

3. This problem appears in David T. Allen, N. Bakshani, and Kirsten Sinclair Rosselot, *Pollution Prevention: Homework and Design Problems for Engineering Curricula*, American Institute of Chemical Engineers and other societies (1992). Used with permission.

4. This problem first appeared in A. Dorner, *Mathematics in the Service of National-Political Association: A Handbook for Teachers*, and is reprinted in Elie A. Cohen, *Human Behaviour in the Concentration Camp* (1988). The Dorner book was intended to make students familiar and "comfortable" with mass murder during the Nazi era in Germany.

4

REACTIONS

$\mathbf{M}$any reactions, whether in nature or in human-created environments, are predictable. Engineers and scientists first strive to understand these reactions and then create mathematical models to describe the behavior of the phenomena. With these models, they can try to predict what would happen if . . .

Such models can be simple or highly complex. If a pipeline under pressure discharges a certain flow rate, and if the pressure is doubled, it is possible to predict fairly accurately the new flow rate. Such simple hydraulic models are the bread and butter of civil engineering.

It is considerably more difficult to try to predict the future of some global phenomenon such as the presence of the hole in the ozone layer. What we know is that chlorofluorocarbons (CFCs) emitted from various sources such as refrigerators and air conditioners are highly stable, travel to the upper atmosphere, and once there, probably react with the ozone. But nobody knows this for sure. It is possible to perform experiments in the laboratory that simulate upper atmosphere conditions and indicate that the effect of the CFC indeed is to decrease ozone concentrations, and it is possible to measure the concentration of ozone in the stratosphere, but there still is no proof that this is what is actually happening. The models that would predict this reaction have not been empirically proven and the drop in the ozone concentration could be a purely natural phenomenon, independent of the CFC emissions. Is it fair to require the manufacturers of CFCs to cut back on production, and for the users of CFCs to use other, less-efficient gases, all based on unprovable mathematical models?

Any large and complex model such as ozone depletion or photochemical smog formation can be in error. What has to be evaluated are the consequences of not believing the model contrasted with the degree of uncertainty. Although the ozone depletion model cannot be proven conclusively, it is mathematically and chemically reasonable and logical and is therefore most probably correct. Ignoring the potential for global disaster predicted by such models is simply foolhardy.

Because the processes are complex, global models are intricate and highly interconnected. Fortunately, these powerful mathematical models have humble beginnings. We begin by assuming that some quantity (mass or volume) changes with time, and that the quantity can be predicted using simple rate equations.

In the previous chapter, materials flow is analyzed as steady-state operations. Time is not a variable. In this chapter we consider the case where materials concentrations change with time.

A general mathematical expression describing a rate at which the mass or volume of some material A is changing with time t is

$$\frac{dA}{dt} = r$$

where r = reaction rate. *Zero-order reactions* are defined as those where r is a constant, or

$$r = k$$

where k = reaction rate constant, so that

$$\frac{dA}{dt} = k \tag{4.1}$$

First-order reactions are defined as those where the change of the component A is proportional to the quantity of the component itself, so that

$$r = kA$$

and

$$\frac{dA}{dt} = kA \tag{4.2}$$

Note that the units of the reaction rate constant k in first-order reactions is time^{-1}, such as d^{-1}.

   Second-order reactions are ones where the change is proportional to the square of the component, or

$$r = kA^2$$

and

$$\frac{dA}{dt} = kA^2 \tag{4.3}$$

   In environmental engineering applications, equations 4.1, 4.2, and 4.3 are usually written in terms of the concentration, so that

$$\frac{dC}{dt} = k \tag{4.4}$$

for zero-order reactions,

$$\frac{dC}{dt} = kC \tag{4.5}$$

for first-order reactions, and

$$\frac{dC}{dt} = kC^2 \tag{4.6}$$

for second-order reactions.

4.1 ◆ Zero-Order Reactions

Many changes occur in nature at a constant rate. Consider a simple example where a bucket is being filled from a garden hose. The volume of water in the bucket is changing with time, and this change is constant. If at zero time the bucket has 2 L of water in it, at 4 seconds it has 3 L, at 6 seconds it has 4 L, and so on, the change in the volume of water in the bucket is constant, at a rate of

$$1 \text{ liter}/2 \text{ seconds} = 0.5 \text{ L/sec}$$

This can be expressed mathematically as a zero-order reaction,

$$\frac{dA}{dt} = k$$

where A is the volume of water in the bucket and k is the rate constant. Defining $A = A_0$ at time $t = t_0$, and integrating,

$$\int_{A_0}^{A} dA = k \int_0^t dt$$

$$A - A_0 = -kt \tag{4.7}$$

This equation is of course how the "water-in-the-bucket" problem is solved. Substitute $A_0 = 2$ L at $t = 0$ and $A = 3$ L at $t = 2$ sec,

$$3 = 2 + k(2)$$

$$k = 0.5 \text{ L/sec} \tag{4.8}$$

In this case the units of k are in volume/time. When the equation is written in terms of mass, the units of the rate constant k are mass/time, such as kg/sec. When the equation is in terms of concentration, which it very often is, the rate constant in a zero-order reaction has units of mass/volume-time, or mg/L-sec if C is in mg/L and t in seconds.

The integrated form of the zero-order reaction when the concentration is *increasing* is

$$C = C_0 + kt \tag{4.9}$$

and if the concentration is *decreasing*, the equation is

$$C = C_0 - kt \tag{4.10}$$

This equation can be plotted as shown in Figure 4-1, if the material is being destroyed or consumed and the concentration C is decreasing. If, however, the material is being produced and C is increasing, the slope is positive.

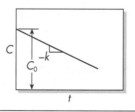

FIGURE 4-1   Plot of a zero-order reaction in which the concentration is decreasing.

e • x • a • m • p • l • e   4.1

Problem   An anteater finds an anthill and starts eating. The ants are so plentiful that all he has to do is flick out his tongue and gobble them up at a rate of 200 per minute. How long will it take him to have a concentration of 1000 ants per anteater in the anteater?

Solution

C = concentration of ants in the anteater at any time t, ants/anteater.

C_0 = initial concentration of ants at time $t = 0$, ants/anteater.

k = reaction rate, the number of ants consumed per minute,

= 200 ants/anteater-minute. (*Note:* k is positive, since the concentration is increasing.)

According to equation 4.9;

$$C - C_0 = kt$$

$$1000 - 0 = 200(t)$$

$$t = 5 \text{ min}$$

$\blacklozenge$

4.2 $\blacklozenge$ First-Order Reactions

The *first-order* reaction can be expressed as

$$\frac{dA}{dt} = r = -kA$$

This equation can be integrated between A_0 and A, and between $t = 0$ and t:

$$\int_{A_0}^{A} \frac{dA}{A} = -k \int_0^t dt$$

$$\ln \frac{A}{A_0} = -kt \tag{4.11}$$

or

$$\frac{A}{A_0} = e^{-kt}$$

or

$$\ln A - \ln A_0 = -kt$$

The logarithms used in this text are the so-called natural or base e logarithms. It is also sometimes useful to use the so-called common or base 10 logarithms. This conversion from natural (base e) to common logarithms (base 10) makes sense because the ratio of one to the other is

$$\frac{\log A}{\ln A} = \frac{k't}{kt} = \frac{k'}{k}$$

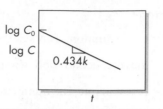

$$\log C_0$$
$$\log C$$
$$0.434k$$
$$t$$

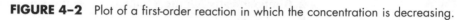

FIGURE 4–2   Plot of a first-order reaction in which the concentration is decreasing.

Suppose $A = 10$, then

$$\frac{\log 10}{\ln 10} = \frac{1}{2.302} = \frac{k'}{k} = 0.434$$

Thus

$$k' = (0.434)k$$

$$k_{\text{base } 10} = (0.434)k_{\text{base } e}$$

The advantage of base 10 logs is that they can be conveniently plotted on ordinary semi-log paper, and Figure 4–2 shows that plotting $\log C$ versus t yields a straight line with a slope $0.434k$. The intercept is $\log C_0$. (Don't be confused between A and C. If the mass of some material is A, in a given volume V, C is the mass A divided by the volume V. If the volume is constant, this term drops out. Thus the previous argument could have been in terms of C as well as A.)

The slope on a semi-log paper is found by recognizing that

$$\text{slope} = \frac{\Delta y}{\Delta x}$$

where Δy is the increment on the y-axis (the ordinate) corresponding to the increment Δx on the x-axis (the abscissa). Choosing the y as one cycle on the log scale, such as $(\log 10 - \log 1)$, and since $\log 10 = 1$ and $\log 1 = 0$,

$$\text{slope} = \frac{1}{\Delta x_{\text{for one cycle}}}$$

Note again that the mass A and the concentration C are interchangeable in the foregoing equations if the volume is constant.

e • x • a • m • p • l • e   **4.2**

Problem   An owl eats frogs as a delicacy, and his intake of frogs is directly dependent on how many frogs are available. Since the rate is a function of the concentration, this can be described as a first-order reaction. $C =$ concentration of frogs at a pond, at any time t, $C_0 =$ initial frog population, 200 frogs per pond, and k = rate constant, 0.1 day^{-1}. How many frogs are left at the end of 10 days?

Solution   Using equation 4.11

$$\ln C - \ln C_0 = -kt$$

$$\ln C - \ln(200) = -0.1(10)$$

$$\ln C = 4.3$$

$$C = 73 \text{ frogs per pond}$$

◆

4.2.1   Deep Ecology

Owls eating frogs is a part of an ecosystem; frogs eat daphnae, daphnae eat algae, and algae grow in the water. As you will see described in Chapter 6, life in such an ecosystem is in balance, allowing all species to survive. Only people seem to be capable of consciously and purposefully destroying such an ecological balance. These considerations prompted a contemporary Norwegian philosopher Arne Naess to try to create a new environmental philosophy, which he called "Deep Ecology." He proposed that environmental ethics should be based on two fundamental values—*self-realization* and *biocentric equality*. These values are intuited and cannot be rationally justified.

Self-realization is the recognition of oneself as a member of the greater universe, and not just as a single individual or even as a member of a restricted community. This self-realization can be achieved, according to Naess, by reflection and contemplation. The tenets of deep ecology are not inherently tied to any particular religious belief, although like Marxism or any other set of beliefs and values, it may function as a religion for the believer.

> Most people in deep ecology have had the feeling—usually, but not always in nature—that they are connected with something greater than their ego. . . . Insofar as these deep feelings are religious, deep ecology has a religious component . . . [a] fundamental intuition that everyone must cultivate if he or she is to have a life based on value and not function like a computer.[1]

The second tenet of deep ecology is biocentric equality. According to Naess, this follows from self-realization, in that once we have understood ourselves to be at one with other creatures and places in the world, we cannot regard ourselves as superior. Everything has an equal right to flourish, and humans are not special. We must eat and use other creatures in nature to survive, of course, but we must not exceed the limits of our "vital" needs. The collecting of material wealth, or goods above the vital necessities, is therefore unethical to a deep ecologist.

If one accepts these fundamental propositions, then one must agree with the following three corollaries:[2]

1. All life, human and nonhuman, has value in itself, independent of purpose, and humans have no right to reduce its richness and diversity except for vital needs.

2. Humans at present are far too numerous and intrusive with respect to other forms and the living earth, with disastrous consequences for all, and must achieve a "substantial decrease" in population to permit the flourishing of both human and nonhuman life.

3. To achieve this requisite balance, significant change in human economic, technical, and ideological structures must be made, stressing not bigness, growth, and higher standards of living, but sustainable societies emphasizing the (nonmaterial) quality of life.

To a deep ecologist, wilderness has a special value and must be preserved for the psychological benefit of humans, as well as providing a place for other species to survive unhindered by human activity. Deep ecologists seek a sense of place in their lives, a part of the earth that they can call home. Deep ecologists also oppose the capitalist/industrial system that seeks to create what they view as unnecessary and detrimental wealth. They disdain the idea of stewardship since it implies human decision making and human intervention in the workings of natural environments. It seeks to regain what it views as the sympathetic and supportive relationship between people and nature held by primitive peoples such as the Native Americans prior to the European invasion.

There is of course no reason why someone who wishes to adopt deep ecology as a personal choice, to adopt it as an ideal, should not do so. Indeed, there may be deep ecologists who consider that the purpose of ethics is to develop a virtuous character. In fact some philosophers have criticized deep ecology precisely because of what they consider to be its excessive focus on the spiritual development of the individual. But most of the tenets of deep ecology require large-scale changes involving the whole population.

These ideas will, obviously, rattle a lot of cages in the capitalist and political mainstream. The precepts of deep ecology offer a challenge to fundamental mainstream ways of doing business. As a result, it has sometimes been characterized as nonhumanist because it does not consider humans to be deserving of special benefits and indicts humans as the exploiters of nature. Especially galling to critics is the call by deep ecologists to depopulate the world, an objective that has been called fascist[3] and dehumanizing. There is no doubt that if we are to have less adverse effect on nature, it would be necessary first to reduce the human population by a significant portion (90% has been suggested by some deep ecologists). "Reduce" is presumably a euphemism for "kill off." In all likelihood there is little popular support for this view. Critics also argue that the acceptance of deep ecology as a global philosophy would revert civilization to the hunter-gatherer societies.

Some critics of deep ecology have argued that it is unacceptably elitist: that is, while its vision of a world largely reverted to wilderness and containing far fewer people may appeal to a small number of the relatively rich and highly educated, it has nothing to offer to the great masses of poor humans trying to survive. Because deep ecology requires voluntary acceptance of a lower material quality of life, it is unlikely that its basic tenets—self-realization and biocentric equality—will be widely and voluntarily accepted.

But back to reactions.

4.3 Second-Order and Noninteger-Order Reactions

The *second-order* reaction is defined as

$$\frac{dA}{dt} = r = -kA^2$$

Integrated

$$\int_{A_0}^{A} \frac{dA}{A^2} = -k \int_0^t dt$$

$$\left.\frac{1}{A}\right]_{A_0}^{A} = kt$$

$$\frac{1}{A} - \frac{1}{A_0} = kt$$

which plots as a straight line as shown in Figure 4-3.

The *noninteger order* (any number) is defined as

$$\frac{dA}{dt} = r = -kA^n$$

where n is any number. Integrated,

$$\left(\frac{A}{A_0}\right)^{1-n} - 1 = \frac{(n-1)kt}{A_0^{(1-n)}}$$

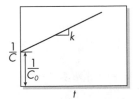

FIGURE 4–3   Plot of a second-order reaction in which the concentration is decreasing.

4.4 ◆ Half-Life

The *half-life* is defined as that time that one-half of the component in question has been converted. At $t = t_{1/2}$, A is 50% of A_0. By substituting $A/A_0 = 0.5$ in the previous expressions, the half-lives for various reaction orders are:

$$\text{First order:} \qquad t_{1/2} = \frac{\ln 2}{k} = \frac{0.693}{k}$$

$$\text{Second order:} \quad t_{1/2} = \frac{1}{k[A]_0} \qquad\qquad (4.12)$$

$$\text{Integer order:} \quad t_{1/2} = \frac{[(1/2)^{1-n} - 1] \cdot [A]}{(n-1)k}$$

e • x • a • m • p • l • e   4.3

Problem   Strontium-90 (^{90}Sr) is a radioactive nuclide found in water and many foods. It has a half-life of 28 years, and decays (as do all other radioactive nuclides) as a first-order reaction. Suppose the concentration of ^{90}Sr in a water sample is to be reduced by 99.9%. How long would this take?

Solution   Let $A_0 = 1$, then $A = 0.001$, or 99.9% reduction. Using equation 4.12, $t_{1/2} = 0.693/k$, and setting $t = 28$, $k = 0.02475$ yr^{-1}. Using equation 4.11,

$$\ln \frac{A}{A_0} = -kt = \ln\left(\frac{0.001}{1}\right) = -0.02475t$$

$$t = \frac{-6.907}{-0.02475} = 279 \text{ yr}$$

4.5 ◆ Consecutive Reactions

Finally, some reactions are *consecutive reactions* such that

$$A \rightarrow B \rightarrow C \rightarrow \cdots$$

If the first reaction is first order, then

$$\frac{dA}{dt} = -k_1 A$$

Likewise, if the second reaction is first order with respect to B, the overall reaction is

$$\frac{dB}{dt} = k_1 A - k_2 B$$

where k_2 is the rate constant for the reaction $B \to C$. Note that some B is being made at a rate of $k_1 A$ while some B is being destroyed, at a rate of $-k_2 B$. Integrated,

$$B = \frac{k_1 A_0}{k_2 - k_1}(e^{-k_1 t} - e^{-k_2 t}) + B_0 e^{-k_2 t} \qquad (4.13)$$

This equation is reintroduced in Chapter 6, where the oxygen level in a stream is analyzed and the oxygen deficit is defined as B, being increased as the result of O_2 consumption by the microorganisms and decreased by the O_2 diffusing into the water from the atmosphere. Before this reaction has much meaning, however, the question of stream ecosystems and gas transfer must be addressed.

Abbreviations

A = amount or any constituent, usually mass or volume
C = concentration, mass/volume
k = rate constant, base e logarithms
k' = rate constant, base 10 logarithms (reaction rate constants have different units depending on the order of the reaction)

r = reaction rate
t = time
V = volume

Problems

4-1   The die-off of coliform organisms below a wastewater discharge point can be described as a first-order reaction. It has been found that 30% of the coliforms die in 8 hours and 55% die in 16 hours. About how long would it take to have 90% die? Use semi-log paper to solve this problem.

4-2   A radioactive nuclide is reduced by 90% in 12 minutes. What is its half-life?

4-3   In a first-order process a blue dye reacts to form a purple dye. The amount of blue at the end of one hour is 480 g and at the end of three hours is 120 g. Graphically estimate the grams of blue dye present initially.

4-4   A reaction of great social significance is the fermentation of sugar with yeast. This is a zero-order (in sugar) reaction, where the yeast is a catalyst (it does not enter the reaction itself). If a 0.5-liter bottle contains 4 grams of sugar, and it takes 30 minutes to convert 50% of the sugar, what is the rate constant?

4-5 Integrate the differential equation where A is being made at a rate k_1 and destroyed simultaneously at a rate k_2.

$$\frac{dA}{dt} = k_1 A - k_2 A$$

4-6 A batch reactor is designed to remove "gobbledygook" by adsorption. The data are as follows:

Time (min)	Concentration of "Gobbledygook" (mg/L)
0	170
5	160
10	98
20	62
30	40
40	27

What order of reaction does this appear to be? Graphically estimate the rate constant.

4-7 There are many everyday processes that can be described in terms of reactors.

a. Give one example of a first-order reaction not already described in the text.

b. Describe *in words* what is going on.

c. Draw a *graph* describing this reaction.

4-8 An oil-storage area was abandoned 19 years ago. Oil spilled on the ground had saturated the soil at a concentration of perhaps 400 mg/kg of soil. A fast-food chain now wants to build a restaurant there, and samples the soil for contaminants, only to discover that the soil still contains oil residues at a concentration of 20 mg/kg. The local engineer concludes that since the oil must have been destroyed by the soil microorganisms at a rate of 20 mg/kg each year, that in one more year the site will be free of all contamination.

a. Is this a good assumption? Why or why not?

b. How long would *you* figure the soil will be below the acceptable contamination of 1 mg/kg?

4-9 A radioactive waste from a clinical laboratory contains 0.2 microcuries of calcium-45 (^{45}Ca) per liter. The half-life of ^{45}Ca is 152 days.

a. How long must this waste be stored before the level of radioactivity falls below the required 0.01 microcuries/liter?

b. Radioactive waste can be stored in many ways, including deep-well injection and above-ground storage. Deep-well injection involves the pumping of the waste thousands of feet below the earth's surface. Above-ground storage is in buildings in isolated areas that are then guarded to prevent people from getting near the waste. What would be the risk factors associated with each of these storage methods? What could go wrong? Which system of storage do you believe is superior for such a waste?

c. How would a deep ecologist view the use of radioactive substances in the health field, considering especially the problem of storage and disposal of radioactive waste? Try to think like a deep ecologist when responding. Do you agree with the conclusion you have drawn?

◆ Endnotes

1. Quoted in K. Sale, "Deep Ecology and Its Critics," *The Nation* 14:670–675 (May 1988).

2. K. Sale, "Deep Ecology and Its Critics," *The Nation* 14:670–675 (May 1988).

3. The term "environmental fascism" was coined by Tom Regan. See Joseph desJardins, *Environmental Ethics*, Wadsworth Publishing Co., Belmont, CA, p. 142 (1995).

chapter
5

REACTORS

Recall from Chapter 3 that processes can be analyzed by using a black box and by writing a materials balance:

$$
\begin{bmatrix} \text{rate of} \\ \text{materials} \\ \text{ACCUMULATED} \end{bmatrix} = \begin{bmatrix} \text{rate of} \\ \text{materials} \\ \text{IN} \end{bmatrix} - \begin{bmatrix} \text{rate of} \\ \text{materials} \\ \text{OUT} \end{bmatrix}
$$

$$
+ \begin{bmatrix} \text{rate of} \\ \text{materials} \\ \text{PRODUCED} \end{bmatrix} - \begin{bmatrix} \text{rate of} \\ \text{materials} \\ \text{CONSUMED} \end{bmatrix}
$$

In the materials flow chapter the first term [materials accumulated] is set to zero, invoking the steady-state assumption.

Reactions, as discussed in Chapter 4, are of course time dependent. If these reactions occur in a black box, the black box becomes a *reactor*, and the first term of the materials balance equation can no longer be assumed to be zero.

Many natural processes as well as engineered systems can be conveniently analyzed using the notion of ideal reactors. A black box can be thought of as a reactor if it has volume and if it is either mixed or materials flow through it.

Three types of ideal reactors are defined on the basis of certain assumptions about their flow and mixing characteristics. The *mixed batch reactor* is fully mixed and does not have a flow into or out of it; in the *plug flow reactor* there is flow but no longitudinal mixing takes place; and the *completely mixed flow reactor*, as the name implies, is a flow reactor with perfect mixing.

First assume that *no reactions* occur inside the reactor; that is, there is nothing being consumed or produced in the reactor. This assumption leads to the so-called *mixing model* of reactors. When reactions occur in the reactors, something is being either produced or consumed, and the *reactor model* describes the operation of these reactors. We focus first on reactors with no reactions, or the mixing model of reactors.

5.1 ◆ Mixing Model

To discuss the mixing model, a device known as a *conservative and instantaneous signal* is used. A signal is simply a tracer placed into the flow as the flow enters the reactor. The signal allows for the characterization of the reactor by measuring the signal or tracer concentration with time. The term *conservative* means that the signal does not itself react. For example, a color dye introduced into water would not react chemically and would neither lose nor gain color. *Instantaneous* means that the signal (again such as a color dye) is introduced to the reactor instantaneously. In real life, this can be pictured as instantaneously dropping a cup of color dye into water, for example.

Conservative and instantaneous signals can be applied to three basic types of reactors: mixed batch, plug flow, and completely mixed flow reactors.

5.1.1 Mixed Batch Reactors

The first type of reactor, illustrated in Figure 5–1, is the mixed batch reactor, which has no flow in or out. If the conservative instantaneous signal is introduced into this reactor, ideally this signal is mixed instantaneously (zero mixing time). The mixing assumption can be illustrated by using a curve shown in Figure 5–1B. The ordinate indicates the concentration at any time t divided by the initial concentration. Before $t = 0$, there is no signal (dye) in the reactor. When the signal is introduced, and since perfect mixing is assumed, the signal is immediately and evenly distributed in the reactor vessel. Thus the concentration instantaneously jumps to C_0, the concentration at $t = 0$ after the signal is introduced. The concentration does not change after that time because there is no

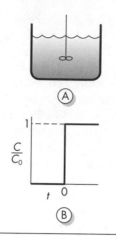

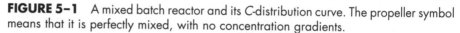

FIGURE 5–1 A mixed batch reactor and its *C*-distribution curve. The propeller symbol means that it is perfectly mixed, with no concentration gradients.

flow in or out of the reactor and because the dye is not destroyed (it is conservative). A plot such as Figure 5–1B is commonly called a *C-distribution* and is a useful way of graphically representing the behavior of reactors. A *C*-distribution curve can be plotted as simply *C* vs. *t*, but is more often normalized and plotted as C/C_0 vs. time. For a mixed batch reactor, at $t = 0$, after the signal has been introduced, $C = C_0$ and $C/C_0 = 1$.

Although mixed batch reactors are useful in a number of industrial and pollution-control applications, a far more common reactor is one in which the flow into and out of the reactor is continuous. Such reactors can be described by considering two ideal reactors—the plug flow and the CMF reactor.

5.1.2   Plug Flow Reactors

Figure 5–2A illustrates the characteristics of a plug flow reactor. Picture a very long tube (like a garden hose) into which is introduced a continuous flow. Assume that while the fluid is in the tube it experiences no longitudinal mixing. If a conservative signal is instantaneously introduced into the reactor at the influent end, any two elements within that signal that enter the reactor together will always exit at the same time. All signal elements have exactly equal *residence times*, defined as the time between entering and exiting the reactor, and calculated as

$$\bar{t} = \frac{V}{Q}$$

where $\bar{t}$ = hydraulic residence time, sec
  V = volume of the reactor, m³
  Q = flow rate to the reactor, m³/sec

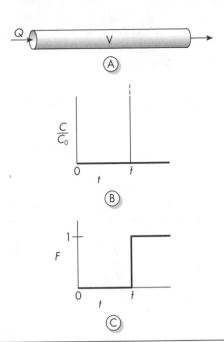

FIGURE 5–2 A plug flow reactor and its *C-* and *F*-distribution curves.

The residence time in an ideal plug flow reactor is the time *any* particle of water spends in the reactor because all particles of water spend exactly the same amount of time in the reactor. As noted in Chapter 2, another way of thinking of residence time is the time necessary to *fill* a reactor. If it's plug flow, it's like turning on an empty garden hose and waiting for water to come out the other end. The same definition for residence time holds for *any* type of reactor, as discussed subsequently.

If a conservative instantaneous signal is now introduced to the reactor, the signal moves as a plug (!) through the reactor, exiting at time $\bar{t}$. Before time $\bar{t}$ none of the signal exits the reactor and the concentration of the signal in the flow is zero ($C = 0$). Immediately after $\bar{t}$, all of the signal has exited, and again $C = 0$. The *C*-distribution curve is thus one instantaneous peak as shown in Figure 5–2B.

Another convenient means of describing a plug flow reactor is to use the *F-distribution*. *F* is defined as the fraction of the signal that has left the reactor at any time t,

$$F = \frac{A_0 - A_R}{A_0}$$

where A_0 = mass of tracer added to the reactor
A_R = mass of tracer remaining in the reactor

The *F*-distribution for a plug flow reactor is shown in Figure 5–2C. At $t = \bar{t}$ all of the signal exits at once, and $F = 1$.

To recap, perfect plug flow occurs when there is no mixing in the reactor. Obviously, this is unrealistic in practice. Equally unrealistic is the third type of ideal reactor, discussed next.

5.1.3 Completely Mixed Flow Reactors

In a completely mixed flow (CMF) reactor, perfect mixing is assumed. There are no concentration gradients at any time, and a signal is mixed perfectly and instantaneously. Figure 5–3A illustrates such a reactor. (Incidentally, chemical engineers call these reactors CSTRs for completely stirred tank reactors.)

A conservative instantaneous signal can be introduced into the CMF reactor feed and the mass balance equation written as:

$$
\begin{bmatrix} \text{rate of} \\ \text{signal} \\ \text{ACCUMULATED} \end{bmatrix} = \begin{bmatrix} \text{rate of} \\ \text{signal} \\ \text{IN} \end{bmatrix} - \begin{bmatrix} \text{rate of} \\ \text{signal} \\ \text{OUT} \end{bmatrix}
$$
$$
+ \begin{bmatrix} \text{rate of} \\ \text{signal} \\ \text{PRODUCED} \end{bmatrix} - \begin{bmatrix} \text{rate of} \\ \text{signal} \\ \text{CONSUMED} \end{bmatrix}
$$

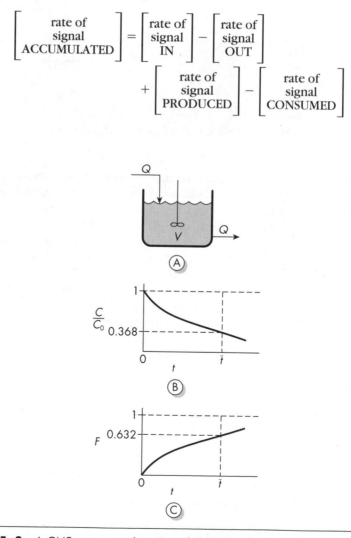

FIGURE 5–3   A CMF reactor and its C- and F-distribution curves.

Since the signal is assumed to be instantaneous, the *rate* at which the signal is introduced is zero. Likewise, the signal is conservative and therefore the rate produced and the rate consumed are both zero. Thus

$$\begin{bmatrix} \text{rate of} \\ \text{signal} \\ \text{ACCUMULATED} \end{bmatrix} = 0 - \begin{bmatrix} \text{rate of} \\ \text{signal} \\ \text{OUT} \end{bmatrix} + 0 - 0$$

If the amount of signal in the reactor at any time t is A, and A_0 is the amount of signal at $t = 0$, the concentration of the signal in the reactor at $t = 0$ is

$$C_0 = \frac{A_0}{V}$$

where C_0 = concentration of signal in the reactor at time zero, in units such as kg/m^3
$\quad A_0$ = mass of signal in the reactor at time zero, kg
$\quad V$ = reactor volume, m^3

After the signal has been instantaneously introduced, the clear liquid (no signal or tracer) continues to flow into the reactor, and the signal in the reactor is progressively diluted. At any time t, the concentration of the signal is

$$C = \frac{A}{V}$$

where C = concentration of signal at any time t, kg/m^3
$\quad A$ = mass of signal at any time t, kg
$\quad V$ = volume of the reactor, m^3

Since the reactor is perfectly mixed, the concentration of the signal in the flow withdrawn from the reactor must also be C. The *rate* at which the signal is withdrawn from the reactor is equal to the concentration times the flow rate. (See Chapter 2, p. 50.) Substituting this into the mass balance equation,

$$\begin{bmatrix} \text{rate of} \\ \text{signal} \\ \text{ACCUMULATED} \end{bmatrix} = -CQ = -\left(\frac{A}{V}\right) Q$$

where Q = flow rate into and out of the reactor

The volume of the reactor, V, is assumed to be constant.
The rate at which signal A is accumulated is dA/dt, and

$$\frac{dA}{dt} = -\left(\frac{A}{V}\right) Q$$

The minus sign comes from the mass balance equation since the concentration is decreasing. This equation can be integrated:

$$\int_{A_0}^{A} \frac{dA}{A} = -\int_{0}^{t} \frac{Q}{V} dt$$

or

$$\ln A - \ln A_0 = -\frac{Q}{V} t$$

$$\frac{A}{A_0} = e^{-\frac{Qt}{V}}$$

The residence time in a CMF reactor is

$$\bar{t} = \frac{V}{Q}$$

Note that although this equation is the same as for the plug flow reactor, now the residence time is defined as the average time a particle of water spends in the reactor. This is still numerically equal to the time necessary to fill the reactor volume V with a flow Q, just as for the plug flow reactor.

Substituting,

$$\frac{A}{A_0} = e^{-(t/\bar{t})}$$

Both A and A_0 are divided by V to obtain C and C_0, the concentration at any time t and the concentration at t_0:

$$\frac{C}{C_0} = e^{-(t/\bar{t})}$$

This equation can now be plotted as the C-distribution shown in Figure 5–3B. At the residence time, $t = \bar{t}$

$$\frac{C}{C_0} = e^{-1} = 0.368$$

meaning that at the residence time, 36.8% of the signal is still in the reactor and 63.2% of the signal has exited. This is best illustrated with the F-distribution, defined as the fraction of the signal that has left the reactor at any time t. The F-distribution curve for a CMF reactor is shown in Figure 5–3C. Recall that F is defined as the fraction of the signal that has left the reactor. As calculated previously, at the hydraulic residence time only 63.2% has left the reactor.

5.1.4   Completely Mixed Flow Reactors in Series

The practicality of the mixing model of reactors can be greatly enhanced by considering one further reactor configuration: a series of CMF reactors, as shown in Figure 5–4. The n reactors each have a volume of V_0, so that $V = n \times V_0$.

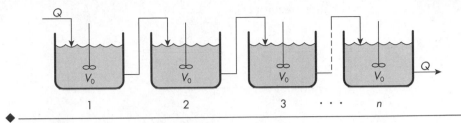

FIGURE 5–4 A series of CMF reactors.

Performing a mass balance on the first reactor, and using an instantaneous, conservative signal as before, results in

$$\frac{dA_1}{dt} = -\left(\frac{Q}{V_0}\right) A_1$$

where A_1 = mass of the signal in the first reactor at any time t, in units such as kg

V_0 = volume of reactor 1, m³

Q = flow rate, m³/sec

and as before,

$$A_1 = A_0 e^{-(Qt/V)} \qquad (5.1)$$

where A_0 = mass of signal in the reactor at $t = 0$. For each reactor the residence time is $\bar{t}_0 = V_0/Q$, so that

$$\frac{A_1}{A_0} = e^{-(t/\bar{t}_0)}$$

For the second reactor, the mass balance reads as before, and the last two terms are again zero. But now the reactor is receiving a signal over time, as well as discharging it. In addition to the signal accumulation, there is also an inflow and outflow. In differential form,

$$\frac{dA_2}{dt} = \left(\frac{Q}{V_0}\right) A_1 - \left(\frac{Q}{V_0}\right) A_2$$

Substituting for A_1 (equation 5.1)

$$\frac{dA_2}{dt} = \left(\frac{Q}{V_0}\right) A_0 e^{-(Qt/V_0)} - \left(\frac{Q}{V_0}\right) A_2$$

and integrating

$$A_2 = \left(\frac{Qt}{V_0}\right) A_0 e^{-(Qt/V_0)}$$

For three reactors,

$$\frac{A_3}{A_0} = \frac{t}{\bar{t}_0}\left(\frac{e^{-(t/\bar{t}_0)}}{2!}\right) = \frac{C_3}{C_0}$$

Generally, for i reactors,

$$\frac{A_i}{A_0} = \left(\frac{t}{\bar{t}_0}\right)^{i-1}\left(\frac{e^{-(t/\bar{t}_0)}}{(i-1)!}\right)$$

For a series of n reactors, and dividing both A_i and A_0 by the reactor volume V_0 to obtain a ratio of concentrations,

$$\frac{C_n}{C_0} = \left(\frac{t}{\bar{t}_0}\right)^{n-1} e^{-(t/\bar{t}_0)}\left(\frac{1}{(n-1)!}\right) \tag{5.2}$$

This equation describes the amount of a conservative, instantaneous signal in any one of a series of n reactors at time t, and can be used again to plot the C-distribution. Also recall that $nV_0 = V$, so that the *total volume* of the entire reactor never changes. The big reactor volume is simply divided up into n equal smaller volumes, or

$$n\bar{t}_0 = \bar{t}$$

Consider now what the concentration would be in the first reactor as the signal is applied. Since the volume of this first reactor of a series of n reactors is $V_0 = V/n$, the concentration at any time must be n times that of only one large reactor (same amount of signal diluted by only one nth of the volume). If it is then necessary to calculate the concentration of the signal in any subsequent reactor, the equation must be

$$\frac{C_n}{C_0} = n\left(\frac{t}{\bar{t}_0}\right)^{(n-1)} e^{-(t/\bar{t}_0)}\left(\frac{1}{(n-1)!}\right) \tag{5.3}$$

where C_0 = concentration of signal in the first reactor

The effect of dividing a reactor volume V into n smaller reactors of volume V_0 each is perhaps more clearly illustrated using the F-distribution.
    Recall that this distribution is defined as

$$F = \frac{A_0 - A_R}{A_0}$$

where A_R = amount of tracer remaining in the reactor at any time

In the case of a series of n reactors,

$$A_R = A_1 + A_2 + A_3 + \cdots + A_n$$

and

$$F = \frac{A_0 - (A_1 + A_2 + \cdots + A_n)}{A_0}$$

Rearranged,

$$F = 1 - \left(\frac{A_1}{A_0} + \frac{A_2}{A_0} + \cdots + \frac{A_n}{A_0}\right)$$

Substituting the equations derived previously,

$$F = 1 - \left[e^{-(t/\bar{t}_0)} + \left(\frac{t}{\bar{t}_0}\right) e^{-(t/\bar{t}_0)} + \left(\frac{t}{\bar{t}_0}\right)^2 e^{-(t/\bar{t}_0)} \left(\frac{1}{2!}\right) + \cdots \right. $$
$$\left. + \left(\frac{t}{\bar{t}_0}\right)^{n-1} e^{-(t/\bar{t}_0)} \left(\frac{1}{(n-1)!}\right) \right]$$

The F-distribution for several reactors in series is shown in Figure 5-5.

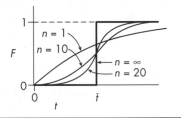

FIGURE 5-5 • F-distribution curve for a series of CMF reactors.

As the number of reactors increases, the F-distribution curve becomes more and more like an S-shaped curve. At $n =$ infinity, it becomes exactly like the F-distribution curve for a plug flow reactor (Figure 5-2C). This can be readily visualized as a great many small CMF reactors in series, so that a plug is moving rapidly from one very small reactor to another, and because each little reactor has a very short residence time, as soon as it enters one reactor, it gets flushed out into the next one. This of course is exactly how the plug moves through the plug flow reactor.

e • x • a • m • p • l • e   5.1

Problem   Estimate the effect of dividing a large completely mixed aeration pond (used for wastewater treatment) into 2, 5, 10, and 20 sections so that the flow enters each section in series. Draw the C- and F-distributions for an instantaneous conservative signal for the single pond and the divided pond.

Solution   For a single pond

$$\frac{C}{C_0} = e^{-(t/\bar{t})}$$

where C = concentration of signal in the effluent at time t
C_0 = concentration at time t_0
$\bar{t}$ = residence time

Substituting various values of $t/\bar{t}$, at

$$t = 0.25\bar{t} \quad C/C_0 = e^{-(0.25)} \quad = 0.779$$

$$t = 0.5\bar{t} \quad C/C_0 = e^{-(0.50)} \quad = 0.607$$

$$t = 0.75\bar{t} \quad C/C_0 = e^{-(0.75)} \quad = 0.472$$

$$t = \bar{t} \quad C/C_0 = e^{-1} \quad = 0.368$$

$$t = 2\bar{t} \quad C/C_0 = e^{-2} \quad = 0.135$$

These results are plotted in Figure 5-6 and describe a single reactor, $n = 1$. Similar calculations can be performed for a series of reactors. For example, for $n = 10$, remember that $10\,\bar{t}_0 = \bar{t}$, and using equation 5.3:

$$t = \bar{t} \quad \frac{C_{10}}{C_0} = (10)(10)^9 e^{-10} \frac{1}{9!} = 1.25$$

$$t = 0.5\bar{t} \quad \frac{C_{10}}{C_0} = (10)(5)^9 e^{-5} \frac{1}{9!} = 0.36$$

These data are also plotted in Figure 5-6. Now for the F-distribution, with one reactor, at

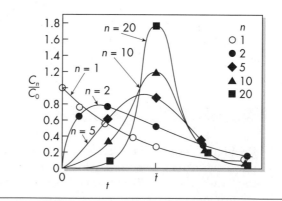

FIGURE 5-6   Actual C-distribution curves for Example 5.1.

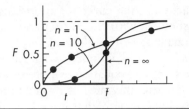

◆──

FIGURE 5-7 Actual *F*-distribution curves for Example 5.1.

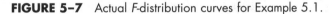

$$t = 0.25\bar{t} \quad F = 1 - e^{-0.25} \quad = 0.221$$

$$t = 0.5\bar{t} \quad F = 1 - e^{-0.50} \quad = 0.394$$

$$t = \bar{t} \quad F = 1 - e^{-1} \quad = 0.632$$

$$t = 2\bar{t} \quad F = 1 - e^{-2} \quad = 0.865$$

and for the 10 reactors in series

$$F = 1 - e^{-\frac{nt}{\bar{t}}}\left[1 + \frac{nt}{\bar{t}} + \left(\frac{nt}{\bar{t}}\right)^2\left(\frac{1}{2!}\right) + \cdots + \left(\frac{nt}{\bar{t}}\right)^{n-1}\frac{1}{(n-1)!}\right]$$

At $t = 0.5\bar{t}$,

$$F = 1 - e^{-5}[1 + 5 + (5)^2(1/2!) + (5)^3(1/3!) + \cdots + (5)^9(1/9!)]$$
$$= 1 - 0.00674[1 + 5 + 12.5 + 12.5 + 20.83 + 26.04 + 26.04$$
$$+ 21.70 + 15.50 + 9.68 + 5.48] = 1 - 0.969 = 0.032$$

Similarly, at $t/\bar{t} = 1$,

$$F = 1 - e^{-10}\left[1 + 10 + (10)^2\left(\frac{1}{2!}\right) + \cdots + (10)^9\left(\frac{1}{9!}\right)\right]$$

and so on. These results are plotted in Figure 5-7.

Note again, as the *C* and *F* curves clearly show, that as the number of small reactors (*n*) increases, the series of CMF reactors begin to behave increasingly like an ideal plug flow reactor.

5.1.5 Mixing Models with Continuous Signals

Thus far the signals used have been instantaneous and conservative. If the instantaneous constraint is removed, the signal can be considered continuous. A continuous conservative signal applied to an ideal plug flow reactor simply produces a *C*-distribution curve with a discontinuity going from $C = 0$ to $C = C_0$ at time *t*. For a CMF reactor, the *C* curve for when a signal is cut off is exactly like the curve for an instantaneous signal. The reactor is simply being flushed out with clear water and the concentration of the dye decreases exponentially as before. If, however, a continuous signal is introduced to a reactor at $t = 0$ and continued, what does the C curve look like? Start by writing a mass balance equation as before. Remember that the flow of dye *in* is fixed, while again the rates of production and consumption are zero. The equation describing a completely mixed reactor with a continuous (from $t = 0$) conservative signal thus is written in differential form as

$$\frac{dC}{dt} = QC_0 - QC$$

and after integration

$$C = C_0 (1 - e^{Qt})$$

where e is the natural log base.

5.1.6   Arbitrary Flow Reactors

Nothing in this world is ideal, including reactors. In plug flow reactors, there obviously is *some* longitudinal mixing, producing a C-distribution more like that shown in Figure 5-8A instead of like Figure 5-2B. Likewise, a CMF reactor cannot be ideally mixed, so that its C-distribution curve behaves more like that shown in Figure 5-8B instead of Figure 5-3B. These non-ideal reactors are commonly called *arbitrary flow reactors*.

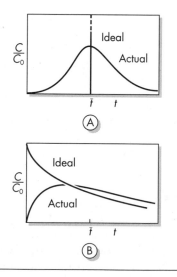

FIGURE 5-8 C-distribution curves for ideal and actual reactors: (A) plug flow reactor, (B) CMF reactor.

It should be apparent that the actual *C*-distribution curve in Figure 5–8B for the non-ideal CMF (arbitrary flow) reactor looks suspiciously like the *C* curve for 10 CMF in series, in Figure 5–5. In fact, looking at Figure 5–5, as the number of CMF in series increases, the *C*-distribution curve approaches the *C* curve for the perfect (ideal) plug flow reactor. Thus it seems reasonable to expect that all real-life (arbitrary flow) reactors really operate, in the mixing mode, as a series of CMF reactors. This observation allows for a quantitative description of reactor flow properties in terms of *n* CMFs in series, where *n* represents the number of CMFs in series and defines the type of reactor.

Further analysis of mixing models is beyond the scope of this brief discussion, and the student is directed to any modern text on reactor theory for more advanced study.

5.2 ◆ Reactor Models

As noted earlier, reactors can be described in two ways: in terms of their mixing properties only, with no reactions taking place, or as true reactors, in which a reaction occurs. In the previous section the mixing concept is introduced, and now it is time to introduce reactions into these reactors. Stated in another way, the constraint that the signal is conservative is now removed.

Again consider three different ideal reactors: the mixed batch reactor, the plug flow reactor, and the CMF reactor.

5.2.1 Mixed Batch Reactors

The same assumptions hold here as in the last section, namely, perfect mixing is assumed (there are no concentration gradients). The mass balance in terms of the material undergoing some reaction is

$$\left[\begin{array}{c}\text{rate}\\\text{ACCUMULATED}\end{array}\right] = \left[\begin{array}{c}\text{rate}\\\text{IN}\end{array}\right] - \left[\begin{array}{c}\text{rate}\\\text{OUT}\end{array}\right] + \left[\begin{array}{c}\text{rate}\\\text{PRODUCED}\end{array}\right] - \left[\begin{array}{c}\text{rate}\\\text{CONSUMED}\end{array}\right]$$

Because it is a batch reactor, there is no inflow or outflow, and thus

$$\left[\begin{array}{c}\text{rate}\\\text{ACCUMULATED}\end{array}\right] = \left[\begin{array}{c}\text{rate}\\\text{PRODUCED}\end{array}\right] - \left[\begin{array}{c}\text{rate}\\\text{CONSUMED}\end{array}\right]$$

If the material is being produced and there is no consumption, this equation can be written as

$$\frac{dC}{dt}V = rV$$

where C = concentration of material at any time t, mg/L
$\quad\quad V$ = volume of the reactor, L
$\quad\quad r$ = reaction rate, mg/L-sec
$\quad\quad t$ = time, sec

The volume term, V, appears because this is a mass balance, and recall that

$$\begin{bmatrix} \text{mass} \\ \text{flow} \end{bmatrix} = [\text{concentration}] \times \begin{bmatrix} \text{volume} \\ \text{flow} \end{bmatrix}$$

The left (accumulation) term units are

$$\frac{(\text{mg/L})}{\text{sec}} \text{L} = \frac{\text{mg}}{\text{sec}}$$

and the right-side units are

$$(\text{mg/L} \cdot \text{sec}) \times (\text{L}) = \frac{\text{mg}}{\text{sec}}$$

The volume term of course can be canceled out, so that

$$\frac{dC}{dt} = r$$

Integrated,

$$\int_{C_0}^{C} dC = r \int_{0}^{t} dt$$

For a zero-order reaction, $r = k$, where k = reaction rate constant, and thus

$$C - C_0 = kt \qquad (5.4)$$

The equation 5.4 holds when the material in question is being produced. In the situation where the reactor destroys the component, the reaction rate is negative, and

$$C - C_0 = -kt \qquad (5.5)$$

In both cases,

$$C = \text{concentration of the material at any time } t$$

$$C_0 = \text{concentration of the material at } t = 0$$

$$k = \text{reaction rate constant}$$

If the reaction is first order,

$$r = kC$$

and

$$\frac{dC}{dt} = kC$$

with the rate constant k having units of time^{-1}. Integrated,

$$\int_{C_0}^{C} \frac{dC}{C} = k \int_0^t dt$$

$$\ln \frac{C}{C_0} = kt \qquad (5.6)$$

$$C = C_0 e^{kt}$$

If the material is being consumed, the reaction rate is negative, and

$$\ln \frac{C}{C_0} = -kt$$

$$\ln \frac{C_0}{C} = kt \qquad (5.7)$$

$$C = C_0 e^{-kt}$$

e • x • a • m • p • l • e **5.2**

Problem   An industrial wastewater treatment process uses activated carbon to remove color from the water. The color is reduced as a first-order reaction in a batch-adsorption system. If the rate constant (k) is 0.35 days^{-1}, how long will it take to remove 90% of the color?

Solution   Let C_0 = initial concentration of the color, and C = concentration of the color at any time t. It is necessary to reach $0.1 C_0$.

$$\ln \left(\frac{C_0}{C} \right) = kt$$

$$\ln \left(\frac{C_0}{0.1 C_0} \right) = 0.35t$$

$$\ln \left(\frac{1}{0.1} \right) = 0.35t$$

$$t = \frac{2.30}{0.35} = 6.58 \text{ days}$$

5.2.2 Plug Flow Reactors

The equations for mixed batch reactors apply equally well to plug flow reactors, since in perfect plug flow reactors a plug of reacting materials flows through the reactor and this plug is itself like a miniature batch reactor. Thus for a zero-order reaction occurring in a plug flow reactor, with the material being made,

$$C = C_0 + k\bar{t} \tag{5.8}$$

where C = effluent concentration
       C_0 = influent concentration
       $\bar{t}$ = residence time of the reactor = V/Q
       V = volume of the reactor
       Q = flow rate through the reactor

If the material is consumed as a zero-order reaction,

$$C = C_0 - k\frac{V}{Q} \tag{5.9}$$

If the material is being produced as a first-order reaction,

$$\ln\frac{C}{C_0} = k\bar{t}$$

$$\ln\frac{C}{C_0} = k\left(\frac{V}{Q}\right) \tag{5.10}$$

$$V = \left(\frac{Q}{k}\right)\ln\frac{C}{C_0}$$

and if it is being consumed,

$$V = \left(-\frac{Q}{k}\right)\ln\frac{C}{C_0} \tag{5.11}$$

$$V = \frac{Q}{k}\ln\frac{C_0}{C}$$

e • x • a • m • p • l • e   **5.3**

Problem   An industry wants to use a long drainage ditch that can be assumed to act as a plug flow reactor in removing odor from their waste. The odor reduction behaves as a first-order reaction, with the rate constant $k = 0.35$ days^{-1}. The flow rate is 1600 L/day. How long must the ditch be if the velocity of the flow is 0.5 m/sec and 90% odor reduction is desired?

Solution

$$\ln\frac{C}{C_0} = -\overline{k}t$$

$$\ln\frac{0.1C_0}{C_0} = (0.35)\bar{t}$$

$$\ln(0.1) = (0.35)\bar{t}$$

$$\bar{t} = 6.57 \text{ days}$$

$$\text{length of ditch} = 0.5 \frac{m}{sec} \times 6.57 \text{ days} \times 86{,}400 \frac{sec}{day}$$

$$= 2.84 \times 10^5 \text{ m (!)}$$

◆

5.2.3 Completely Mixed Flow Reactors

The mass balance for the CMF reactor can be written in terms of the material in question as:

$$\begin{bmatrix} \text{rate} \\ \text{ACCUMULATED} \end{bmatrix} = \begin{bmatrix} \text{rate} \\ \text{IN} \end{bmatrix} - \begin{bmatrix} \text{rate} \\ \text{OUT} \end{bmatrix} + \begin{bmatrix} \text{rate} \\ \text{PRODUCED} \end{bmatrix} - \begin{bmatrix} \text{rate} \\ \text{CONSUMED} \end{bmatrix}$$

$$\frac{dC}{dt} V = QC_0 - QC + rV - rV$$

where C_0 = concentration of the material in the influent, mg/L

$\quad\ C$ = concentration of material in the effluent, and at any place and time in the reactor, mg/L

$\quad\ r$ = reaction rate

$\quad\ V$ = volume of reactor, L

It is now necessary to again assume a steady-state operation. Although a reaction is taking place within the reactor, the effluent concentration (and hence the concentration in the reactor) is not changing with time. That is,

$$\frac{dC}{dt} V = 0$$

If the reaction is of a zero order and the material is being produced, $r = k$, the consumption term is zero and

$$0 = QC_0 - QC + kV$$

$$C = C_0 + k \frac{V}{Q} \tag{5.12}$$

$$C = C_0 + k\bar{t}$$

Or if the material is being consumed, the zero-order reaction occurring in a CMF is

$$C = C_0 - k\bar{t} \tag{5.13}$$

If the reaction is first order and the material is produced, $r = kC$ and

$$0 = QC_0 - QC + kCV$$

$$\frac{C_0}{C} = 1 - k \frac{V}{Q} \tag{5.14}$$

$$C = \frac{QC_0}{Q - kV}$$

Or if the reaction is first order and the material is being destroyed in a CMF reactor,

$$\frac{C_0}{C} = 1 + k\left(\frac{V}{Q}\right)$$

$$C = \frac{QC_0}{Q + kV} \tag{5.15}$$

$$V = \frac{Q}{k}\left[\frac{C_0}{C} - 1\right]$$

Equation 5.15 makes common sense. Suppose it is necessary to maximize the performance of a reactor that destroys a component; that is, we want to *increase* (C_0/C) (C is to be small). This can be accomplished by

a. Increase volume of the reactor, V
b. Decrease the flow rate to the reactor, Q
c. Increase the rate constant, k

The rate constant k is dependent on numerous variables such as temperature and the intensity of mixing. The variability of k with temperature is commonly expressed in exponential form as

$$k_T = k_0\, e^{\Phi(T-T_0)}$$

where Φ = constant
k_0 = the rate constant at temperature T_0
k_T = the rate constant at temperature T

e • x • a • m • p • l • e **5.4**

Problem   A new disinfection process destroys coliform (coli) organisms in water. The reaction is first order, with $k = 1.0$ day^{-1}. The influent concentration, $C_0 = 100$ coli/mL. The reactor volume, $V = 400$ L, and the flow rate, $Q = 1600$ L/day. What is the effluent concentration of coliforms?

Solution

$$\left[\begin{array}{c}\text{rate}\\\text{ACCUMULATED}\end{array}\right] = \left[\begin{array}{c}\text{rate}\\\text{IN}\end{array}\right] - \left[\begin{array}{c}\text{rate}\\\text{OUT}\end{array}\right] + \left[\begin{array}{c}\text{rate}\\\text{PRODUCED}\end{array}\right] - \left[\begin{array}{c}\text{rate}\\\text{CONSUMED}\end{array}\right]$$

$$0 = QC_0 - QC + 0 - rV$$

where

$$r = kC$$

$$0 = 1600(100) - 1600(C) - 1.0(400)\,C$$

$$C = 80 \text{ coli/mL}$$

◆

5.2.4   Completely Mixed Flow Reactors in Series

For two CMF reactors in series, the effluent from the first is C_1 and assuming a first-order reaction where the material is being destroyed,

$$\frac{C_0}{C_1} = 1 + k\left(\frac{V_0}{Q}\right)$$

$$\frac{C_1}{C_2} = 1 + k\left(\frac{V_0}{Q}\right)$$

where V_0 is the volume of the individual reactor. Similarly, for the second reactor, the influent is C_1, and C_2 is its effluent, and for the two reactors,

$$\frac{C_0}{C_1} \cdot \frac{C_1}{C_2} = \frac{C_0}{C_2} = \left[1 + k\left(\frac{V_0}{Q}\right)\right]^2$$

For any number of reactors in series,

$$\frac{C_0}{C_n} = \left[1 + k\left(\frac{V_0}{Q}\right)\right]^n$$

$$\left(\frac{C_0}{C_n}\right)^{1/n} = 1 + k\,\frac{V}{nQ}$$

where V = volume of all the reactors, equal to nV_0
$\quad\quad\ n$ = number of reactors
$\quad\quad V_0$ = volume of each reactor

The equations for CMF reactors are summarized in Table 5-1.

5.2.5   Comparison of Reactor Performance

The efficiency of ideal reactors can now be compared by first solving all the descriptive equations in terms of reactor volume, as shown in Table 5-2 and illustrated in Example 5.5.

TABLE 5-1   Performance Characteristics of Completely Mixed Flow Reactors

Zero-order reaction, material produced:

$$C = C_0 + k\bar{t}$$

Zero-order reaction, material destroyed:

$$C = C_0 - k\bar{t}$$

First-order reaction, material produced:

$$\frac{C_0}{C} = 1 - k\bar{t}$$

First-order reaction, material destroyed:

$$\frac{C_0}{C} = 1 + k\bar{t}$$

Second-order reaction, material destroyed:

$$C = \frac{-1 + [1 + 4k\bar{t}C_0]^{1/2}}{2\,k\bar{t}}$$

Series of n CMF reactors, zero-order reaction, material destroyed:

$$\frac{C_0}{C_n} = \left(1 + \frac{k\bar{t}}{C_0}\right)^n$$

Series of n CMF reactors, first-order reaction, material destroyed:

$$\frac{C_0}{C_n} = (1 + k\bar{t}_0)^n$$

Note: C = effluent concentration; C_0 = influent concentration; $\bar{t}$ = residence time = V/Q; k = rate constant; C_n = concentration of the nth reactor; $\bar{t}_0$ = residence time in each of n reactors.

e • x • a • m • p • l • e   **5.5**

Problem   Consider a first-order reaction, requiring 50% reduction in the concentration. Would a plug flow or a CMF require the least reactor volume?

Solution

$$\frac{V_{\mathrm{CMF}}}{V_{\mathrm{PF}}} = \frac{\dfrac{Q}{k}\left(\dfrac{C_0}{C} - 1\right)}{\dfrac{Q}{k}\left(\ln\dfrac{C_0}{C}\right)}$$

	CMF		
Reaction Order	**Single Reactor**	**n Reactors**	**Plug Flow Reactor**
Zero	$V = \dfrac{Q}{k}(C_0 - C)$	$V = \dfrac{Q}{k}(C_0 - C_n)$	$V = \dfrac{Q}{k}(C_0 - C)$
First	$V = \dfrac{Q}{k}\left(\dfrac{C_0}{C} - 1\right)$	$V = \dfrac{Qn}{k}\left[\left(\dfrac{C_0}{C}\right)^{1/n} - 1\right]$	$V = \dfrac{Q}{k}\ln\dfrac{C_0}{C}$
Second	$V = \dfrac{Q}{k}\left(\dfrac{C_0}{C} - 1\right)\dfrac{1}{C}$	complex	$V = \dfrac{Q}{k}\left(\dfrac{1}{C} - \dfrac{1}{C_0}\right)$

TABLE 5-2 Summary of Ideal Reactor Performance

Note: Material destroyed; V = reactor volume; Q = flow rate; k = reaction constant; C_0 = influent concentration; C = effluent concentration; n = number of CMF reactors in series.

For 50% conversion,

$$\frac{C_0}{C} = 2$$

$$\frac{V_{CMF}}{V_{PF}} = \frac{(2 - 1)}{\ln 2} = 1.44$$

Conclusion A CMF reaction would require 44% more volume than a plug flow reactor.

◆

The conclusion reached in Example 5.5 is a very important concept used in many environmental engineering systems. Stated generally,

> For reaction orders of greater than or equal to one, the ideal plug flow reactor will always outperform the ideal CMF reactor.

This fact is a powerful tool in the design and operation of treatment systems.

◆ Abbreviations

A = amount (mass) of a signal in a reactor at any time t
C = concentration at any time t
C_0 = concentration at $t = 0$
F = fraction of a signal that has left the reactor at any time t

k = reaction rate constant
Q = flow rate
t = time
$\bar{t}$ = residence time
V = reactor volume

Problems

5–1   A dye mill has a highly colored 8-mgd effluent. A pilot study (small-scale experiment) is performed to evaluate a biological means of treating this wastewater and removing the coloration. Using a mixed batch reactor, the following data result:

Time (hr)	Dye Concentration (mg/L)
0	900
10	720
20	570
40	360
80	230

A completely mixed aerated lagoon (batch reactor) is to be used. How large must the lagoon be to achieve an effluent of 50 mg/L? (HINT: Graph paper is very useful here.)

5–2   A settling tank has an influent rate of 0.6 mgd. It is 12-ft deep and has a surface area of 8000 ft². What is the hydraulic residence time?

5–3   An activated sludge tank, 30 × 30 × 200 ft, is designed as a plug flow reactor, with an influent BOD of 200 mg/L and a flow rate of 1 million gallons per day.

   a.   If BOD removal is a first-order reaction, and the rate constant is 2.5 days⁻¹, what is the effluent BOD concentration?

   b.   If this system operates as a completely mixed reactor, what must its volume be (for the same BOD reaction)? How much bigger is this, as a percentage of the plug flow volume?

   c.   If the plug flow system is constructed and found to have an effluent concentration of 27.6 mg/L, the system could be theoretically characterized as a series of CMF reactors. How many? ($n = ?$)

5–4   The plant manager has a decision to make. She needs to reduce the concentration of salt in the 8000-gallon tank from 30,000 mg/L to 1000 mg/L. She can do it in one of two ways.

   1.   She can start flushing it out by keeping the tank well mixed while running in a hose with clean water (zero salt) at a flow rate of 60 gallons per minute (with an effluent of 60 gallons per minute, obviously).

   2.   She can empty out some of the saline water and fill it up again with enough clean water to get the concentration down to 1000 mg/L. The maximum rate at which the tank will empty is 60 gallons per minute and the maximum flow of clean water is 100 gallons per minute.

   a.   If she intends to do this job at the shortest time possible, which alternative will she choose?

   b.   Jeremy Rifkin points out that the most pervasive concept of modern times is *efficiency*. Everything has to be done so as to expend the least energy, effort, and especially time. He notes that we are losing our perspective on time, especially if we think of time in a digital way (as digital numbers on a watch) instead of in the analog (hands on a conventional watch). With the digits we cannot see where we have been, and we cannot see where we are going, and we lose all perspective of time. Why indeed would the plant manager want to empty out the tank in the shortest time? Why is she so hung up with time? Having time (and not wasting it) seems to have become a pervasive value in our lives, and it sometimes overwhelms our other values. Write a one-page paper on how you value time in your life and how this value influences your other values.

5-5 A first-order reaction is employed in the destruction of a certain kind of microorganism. Ozone is used as the disinfectant, and the reaction is found to be

$$\frac{dC}{dt} = -kC$$

where C = concentration of microorganisms, microbes/mL
k = rate constant, 0.1 min^{-1}
t = time, min

The present system employs a completely mixed tank and there is some thought of baffling it to create a series of completely mixed tanks.

a. If the objective is to increase the percentage microorganism destruction from 80% to 95%, how many CMF reactors are needed in series? (The volume will not be changed.)

b. We routinely kill microorganisms and think nothing of it. But do microorganisms have the same right to exist as larger organisms, such as whales for example? or as people? Should we afford moral protection to microorganisms? Can a microorganism ever become an endangered species? As a part of this assignment, write a letter to the editor of your school newspaper on behalf of Microbe Coliform, a typical microorganism who is fed up with not being given equal protection and consideration within the human society, and who is demanding microbe-rights. What philosophical arguments can be mounted to argue for microbe rights? Do not make this a silly letter. Consider the question seriously, because it reflects on the entire problem of environmental ethics.

5-6 A CMF reactor, with a volume of V and a flow rate of Q, has a zero-order reaction, $dA/dt = k$, where A is the material being produced and k is the rate constant. Derive an equation that would allow for the direct calculation of the required reactor volume.

5-7 A completely mixed continuous bioreactor used for growing penicillin operates as a zero-order system. The input, glucose, is converted to various organic yeasts. The flow rate to this system is 20 liters per minute and the conversion rate constant is 4 mg/(min-L). The influent glucose concentration is 800 mg/L and the effluent must be less than 100 mg/L. What is the smallest reactor capable of producing this conversion?

5–8   Suppose you are to design a chlorination tank for killing microorganisms in the effluent from a wastewater treatment plant. You must achieve 99.99% kill in a wastewater flow of 100 m³/hr. Assume the disinfection is a first-order reaction with a rate constant of 0.2 min⁻¹.

    a.   Calculate the tank volume if the contact tank is a CMF reactor.

    b.   Calculate the tank volume if the contact tank is a plug-flow reactor.

    c.   What is the residence time of both reactors?

5–9   A CMF reactor, operating at steady state, has an inflow of 4 L/min and an inflow "gloop" concentration of 400 mg/L. The volume is 60 liters. The reaction is zero order. The "gloop" concentration in the reactor is 100 mg/L.

    a.   What is the reaction rate constant?

    b.   What is the hydraulic residence time?

    c.   What is the outflow (effluent) "gloop" concentration?

    (HINT: Graph paper would be useful.)

chapter 6

ENERGY FLOWS & BALANCES

An energetic person is one who is continually in motion; someone who has a lot of energy and is always active. But some energetic people never seem to get anything *done*—they expend a lot of effort but have little to show for it. Obviously, it is not enough to be energetic; one is also expected to be *efficient*. Available energy, to be useful, must be funneled efficiently into productive use.

In this chapter we look at quantities of energy, how energy flows and is put to use, and the efficiencies of such use.

6.1 ◆ Units of Measure

One of the earliest measures of energy, still widely used by American engineers, is the British Thermal Unit (BTU), defined as that amount of energy necessary to heat one pound of water one degree Fahrenheit. The internationally accepted unit of energy is the joule. Other common units for energy are the calorie and kilowatt-hour (kWh); the former used in natural sciences, the latter in engineering. Table 6–1 shows the conversion factors for all these units, emphasizing the fact that *all* are measures of energy and are interchangeable.

e • x • a • m • p • l • e **6.1**

Problem   One gallon of gasoline has an energy value of 126,000 BTU. Express this in (a) calories, (b) joules, (c) kWh.

Solution

   a.   126,000 BTU $\times$ 252 cal/BTU = 3.18 $\times$ 10^7 cal

   b.   126,000 BTU $\times$ 1054 J/BTU = 1.33 $\times$ 10^8 J

   c.   126,000 BTU $\times$ 2.93 $\times$ 10^4 kWh/BTU = 37 kWh

TABLE 6–1	Energy Conversion Factors	
To Convert	*to*	*Multiply by*
BTU	calories	252
	joules	1054
	kWh	0.000293
Calories	BTU	0.00397
	joules	4.18
	kWh	0.00116
Joules	BTU	0.000949
	calories	0.239
	kWh	2.78 $\times$ 10^{-7}
Kilowatt-hours	BTU	3413
	calories	862
	joules	3.6 $\times$ 10^6

6.2 ◆ Energy Balances and Conversion

There are of course many forms of energy, such as chemical, heat, potential energy due to elevation, and so on. Often the form of energy available is not the form that is most useful, and one form of energy must be converted to another form. For example, the water in a mountain lake has potential energy and can be run through a turbine to convert this potential to electrical energy that can be converted to heat or light, both forms of useful energy. Chemical energy in organic matter, stored in the carbon–carbon and carbon–hydrogen bonds formed by plants, can be severed by a process such as combustion, which liberates heat energy that can be used directly, or indirectly to produce steam to drive electrical generators. Wind has kinetic energy and this can be converted to mechanical energy with a windmill, and this energy can be converted to electrical energy to produce heat energy to warm your house. Energy conversion is thus an important and ancient engineering process. Unfortunately, energy conversions are *always* less than 100% efficient.

Energy (of whatever kind) when expressed in common units can be pictured as a quantity that flows, and thus it is possible to analyze energy flows using the same concepts used for materials flows and balances. As before, a "black box" is any process or operation into which certain flows enter and others leave. If all the flows can be correctly accounted for, then there must be a balance.

Looking at Figure 6-1, note that in a black box the "energy in" has to equal the "energy out" (energy wasted in the conversion + useful energy) plus the energy accumulated in the box. This can be expressed as an equation:

$$\begin{bmatrix} \text{rate of} \\ \text{energy} \\ \text{ACCUMULATED} \end{bmatrix} = \begin{bmatrix} \text{rate of} \\ \text{energy} \\ \text{IN} \end{bmatrix} - \begin{bmatrix} \text{rate of} \\ \text{energy} \\ \text{OUT} \end{bmatrix}$$

$$+ \begin{bmatrix} \text{rate of} \\ \text{energy} \\ \text{PRODUCED} \end{bmatrix} - \begin{bmatrix} \text{rate of} \\ \text{energy} \\ \text{CONSUMED} \end{bmatrix}$$

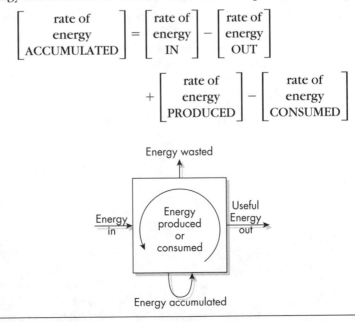

FIGURE 6-1   Black box for energy flows.

Of course, energy is never produced or consumed in the strict sense; it is simply changed in form.

Just as processes involving materials can be studied in their "steady-state" condition, defined as no change occurring over time, energy systems can also be in steady state. Obviously, if there is no change over time, there cannot be a continuous accumulation of energy, and the equation must read

$$[\text{rate of energy IN}] = [\text{rate of energy OUT}]$$

or if energy wasted is considered a type of consumption,

$$\begin{bmatrix} \text{rate of} \\ \text{energy} \\ \text{IN} \end{bmatrix} = \begin{bmatrix} \text{rate of} \\ \text{energy} \\ \text{OUT} \end{bmatrix} + \begin{bmatrix} \text{rate of} \\ \text{energy} \\ \text{WASTED} \end{bmatrix}$$

If the input and useful output from a black box are known, the efficiency of the process can be calculated as

$$\text{efficiency (\%)} = \frac{\text{useful energy OUT}}{\text{energy IN}} \times 100$$

e • x • a • m • p • l • e 6.2

Problem   A coal-fired power plant uses 1000 Mg (mega-grams, or 1000 kg, commonly called a metric ton) of coal per day. The energy value of the coal is 28,000 kJ/kg (kilojoules/kilogram). The plant produces 2.8×10^6 kWh of electricity each day. What is the efficiency of the power plant?

Solution

$$\text{energy IN} = 28{,}000 \text{ kJ/kg} \times 1000 \text{ Mg/day} (1 \times 10^3) \text{ kg/Mg.}$$

$$= 28 \times 10^9 \text{ kJ/day}$$

useful energy output

$$= 2.8 \times 10^6 \text{ kWh/day} \times (3.6 \times 10^6) \text{ J/kWh} \times 10^{-3} \text{ kJ/J}$$

$$= 10.1 \times 10^9 \text{ kJ/day}$$

$$\text{efficiency (\%)} = [10.1 \times 10^9]/[28 \times 10^9] \times 100 = 36\%$$

Another example of how one form of energy can be converted to another form uses the *calorimeter*, the standard means of measuring the heat energy

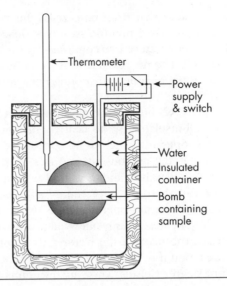

- Thermometer
- Power supply & switch
- Water
- Insulated container
- Bomb containing sample

FIGURE 6-2   Simplified drawing of a bomb calorimeter.

value of materials when they combust. Figure 6-2 shows a schematic sketch of a *bomb calorimeter*. The bomb is a stainless steel ball that screws apart. The ball has an empty space inside into which the sample to be combusted is placed. A sample of known weight, such as a small piece of coal, is placed into the bomb and the two halves screwed shut. Oxygen under high pressure is then injected into the bomb and the bomb is placed in an adiabatic water bath, with wires leading from the bomb to a source of electrical current. By means of a spark from the wires, the material in the steel ball combusts and heats the bomb, which in turn heats the water. The temperature rise in the water is measured with a thermometer and recorded as a function of time. Figure 6-3 shows the trace of a typical calorimeter curve.

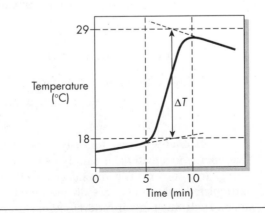

FIGURE 6-3   Results of a bomb calorimeter test.

Note that from time zero, the water is heating due to the heat in the room. At $t = 5$ min the switch is closed and combustion in the bomb occurs. The temperature rise continues to $t = 10$ min, at which time the water starts to cool. The net rise due to the combustion is calculated by extrapolating both the initial heating and cooling lines and determining the difference in temperature.

Now consider the calorimeter as a black box and assume that the container is well insulated, so no heat energy escapes the system. Since this is a simple batch operation, there is no accumulation. The [energy in] is due entirely from the material combusted, and this must equal the [energy out] or the energy expressed as heat and measured as temperature with the thermometer. Remember also that only *heat* energy is considered, so by assuming that no heat is lost to the atmosphere, there is no "wasted energy."

The heat energy out is calculated as the temperature increase of the water times the mass of the water plus bomb. Recall that one calorie is defined as the amount of energy necessary to raise the temperature of one gram of water one degree Celsius. Knowing the grams of water in the calorimeter, it is possible to calculate the energy accumulation in calories. An equal amount of energy must have been liberated by the combustion of the sample, and knowing the weight of the sample, its energy value can be calculated.[1]

e • x • a • m • p • l • e **6.3**

Problem   A calorimeter holds 4 liters of water. Ignition of a 10-gram sample of a waste-derived fuel of unknown energy value yields a temperature rise of 12.5°C. What is the energy value of this fuel? Ignore the mass of the bomb.

Solution   [energy IN] = [energy OUT]

$$\text{energy out} = 12.5° \times 4\,L \times 10^3\,mL/L \times 1\,g/mL$$

$$= 50 \times 10^3 (°C)\text{-g, or calories}$$

$$= 50 \times 10^3\,\text{calories} \times 4.18\,(J/cal) = 209 \times 10^3\,J$$

$$\text{energy in} - \text{energy out} = 209 \times 10^3\,J$$

$$\text{energy value of the fuel} = [209 \times 10^3\,J/g]/[10\,g] = 20{,}900\,J/g$$

◆

Heat energy is easy to analyze by energy balances since the quantity of heat energy in a material is simply its mass times its absolute temperature. This is only true, however, if the heat capacity is independent of temperature. In particular, if phase changes do not occur, as in the conversion of water to steam. Such situations are addressed in a thermodynamics course, highly recommended for all environmental engineers.

An energy balance for heat energy would then be in terms of the quantity of heat, or

$$\begin{bmatrix} \text{heat} \\ \text{energy} \end{bmatrix} = \begin{bmatrix} \text{mass of} \\ \text{material} \end{bmatrix} \times \begin{bmatrix} \text{absolute temperature} \\ \text{of the material} \end{bmatrix}$$

This is analogous to mass flows discussed earlier, except now the flow is energy flow. When two heat energy flows are combined, for example, the temperature of the resulting flow at steady state is calculated using the black box technique,

$$0 = [\text{heat energy IN}] - [\text{heat energy OUT}] + 0 - 0$$

or stated another way,

$$0 = [T_1 Q_1 + T_2 Q_2] - [T_3 Q_3]$$

or

$$T_3 = \frac{T_1 Q_1 + T_2 Q_2}{Q_3} \qquad (6.1)$$

where T = absolute temperature
Q = flow, mass/unit time (or volume constant density)
1 and 2 = input streams
3   = output stream

The mass/volume balance is

$$Q_3 = Q_1 + Q_2$$

Although strict thermodynamics requires that the temperature be expressed in *absolute* terms, in the conversion from Celsius (C) to Kelvin (K) the conversion simply cancels out and T can be conveniently expressed in degrees C. Recall that $0°C = 273°K$.

e • x • a • m • p • l • e **6.4**

Problem   A coal-fired power plant discharges 3 m³/sec of cooling water at 80°C into a river that has a flow of 15 m³/sec and a temperature of 20°C. What will be the temperature in the river immediately below the discharge?

Solution   Using equation 6.1 and considering the confluence of the river and cooling water as a black box,

$$T_3 = \frac{T_1 Q_1 + T_2 Q_2}{Q_3}$$

$$T_3 = \frac{[(80 + 273)(3)] + [(20 + 273)(15)]}{(3 + 15)} = 303°K$$

or $303 - 273 = 30°C$. Note that the use of absolute temperatures is not necessary since the 273 cancels out.

◆

6.3 ◆ Energy Sources and Availability

Power utilities use the most efficient fuels possible since these will produce least ash for disposal and will most likely be the cheapest to use in terms of kilowatt-hour of electricity produced per dollar of fuel cost. But the best fuels, natural gas and oil, are in finite supply. Estimates vary as to how much natural gaseous and liquid fuels remain in the earth's crust within our reach, but most experts agree that if they continue to be used at the present expanding rate, the existing supplies will be depleted within 50 years. Others argue that as the supplies begin to run low and the price of these fuels increases, other fuels will become less expensive by comparison and market forces will limit the use of the resources.

If the pessimists are correct, and the world runs out of oil and natural gas within 50 years, will the next generation blame us for our unwise use of natural resources? Or should we, indeed, even be worried about the next generations?

There is a strong argument to be made that the most important thing we can do for the coming generations is not to plan for them. The world will be so different, and the state of technology would have changed so markedly, that it is impossible to estimate what future generations will need. For us to deprive ourselves of needed resources today just so some future person will have the benefit of these resources is simply ludicrous.

However, are there not some things we can be fairly certain of in terms of the future generations? We can assume that they will appreciate and value many of the same things we do. They will like to have clean air and plentiful and safe water. They will like to have open spaces and wilderness. They will appreciate the absence of hazardous waste time bombs. And (this is moot) they will appreciate that some of what we regard as natural resources are there for them to use and manage.

But the most important question is why we should care at all about the future generations. What have they ever done for us? Where is the quid pro quo in all this?

If we ignore the problem of reciprocity and decide to conserve energy resources, what are the available sources of renewable energy? The sources in present use include

- Hydropower from rivers
- Hydropower from tidal estuaries
- Wood and other biomass such as sugarcane and rice hulls
- Solar power
- Refuse and other waste materials

The nonrenewable energy sources include

- Nuclear power
- Coal, peat, and similar materials

- Natural gas
- Oil

The nonrenewable sources are our "energy capital," the amount of energy good that we have to spend. The renewable sources are analogous to our "energy income"; resources that we can continue to use as long as the sun shines and the wind blows. Energy use in the United States is illustrated in Figure 6-4. Note that most of the renewable sources are so small that they don't even make the chart. We are therefore rapidly depleting our energy capital and rely almost not at all on the renewable energy income.

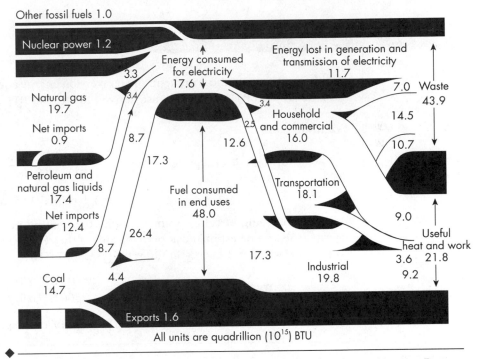

FIGURE 6-4 Energy flow in the United States. (Adapted from R. H. Wagner, *Environment and Man*, New York, Norton (1978), from data by E. Cook.)

6.3.1 Energy Equivalence

There is a big difference between potentially available energy and energy that can be efficiently harnessed. For example, one of the greatest sources of potential energy is tidal energy. The difficulty, however, is how to convert this potential to a useful form such as electrical energy. With a few notable exceptions, tidal systems have not proven cost efficient. That is, the electrical energy from tidal power costs much more than the electrical energy produced by other means.

Further, some types of conversions may not be *energy* efficient, in that it takes more energy to produce the marketable form of energy than the final energy produced. For example, the energy necessary to collect and process household refuse may be more (in terms of BTU) than the energy produced by burning the refuse-derived fuel and producing electricity.

An important distinction must be made between *arithmetic energy equivalence* and *conversion energy equivalence*. The former is calculated simply on the basis of energy, while the latter takes into account the energy loss in conversion. Example 6.5 illustrates this point.

e • x • a • m • p • l • e **6.5**

Problem   What are the arithmetic and conversion energy equivalents between gasoline (20,000 BTU/lb) and refuse-derived fuel (5000 BTU/lb)?

Solution

$$\text{arithmetic energy equivalence} = \frac{20{,}000 \text{ BTU/lb gasoline}}{5000 \text{ BTU/lb refuse}}$$

$$= 4 \text{ lb refuse/1 lb gasoline}$$

But the processing of refuse to make a fuel that can be burned also requires energy. This can be estimated at perhaps 50% of the refuse-derived fuel energy, so that the actual *net* energy in the refuse is 2500 BTU/lb, and the

$$\text{conversion energy equivalence} = \frac{20{,}000 \text{ BTU/lb gasoline}}{2500 \text{ BTU/lb refuse}}$$

$$= 8 \text{ lb refuse/1 lb gasoline}$$

Finally, there is the very practical problem of running an automobile on refuse-derived fuel. If this were a *true* equivalence, it would be a simple matter to substitute one fuel for another, without penalty. Obviously this is not possible, and a conversion equivalence does not mean that one fuel can be substituted for another. In addition, the measured energy value of a fuel such as gasoline is not the net energy of that fuel. This is the energy value as measured by a calorimeter, but net or true value must be calculated by subtracting the energy cost of the surveys, drilling, production, and transport necessary to produce the gasoline. Suffice it to say that energy equivalence calculations are not simple, and thus grandiose pronouncements by politicians such as "We can save 15 zillion barrels of gasoline a year if we would only start burning all our cow pies" should be treated with proper skepticism.

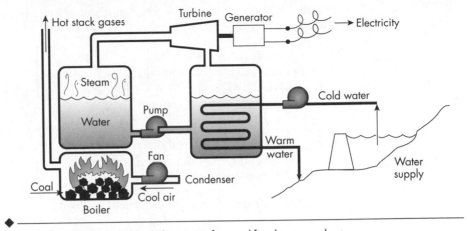

FIGURE 6-5 Simplified drawing of a coal-fired power plant.

6.3.2 Electric Power Production

One of the most distressing conversion problems is the production of electricity from fossil fuels. The present power plants are less than 40% efficient. Why is this?

First, consider how a power plant operates. Figure 6-5 shows that the water is heated to steam in a boiler, and the steam is used to run a turbine, which drives a generator. The waste steam must be condensed to water before it can again be converted to high-pressure steam. This system can be simplified as in Figure 6-6; the resulting schematic is called a *heat engine*. If the work

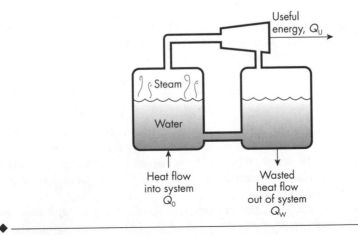

FIGURE 6-6 The heat engine.

performed is also expressed as energy, and steady state is assumed, an energy balance on this heat engine yields

$$
\begin{bmatrix} \text{rate of} \\ \text{energy} \\ \text{ACCUMULATION} \end{bmatrix} = \begin{bmatrix} \text{rate of} \\ \text{energy} \\ \text{IN} \end{bmatrix} - \begin{bmatrix} \text{rate of} \\ \text{useful energy} \\ \text{OUT} \end{bmatrix} - \begin{bmatrix} \text{rate of} \\ \text{wasted energy} \\ \text{OUT} \end{bmatrix}
$$

$$0 = Q_0 - Q_U - Q_W$$

where Q_0 = energy flow into the black box
Q_U = useful energy out of the black box
Q_W = wasted energy out of the black box

The efficiency of this system, as previously defined, is

$$\text{efficiency (\%)} = [Q_U]/[Q_0] \times 100$$

From thermodynamics, it is possible to prove that the most efficient engine (least wasted energy) is called the Carnot engine and that its efficiency is determined by the absolute temperature of the surroundings. The efficiency of the Carnot engine is defined as

$$E_C(\%) = \frac{T_1 - T_0}{T_1} \times 100$$

where T_1 = absolute temperature of the boiler, $°K = °C + 273$
T_0 = absolute temperature of the condenser (cooling water)

Since this is the best possible, any real-world system must be less efficient, or

$$\frac{Q_U}{Q_0} \leq \frac{T_1 - T_0}{T_1}$$

Modern boilers can run at temperatures as high as 600°C, while environmental restrictions limit condenser water temperature to about 20°C. Thus the best expected efficiency is

$$E_C(\%) = \frac{(600 + 273) - (20 + 273)}{(600 + 273)} \times 100 = 66\%$$

A real power plant also has losses in energy due to hot stack gases, evaporation, friction losses, and so on. The best plants so far have been hovering around 40% efficiency. When various energy losses are subtracted from nuclear-powered plants, the efficiencies for these plants seem to be even lower than for fossil fuel plants.

If 60% of the heat energy in coal is not used, it must be wasted, and this energy must somehow be dissipated into the environment. Waste heat energy is emitted from the power plant in two primary ways: stack gases and cooling water. The schematic of the power plant in Figure 6–5 can be reduced to a

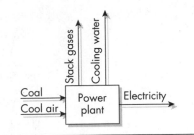

FIGURE 6-7 The power plant as a black box.

black box, and a heat balance performed, as shown in Figure 6-7. Assume for the sake of simplicity that the heat in the cool air is negligible. So the energy balance, at steady state, is

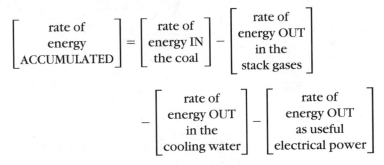

Commonly the energy lost in the stack gases accounts for 15% of the energy in the coal, while the cooling water accounts for the remaining 45%. This large fraction illustrates the problems associated with what is known as *thermal pollution*, the increase in temperature of lakes and rivers due to cooling water discharges.

Most states restrict thermal discharges to less than 1°C rise above ambient stream temperature. The heat in the cooling water must therefore be dissipated into the atmosphere before the cooling water is discharged. Various means are used for dissipating this energy, including large shallow ponds and cooling towers. A cutaway drawing of a typical cooling tower is shown in Figure 6-8. Cooling towers represent a substantial additional cost to the generation of electricity, estimated as doubling the cost of power production for fossil-fuel plant. Cooling water towers may increase the costs for nuclear power plants by as much as 250%.

Even with this expense, watercourses immediately below cooling-water discharges are often significantly warmer than normal. This results in the absence of ice during hard winters and the growth of immense fish. Stories about the size of fish caught in artificially warmed streams and lakes abound, and these places become not only favorite fishing sites for people, but winter roosting places for birds. Wild animals similarly use the unfrozen water during winter when other surface waters are frozen.

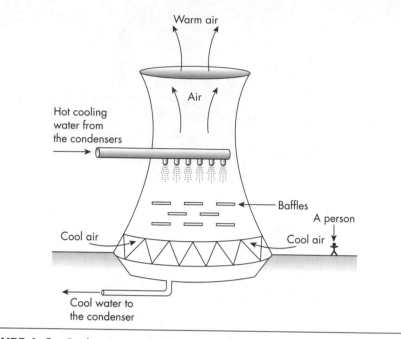

Warm air

Air

Hot cooling
water from
the condensers

Baffles

A person

Cool air

Cool air

Cool water to
the condenser

FIGURE 6-8 Cooling tower used in power plants.

This all sounds pretty good, for people as well as wildlife. The heat clearly changes the aquatic ecosystem, but some would claim that this change is for the better. Everyone seems to benefit from having the warm water. Yet changes in aquatic ecosystems are often unpredictable and potentially disastrous. Heat can increase the chances of various types of disease in fish, and heat will certainly restrict the types of fish that can exist in the warm water. Many cold-water fish such as trout cannot spawn in warmer water and they will die out, their place taken by fish that can survive, such as catfish and carp.

It is unclear what values are involved in governmental restrictions on thermal discharges such as "no more than 1°C rise in temperature." Is it our intent to protect the trout, or would it be acceptable to have the stream populated by other fish? What about the advantage to other life by having ice-free water during winter? And what about the people who like to fish? On a local level, thermal discharges do not seem to produce lasting effects, so why are we paying so much more for cooling the heated water?

Hot water can also be useful for heating buildings. In some parts of the world, cooling water from power generation is used for space heating, piping the hot water through underground lines to homes and businesses. In other cities such as Nashville, Tennessee, high-pressure steam is used for power production and the leftover low-pressure steam becomes a valuable resource for heating buildings in the central business district.

With this much potential good resulting from the discharge of waste heat, why is it necessary to spend so much money cooling the water down before discharge? In the United States, waste heat is not used for space heating because the power plants have been intentionally built as far away from civilization as possible. Our unwillingness to have power plants as neighbors also deprives us of the opportunity to obtain "free" heat. We are too rich to worry about such savings in energy. What will historians 100 years from now say about our values?

Abbreviations

cal	= calories	kWh	= kilowatt hours
E	= efficiency	Q	= heat flow
J	= joules	T	= temperature

Problems

6-1 How many pounds of coal must be burned to keep one 100-watt light bulb lit for one hour? Assume the efficiency of the power plant is 35%, the transmission losses are 10% of the delivered power, and the heating value of coal is 12,000 BTU/lb.

6-2 Coal is a nonreplenishable resource, and a valuable source of carbon for the manufacture of plastics, tires, and the like. Once the stores of coal are depleted, there will not be any more coal to be mined. Is it our responsibility to make sure that there are coal resources left in say 200 years, to be used by people alive then? Should we even worry about this, or should we use up coal as fast as necessary and prudent for our own needs, figuring that future generations can take care of themselves? Write a one-page argument for either not conserving coal or for conserving coal for use by future generations.

6-3 A nuclear power station with a life of 25 years produces 750 MW per year as useful energy. The energy cost is as follows:

150 MW lost in distribution

20 MW needed to mine the fuel

50 MW needed to enrich the fuel

80 MW (spread over 25 years) to build the plant, or 80/25 MW per year

290 MW lost as heat

What is the efficiency of the plant? What is the efficiency of the system (including distribution losses)?

6-4 One of the greatest problems with nuclear power is the disposal of radioactive wastes. If we all received electrical energy from a nuclear power plant, our personal contribution to the nuclear waste is about 1/2 pint, a small milk carton.

This does not seem like a lot to worry about. Suppose, however, that the entire New York City area (population 10 million) receives its electricity from only nuclear power plants. How much waste would be generated each year? What should be done with it? Devise a novel (?) method for high-level nuclear-waste disposal and defend your selection with a one-page discussion. Consider both present and future human generations as well as environmental quality, ecosystems, and future use of resources.

6-5   A one-gram sample of an unknown fuel is tested in a 2-liter (equivalent) calorimeter, with the following results:

Time (min)	Temp. (°C)
0	18.5
5	19.0
6	19.8
7	19.9
8	20.0
9	19.9
10	19.8

What is the heating value of this fuel in kJ/kg?

6-6   One of the cleanest forms of energy is hydroelectric power. Unfortunately, most of our rivers have already been dammed up as much as is feasible, and it is unlikely that we will be able to obtain much more hydroelectric power. In Canada, however, the James Bay area is an ideal location for massive new dams that would provide clean and inexpensive electrical power to the Northeast. The so-called Hydro-Quebec project is already underway, and the Canadians are seeking new customers for their power.

The dams will, however, create lakes that will flood Native American ancestral lands, and the Native Americans are quite upset by this. Because of these and other environmental concerns, New York State and other possible customers have backed out of purchase arrangements, casting a shadow over the project expansion.

Discuss in a two-page paper the conflict of values as you perceive them. Does the Canadian government have legitimate right to expropriate the lands? Recognize that if this project will not be built, other power plants will be constructed. How should the Canadian government resolve this issue?

6-7   The balance of light hitting the planet earth is shown in Figure 6-9. What fraction of the light is actually useful energy absorbed by the earth's surface?

6-8   Figure 6-4 on page 159 shows an energy balance for the United States. All values are in quadrillion BTU (10^{15} BTU). The chart shows how various sources of energy are used, and a large part of our energy budget is wasted. Check the numbers on this figure using an energy balance.

6-9   The Dickey and Lincoln hydroelectric dams on the St. John River in Maine were planned with careful consideration to the Furbish lousewort (*Pedicularis furbishiae*), an endangered plant species. An engineer writing in *World Oil* (January 1977, p. 5) termed this concern for the "lousy lousewort" to be

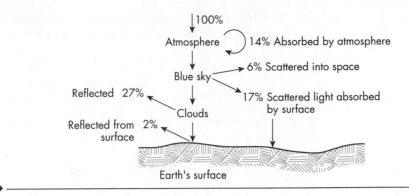

FIGURE 6-9 Global energy flow. See Problem 6-7.

"total stupidity." Based on only this information, construct an ethical profile of this engineer.

6-10[2] Disposable diapers, manufactured from paper and petroleum products, are one of the most convenient diapering systems available. Disposable diapers are also considered by many to be anti-environment, but the truth is not so clear-cut.

In this problem, three diapering systems are considered: home-laundered cloth diapers, commercially laundered cloth diapers, and disposable diapers containing a superabsorbent gel. Energy and materials balances are used to determine the relative merits of each system.

The energy and waste inventory data for each system are shown below. The data are for 1000 diapers.

		Cloth Diapers	
	Disposable Diapers	Commercially Laundered	Home Laundered
Energy requirements, 10^6 BTU	1.9	2.1	3.8
Solid waste, ft^3	17	2.3	2.3
Atmospheric emissions, lb	8.3	4.5	9.6
Waterborne wastes, lb	1.5	5.8	6.1
Water volume requirements, gal	1300	3400	2700

An average of 68 cloth diapers are used per week per baby. Since disposable diapers last longer and never need double diapering, the number of disposable diapers can be expected to be less.

a. Determine the number of disposable diapers required to match the 68 cloth diapers per week. Assume the following:

- 15.8 billion disposable diapers are sold annually.
- 3,787,000 babies are born each year.
- Children wear diapers for the first 30 months.
- Disposable diapers are worn by only 85% of the babies.

b. Complete the table below showing the ratio of impact relative to the home-laundered diapers. The first line is already completed.

	Disposable Diapers	Cloth Diapers Commercially Laundered	Home Laundered
Energy requirements	0.5	0.55	1.0
Solid waste			1.0
Atmospheric emissions			1.0
Waterborne wastes			1.0
Water volume requirements			1.0

c. Using the data below, determine the percentage of disposable diapers that would have to be recycled to make the solid landfill requirements equal for cloth and disposable systems. The table shows the ratios of their impact of solid waste for disposable diapers.

Percentage of Diapers Recycled	Solid Waste per 1000 Diapers (ft³)
0	17
25	13
50	9.0
75	4.9
100	0.80

d. Based on the given data, what is the relative adverse environmental impact of the three different diaper systems?

e. In your opinion, are all the factors considered fairly in this exercise? What is not? Is anything left out?

6–11 You are having lunch with an economist and a lawyer. They are discussing ways to reduce the use of gasoline by private individuals.

Economist: The way to solve this problem is to place such a high tax on gasoline and electric power that people will start to conserve. Imagine if each gallon of gas costs $20? You'd fill your tank for $300! You'd think twice about driving somewhere when you can walk instead.

Lawyer: That would be effective, but a law would be better. Imagine a federal law that forbade the use of a car except when at least four people were passengers? Or we could make a law that required cars to be the size and weight of small motorcycles, with only one passenger seat. That way we could still travel wherever we wanted to but we would triple the gas mileage.

Economist: We could also levy a huge tax on spare parts. It would be more likely then that older cars would be junked sooner, and cleaner, more efficient, newer cars would take their place.

Lawyer: I still like the idea of a 45 mph speed limit on our interstates. This is the most efficient speed for gasoline consumption, and it would also hold down accidents. Besides, the state could raise a lot of revenue by ticketing everyone exceeding the limit. It would be a money-maker!

They continue in this vein until they realize that you have been keeping out of the conversation. They ask you for your opinion about reducing the use of gasoline. How would you respond to their ideas, and what ideas (if any) would you have to offer? Write a one-page response.

6-12   Large consulting firms commonly have many offices, and often communication among the offices is less than efficient. Engineer Stan, in the Atlanta office, is retained by a neighborhood association to write an environmental impact study, which concludes that the plans by a private oil company to build a petrochemical complex would harm the habitat of several endangered species. The client, the neighborhood association, has already reviewed draft copies of the report and is planning to hold a press conference when the final report is delivered. Stan is asked to attend the news conference, in his professional capacity, and charging time to the project.

Engineer Bruce, a partner in the firm and working out of the New York office, receives a phone call.

"Bruce, this is J. C. Octane, president of Bigness Oil Company. As you well know, we have retained your firm for all of our business and have been quite satisfied with your work. There is, however, a minor problem. We are intending to build a refinery in the Atlanta area, and hope to use you as the design engineers."

"We would be pleased to work with you again," replies Bruce, already counting the $1 million design fee.

"There is, however, a small problem," continues J. C. Octane. "It seems that one of your engineers in the Atlanta office has conducted a study for a neighborhood group opposing our refinery. I have received a draft copy of the study, and my understanding is that the engineer and leaders of the neighborhood organization are to hold a press conference in a few days and conclude unfavorable environmental impact as a result of the refinery. I need not tell you how disappointed we will be if this occurs."

As soon as Bruce hangs up the phone with J. C. Octane, he calls Stan in Atlanta.

"Stan, you must postpone the press conference at all costs," Bruce yells into the phone.

"Why?" asks Stan. "It's all ready to go."

"Here's why. You had no way of knowing this, but Bigness Oil is one of the firm's most valued clients, and the president of the oil company has found out about your report and threatens to pull all of their business should the report be delivered to the neighborhood association. You have to rewrite the report in such a way as to show that there would be no significant damage to the environment."

"I can't do that!" pleads Stan.

"Let me see if I can make it clear to you, then," replies Bruce. "You either rewrite the report or withdraw from the project and write a letter to the neighborhood association stating that the draft report was in error, and offer to refund all of their money. You have no other choice!"

Is a consulting engineer an employee of a firm and subject to the dictates and orders of his or her superior, or is he or she functioning as an independent professional who happens to be cooperating with other engineers in the firm? What is the price of loyalty?

State all Stan's alternatives and the probable ramifications of these courses of action, then recommend what action he should take. Make sure you can justify this action. Finally, what would *you* do if you were Stan?

◆ Endnotes

1. This discussion is substantially simplified, and anyone interested in the details of calorimetry should consult any modern thermodynamics text for a thorough discussion.
2. This problem is by David R. Allen, N. Bakshani, and Kirsten Sinclair Rosselot, *Pollution Prevention: Homework and Design Problems for Engineering Curricula*, American Institute of Chemical Engineers and other societies, 1991. The data are from Franklin Associates Ltd., *Energy and Environmental Profile Analysis of Children's Disposable and Cloth Diapers* (report prepared for the American Paper Institute's Diaper Manufacturing Group) Prairie Village, KS (1990). Used with permission.

Some of the most fascinating reactors imaginable are called *ecosystems*. Through these systems flow both energy and materials. *Ecology*, the topic of this chapter, is the study of plants, animals, and their physical environment; that is, the study of ecosystems, and how energy and materials behave in ecosystems.

Specific ecosystems are difficult to define since all plants and animals are in some way related to each other. Because of its shear complexity, we cannot study the earth as a single ecosystem (except in a very crude way) and we must select functionally simpler and spatially smaller systems such as ponds, forests, or even gardens. When the system is narrowed down too far, however, too many external processes affect the system and a meaningful model is difficult to construct. The interaction of squirrels and bluejays at a birdfeeder may be fun to watch, but is not very interesting scientifically to an ecologist because the ecosystem (birdfeeder) is too limited in scope. There are many more organisms and environmental factors that become important in the functioning of the birdfeeder, and these must be taken into account to make this ecosystem meaningful. The problem is deciding where to stop. Is the backyard large enough to study, or must the entire neighborhood be included in the ecosystem? If this is still too limited, where are the boundaries? There are none, of course, and everything truly *is* connected to everything else.

With that caveat, what follows is a brief introduction to the study of ecosystems, however they may be defined.

7.1 ◆ Energy and Materials Flows in Ecosystems

Energy and materials both flow inside ecosystems, but with a fundamental difference: energy flow is only in one direction, whereas materials flow is cyclical.

All energy on earth originates from the sun as light energy. Plants trap this energy through a process called *photosynthesis* and, using nutrients and carbon dioxide, convert the light energy to chemical energy by building high-energy molecules of starch, sugar, proteins, fats, and vitamins. In a crude way, photosynthesis can be pictured as

$$[\text{nutrients}] + CO_2 + \xrightarrow{\text{sunlight}} O_2 + [\text{high-energy molecules}]$$

All other organisms must use this energy for nourishment and growth, using a process called *respiration*:

$$[\text{high-energy molecules}] + O_2 \rightarrow CO_2 + [\text{nutrients}]$$

This conversion process is highly inefficient, with only about 1.6% of the total energy available being converted into carbohydrates through photosynthesis.

Plants, because they manufacture the high-energy molecules, are called *producers*, and the animals using these molecules are called *consumers*.

Both plants and animals produce wastes and eventually die, forming a pool of dead organic matter known as *detritus*, which still contains considerable energy. (That's why we need wastewater treatment plants!) In the ecosystem, the organisms that use this detritus are known as *decomposers*.

In summary, there are three main groups of organisms within an ecosystem:

- Producers receive energy from the sun and produce high-energy molecules.

- Consumers use these molecules as a source of energy.

- Decomposers use the residual energy remaining in dead producers and consumers, and as well as the residual energy in the waste products of producers and consumers.

This one-way flow is illustrated in Figure 7–1. Note that the rate at which this energy is extracted (symbolized by the slope of the line) slows considerably as the energy level decreases, a concept important in wastewater treatment (Chapter 10).

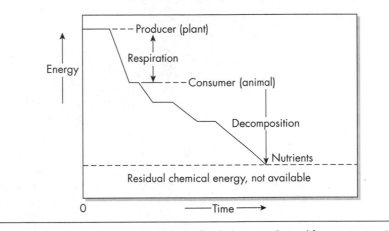

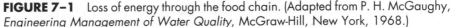

FIGURE 7–1   Loss of energy through the food chain. (Adapted from P. H. McGaughy, *Engineering Management of Water Quality*, McGraw-Hill, New York, 1968.)

Energy flow is one-way flow; from the sun to the plants, to be used by the consumers and decomposers for making new cellular material and for maintenance. Energy is not recycled within an ecosystem, as illustrated by the following argument.

Suppose a plant receives 1000 J of energy from the sun. Of that amount, 760 J is rejected (not absorbed), and only 240 J absorbed. Most of this is released as heat, and only 12 J used for production, 7 of which must go for respiration (maintenance) and the remaining 5 J will go toward building new tissue. If the plant is eaten by a consumer, 90% of the 5 J will go toward the animal's maintenance, and only 10% (or 0.5 J) to new tissue. If this animal is in turn eaten, then again only 10% (or 0.05 J) will be used for new tissue, and the remaining energy is used for maintenance. If the second animal is a human being, then of the 1000 J coming from the sun, only 0.05 J or 0.005% is used for tissue building—a highly inefficient system.

While energy flow is in one direction only, nutrient flow through an ecosystem is cyclical, as represented by Figure 7–2. Starting with the dead organics, or the detritus, the initial decomposition by microorganisms is to some initial products such as ammonia (NH_3), carbon dioxide, and hydrogen

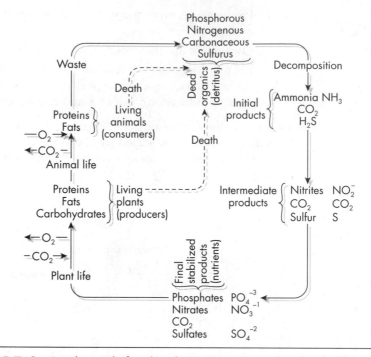

FIGURE 7–2 Aerobic cycle for phosphorus, nitrogen, carbon, and sulfur. (Adapted from P. H. McGaughy, *Engineering Management of Water Quality*, McGraw-Hill, New York, 1968.)

sulfide (H_2S) for nitrogenous, carbonaceous, and sulfurous matter, respectively. These products are in turn decomposed further, until the final stabilized or fully oxidized forms are nitrate (NO_3^-), carbon dioxide (CO_2), sulfate (SO_4^{-2}), and phosphates (PO_4^{-3}). The carbon dioxide is of course used by the plants as a source of carbon, while the nitrates, phosphates, and sulfates are used as nutrients, or the building blocks for the formation of new plant tissue.

There are two types of microbial decomposers, aerobic and anaerobic microorganisms, classified according to whether they require molecular oxygen in their metabolic activity. The equations for aerobic and anaerobic decomposers are

Aerobic: $[\text{detritus}] + O_2 \rightarrow CO_2 + H_2O + [\text{nutrients}]$

Anaerobic: $[\text{detritus}] \rightarrow CO_2 + CH_4 + H_2S + NH_3 + \cdot\ \cdot\ \cdot + [\text{nutrients}]$

The decomposition carried out by the aerobic organisms is much more complete, since some of the end products of anaerobic decomposition (e.g., ammonia nitrogen) are not in their final fully oxidized state. Ammonia, as shown in the aerobic decomposition cycle in Figure 7–2, is an intermediate product of decomposition. Anaerobic decomposition is able to get only part way to the fully oxidized nutrients. Aerobic decomposition is required to oxidize ammonia nitrogen to the fully oxidized form nitrate nitrogen, hydrogen sulfide to sulfate, and so on. Both aerobic and anaerobic microorganisms are used in wastewater treatment, as discussed in Chapter 10.

Decomposition in biochemistry is classified according to whether the microorganisms have the ability to use dissolved oxygen (O_2) as the *hydrogen acceptor* in the decomposition reaction. As the organic molecules decompose, the hydrogen from these molecules hooks on to some other chemical. *Obligate aerobes* are microorganisms that must have dissolved oxygen to survive, and they use oxygen as the hydrogen acceptor, so that in fairly simple terms, the hydrogen from the organic compounds ends up as water, as in the aerobic equation on page 174. When no dissolved oxygen is available, some organisms, called *facultative* microorganisms, will be able to use oxygen from other sources, such as nitrates and sulfates. The nitrates are converted to ammonia (NH_3) and the sulfates to hydrogen sulfide (H_2S), as shown in the anaerobic equation on page 174. The microorganisms find it easier obtaining the oxygen from nitrates, so this process occurs in preference to the use of sulfates. When oxygen is no longer available from nitrates, the next source of oxygen is sulfates. Some microorganisms, for which dissolved oxygen is in fact toxic, are called *obligate anaerobes*, and they also participate in anaerobic decomposition.

Plants, of course, form the proteins, fats, and carbohydrates (the high-energy molecules) and, in the process of photosynthesis, produce oxygen. Some of these plants die, forming the detritus or dead organic matter that must be decomposed by the microorganisms. Some plants are consumed by animals and the nutrients are recycled back to detritus by the wastes produced by the animals or as a result of the death of the animals.

Because nutrient flow in ecosystems is cyclical, it is possible to analyze these flows using the techniques already introduced for materials flow analysis. As introduced earlier,

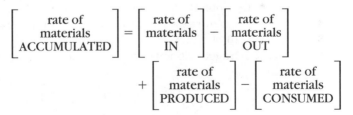

The materials in an ecosystem, as in any other reactor, can be analyzed by the procedures already introduced.

e • x • a • m • p • l • e   **7.1**

Problem   A major concern with the wide use of fertilizers is the leaching of nitrates into the groundwater. Such leaching is difficult to measure, unless a nitrogen balance is constructed for a given ecosystem. Consider, for example, the diagram shown as Figure 7–3, which represents nitrogen transfer in a meadow fertilized with 34 g/m²-yr of ammonia + nitrate nitrogen (17 + 17), both expressed as nitrogen. [Recall that the atomic weight of N is 14 and H is

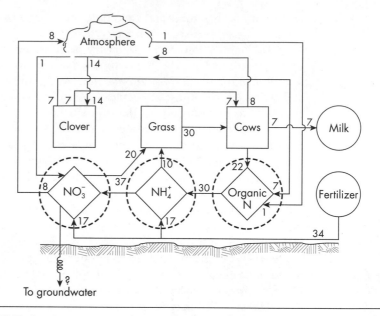

FIGURE 7–3 An ecosystem showing a nitrogen balance.

1.0, so that 17 g/m²-yr of nitrogen requires the application of $17 \times (17/14) =$ 20.6 g/m²-yr of NH_3.] What is the rate of nitrogen leaching into the soil?

Solution   Flows into all black boxes should balance. Note, for example, that organic nitrogen originates from three sources; atmosphere, clover fixing N from the air, and cow manure. The total output of organic N must therefore equal the input, or $22 + 7 + 1 = 30$ g/m²-yr. The organic nitrogen is converted to ammonia nitrogen, and the output of ammonia N must be $17 + 30 = 47$ g/m²-yr, 10 of which is used by the grass, leaving 37 g/m²-yr to be oxidized to nitrate nitrogen. Conducting a mass balance of nitrate nitrogen, and assuming steady state,

$$[IN] = [OUT]$$

$$37 + 17 + 1 = 8 + 20 + \text{leachate}$$

$$\text{leachate} = 27 \text{ g/m}^2\text{-yr}$$

The process by which an ecosystem remains in a steady-state condition is called *homeostasis*. Fluctuations of course occur within a system, but the overall effect is steady state. To illustrate this idea consider a very simple ecosystem consisting of grass, field mice, and owls, as shown in Figure 7–4.

The grass receives energy from the sun, the mouse eats the seeds from the grass, and the owl eats the mice. This progression is known as a *food chain*, and the interaction among the various organisms is a *food web*. Each organism is said to occupy a *trophic level*, depending on its proximity to the producers.

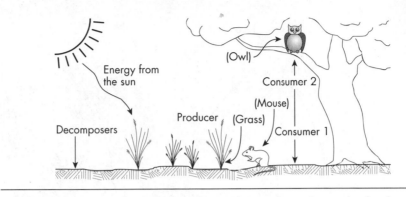

FIGURE 7–4 A simple ecosystem. The numbers indicate the trophic level.

Since the mouse eats the plants, it is at trophic level 1, the owl is at trophic level 2, and so on. It's also possible that a grasshopper eats the grass (trophic level 1) and a praying mantis eats the grasshopper (trophic level 2) and a shrew eats the praying mantis (trophic level 3). If the owl now gobbles up the shrew, it is performing at trophic level 4. Figure 7–5 illustrates such a land-based food web. The arrows show how energy is received from the sun and flows through the system.

If a species is free to grow, unconstrained by food, space, or predators, its growth is described as a first-order reaction,

$$\frac{dN}{dt} = kN$$

where N = number of organisms of a species
k = rate constant
t = time

Fortunately, populations (except for human populations!) within an ecosystem are constrained by food availability, space, and predators. Considering the first two constraints, the maximum population that can exist can be described in mathematical terms as

$$\frac{dN}{dt} = kN - \frac{k}{K}N^2$$

where K = maximum possible population in the ecosystem

Note that at steady state, where $dN/dt = 0$,

$$0 = kN - \frac{k}{K}N^2$$

and canceling,

$$K = N$$

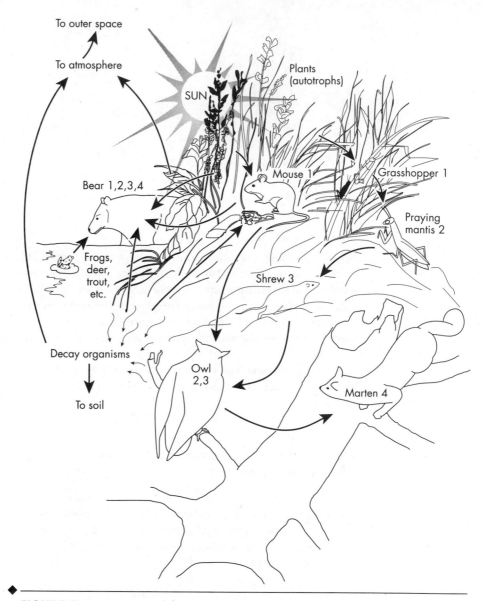

To outer space

To atmosphere

SUN

Plants (autotrophs)

Mouse 1

Grasshopper 1

Bear 1,2,3,4

Praying mantis 2

Frogs, deer, trout, etc.

Shrew 3

Decay organisms

Owl 2,3

Marten 4

To soil

FIGURE 7-5 A terrestrial food web. The numbers indicate the trophic level. (From A. Turk, et al., *Ecosystems, Energy, Population*, W. B. Saunders Company, Philadelphia, 1975. Used with permission.)

or the population is the maximum population possible in that system. If for whatever reason N is reduced to below K, the population increases to eventually again attain the level K.

Other species can, of course, affect the population level of any other species. Suppose there are M organisms of another species in competition with the species of N organisms; then the growth rate of the original species is expressed as

$$\frac{dN}{dt} = kN - \frac{k}{K}N^2 - sMN$$

where s = growth rate constant for the competitive species
M = population of the competitive species

In words, this expression reads

$$\begin{bmatrix} \text{growth} \\ \text{rate} \end{bmatrix} = \begin{bmatrix} \text{unlimited} \\ \text{growth rate} \end{bmatrix} - \begin{bmatrix} \text{self-crowding} \\ \text{effects} \end{bmatrix} - \begin{bmatrix} \text{competitive} \\ \text{effects} \end{bmatrix}$$

If s is small, then both species should be able to exist. If the first species of population N is not influenced by overcrowding, then the competition from the competitive species will be able to keep the population in check.

Note again that in steady state, if $dN/dt = 0$,

$$N = K\left[1 - \frac{Ms}{k} \right]$$

so that if $M = 0$, $N = K$.

Competition in an ecosystem occurs in niches. A *niche* is an animal's or plant's best accommodation with its environment, where it can best exist in the food web. Returning to the simple grass–mouse–owl example, each participant occupies a niche in the food web. If there is more than one kind of grass, they may occupy very similar niches, but there will always be seemingly minor, yet extremely important, differences. Two kinds of clover, for example, may seem at first to occupy the same niche, until we recognize that one species blossoms early while the other blooms later in the summer, thus not competing directly. In a New England hay field the dandelions blossom early, before the hay is able to grow very tall, generating a proliferous yellow sea that produces multitudinous new seeds just days ahead of the rapidly growing hay.

The greater the number of organisms available to occupy various niches within the food web, the more stable is the system. If, in the above example, the grass dies due to a drought or disease, the mice would have no food and they, as well as the owls, would die of starvation. If, however, there are *two* grasses, each of which the mouse could use as a food source (they both fill almost the same niche relative to the mouse's needs), the death of one grass would not result in the collapse of the system. This system is therefore considered to be more stable because it can withstand perturbations without collapsing. Examples of very stable ecosystems are tropical forests and estuaries, while unstable ecosystems include the northern tundra and the deep oceans.

Some perturbations to ecosystems are natural (witness the destruction caused by the eruption of Mount St. Helens in Washington State), whereas many more are caused by human activities. Humans have, of course, adversely affected ecosystems on a small scale (like streams and lakes) and on a large scale (global). The actions of humans can be thought of as *domination* of nature (as in Genesis I) and this idea has spawned a new philosophical approach to environmental ethics—*ecofeminism*.

Ecofeminism is "the position that there are important connections—historical, symbolic, theoretical—between the domination of women and domination of nonhuman nature."[1] The basic premise is that it should be possible to construct a better environmental ethic by incorporating into it, by analogy, the problems of domination of women by men.

If science is the dominant means of understanding the environment, ecofeminists argue that science needs to be transformed by feminism. Because women tend to be more nurturing and caring, this experience can be brought to bear on our understanding of the environment. Ecofeminists argue that the whole idea of domination of nature and the detachment from nature has to change. Ecofeminism concentrates on the ethics of care, as opposed to the ethics of justice, and by that means avoids many of the problems inherent in the application of classical ethical thinking to the environment. Animal suffering, for example, is not a philosophical problem to ecofeminists since the care of all creatures is what is important.

Ecofeminists often invoke the spiritual dimension, asserting that the earth has always been thought of as feminine (mother earth and mother nature) and that the domination of women by men has a striking parallel with the domination of the earth.

But such a similarity does not prove a causal connection. Besides, over the years women have often participated in the destruction of nature with the same vigor as men. Such destruction seems to be a human trait and not necessarily a masculine characteristic.

7.2 ◆ Human Influence on Ecosystems

Probably the greatest difference between people and all other living and dead parts of the global ecosystem is that people are unpredictable. All other creatures play according to well-established rules. Ants build ant colonies, for example. They will always do that. The thought of an ant suddenly deciding that it wants to go to the moon, or write a novel, is ludicrous. Yet people, creatures who are just as much part of the global ecosystem as ants, can make unpredictable decisions such as to develop chlorinated pesticides, or to drop nuclear bombs, or even to change the global climate. From a purely ecological perspective, we are indeed different—and frightening.

Below are several of the more evident ways unpredictable humans can affect ecosystems.

7.2.1   Effect of Pesticides on an Ecosystem

The large-scale use of pesticides for the control of undesired organisms began during World War II with the invention and wide use of the first effective organic pesticide, DDT. Before that time, arsenic and other chemicals had been employed as agricultural pesticides, but the high cost of these chemicals and their toxicity to humans limited their use. DDT, however, was cheap, lasting, effective, and did not seem to harm human beings.

Many years later it was discovered that DDT decomposes very slowly, is stored in the fatty tissues of animals, and is readily transferred from one organism to another through the food chain. As it moves through the food chain, it is *biomagnified*, or concentrated as the trophic levels increase. For example, the concentration of DDT in one estuarine food chain is shown in Table 7-1. Note that the concentration factor from the DDT level in water to the larger birds is about 500,000.

As the result of these very high concentrations of DDT (and subsequently from other chlorinated hydrocarbon pesticides as well), a number of birds are on the verge of extinction since the DDT affects their calcium metabolism, resulting in the laying of eggs with very thin (and easily broken) shells.

People of course occupy the top of the food chain and would be expected to have very high levels of DDT. Because it is impossible to relate directly DDT to any acute human disease, the increase in human DDT levels for many years was not an area of public health concern. More recently, the subtle effects of chemicals such as DDT on human reproduction systems has come into focus, and public concern finally forced DDT to be banned when it was discovered that human milk fed to infants often contained four to five times the DDT content allowable for the interstate shipment of *cows'* milk!

An interesting example of how DDT can cause a perturbation of an ecosystem occurred in a remote village in Borneo. A World Health Organization worker, attempting to enhance the health of the people, sprayed DDT in the inside of thatch huts to kill flies, which he feared would carry disease. The dying flies were easy pray for small lizards that live inside the thatch and feed on the flies. The large dose of DDT, however, made the lizards ill, and they in turn fell pray to the village cats. As the cats began to die, the village was invaded by rats, which were suspected of carrying bubonic plague. As a result, live cats were parachuted into the village to try to restore the balance upset by a well-meaning health official.[2]

TABLE 7-1 DDT Residues in an Estuarine Food Web

	DDT Residues (ppm)
Water	0.00005
Plankton	0.04
Minnows	0.23
Pickerel (predatory fish)	1.33
Heron (feeds on small animals)	3.57
Herring gull (scavenger)	6.00
Merganser (fish-eating duck)	22.8

Data from G. M. Woodell, et al., *Science*, 156:821–824 (1967).

7.2.2   Effect of Nutrients on a Lake Ecosystem

Lakes represent a second example of an ecosystem affected by people. A model of a lake ecosystem is shown in Figure 7–6. Note that the producers (algae) receive energy from the sun, and through the process of photosynthesis produce biomass and oxygen. Because the producers (the algae) receive energy from the sun, they obviously must be restricted to the surface waters in the lake. Fish and other animals also exist mostly in the surface water since much of the food is there, but some scavengers are on the bottom as well. The decomposers mostly inhabit the bottom waters because this is the source of their food supply (the detritus).

Through the process of photosynthesis, the algae use nutrients and carbon dioxide to produce high-energy molecules and oxygen. The consumers, including fish, plankton, and many other organisms, all use the oxygen, produce CO_2, and transfer the nutrients to the decomposers in the form of dead organic matter. The decomposers, including scavengers such as worms and various microorganisms, reduce the energy level further by the process of respiration, using oxygen and producing carbon dioxide. The nutrients nitrogen and phosphorus (as well as other nutrients, often called micronutrients) are then again used by the producers. Some types of algae are able to fix nitrogen from the atmosphere, while some decomposers produce ammonia nitrogen, which bubbles out.[3]

The only element of major importance that does not enter the system from the atmosphere is phosphorus. A given quantity of phosphorus is recycled from the decomposers back to the producers. The fact that only a certain amount of phosphorus is available to the ecosystem limits the rate of metabolic activity. Were it not for the limited quantity of phosphorus, the ecosystem metabolic activity could accelerate and eventually self-destruct, since all other

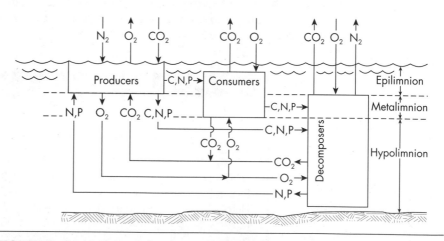

FIGURE 7–6   Movement of carbon, oxygen, and nutrients in a lake. (Adapted from Don Francisco, Department of Environmental Science and Engineering, University of North Carolina, Chapel Hill.)

chemicals (and energy) are in plentiful supply. For the system to remain at homeostasis (steady state), some key component, in this case phosphorus, must limit the rate of metabolic activity by acting as a brake in the process.

Note that in Figure 7-6 the algae must live on or near the surface because this is where the sunlight is, while the decomposition takes place on the bottom. Oxygen, from both the algae and the atmosphere, must travel through the lake to the bottom to supply the aerobic decomposers. If the water is well mixed, this will not present much difficulty, but unfortunately most lakes are thermally stratified and not mixed throughout the full depth.

The reason for such stratification is shown in Figure 7-7. In the winter, with ice on the surface, the temperature of northern lakes is about 4°C, the temperature at which water is the densest. In warmer climates, where ice does not form, the deepest waters are still normally at this temperature. In the spring, the surface water begins to warm, and for a time the water in the entire lake is about the same temperature. For a brief time, wind can produce mixing that can extend to the bottom. As summer comes, however, the deeper, denser water sits on the bottom as the surface water continues to warm, producing a steep gradient and three distinct sections: *epilimnion*, *metalimnion*, and *hypolimnion*. The inflection point is called the *thermocline*. In this condition, the lake is thermally stratified, and there is only mixing between any two adjacent strata. During this season, when the metabolic activity can be expected to be the greatest, oxygen gets to the bottom only by diffusion. As winter approaches and the surface water cools, the water on top may cool enough to be denser than the lower water, and a *fall turnover* occurs, resulting in a

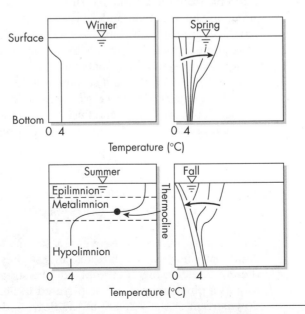

FIGURE 7-7 Temperature profiles in a temperate-climate lake.

thoroughly mixed lake. During the winter, when the water on top is again lighter, the lake is once again stratified.

Consider now what would occur if an external source of phosphorus, such as might originate from farm runoff, would flow into the lake. The brake on the system would be released, and the algae would begin to reproduce at a higher rate, resulting in a greater production of food for the consumers, which in turn would grow at a higher rate. All of this activity would produce ever-increasing quantities of dead organic matter for the decomposers, which would greatly multiply.

Unfortunately, the dead organic matter is distributed throughout the body of water (as shown in Figure 7–6) while the algae, which produce the necessary oxygen for the decomposition to take place, live only near the surface of the lake, where there is sunlight. The other supply of oxygen for the decomposers is from the atmosphere, also at the water's surface. Because both sources of oxygen are at the top of the lake, and because the lake is thermally stratified during much of the year, not enough O_2 can be transported quickly enough to the lake bottom. With an increased food supply, oxygen demand by the decomposers increases and finally outstrips the supply. The aerobic decomposers die out and are replaced by anaerobic forms, which produce large quantities of biomass and result in incomplete decomposition. Eventually, with an ever-increasing supply of phosphorus, the entire lake becomes anaerobic: most fish life dies and the algae is concentrated on the very surface of the water, forming green slimy algae mats, called *algae blooms*. In time, the lake fills with dead and decaying organic matter and a peat bog results.

This process is called *eutrophication*. Eutrophication actually occurs in all lakes, independent of human activity, because all lakes receive some additional nutrients from the air and overland flow. Natural eutrophication is usually very slow, measured in thousands of years before major changes occur. What people have accomplished, of course, is speeding up this process with the introduction of large quantities of nutrients, resulting in *accelerated eutrophication*. Thus an aquatic ecosystem that might have been in essentially steady state (homeostasis) has been disturbed and an undesirable condition results.

Incidentally, the process of eutrophication can be reversed by reducing nutrient flow into a lake and flushing or dredging as much of the phosphorus out as possible, but this is an expensive proposition.

7.2.3 Effect of Organic Wastes on a Stream Ecosystem

The primary difference between a stream and a lake is that the former is continually flushed out. Streams, unless they are exceptionally lethargic, are therefore seldom highly eutrophied. The stream can be hydraulically characterized as a plug-flow reactor, as pictured in Figure 7–8. A plug of water moves downstream at the velocity of the flow, and the reactions that take place in this plug can be analyzed.

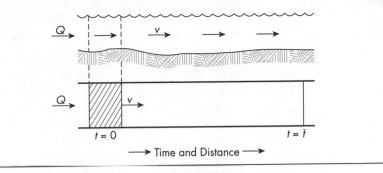

$t = 0$ $t = \bar{t}$

⟶ Time and Distance ⟶

FIGURE 7-8 A stream acting as a plug flow reactor.

The reaction of greatest concern in stream pollution is the depletion of oxygen. Thinking of the plug as a black box and a mixed batch reactor, we can write a materials balance in terms of oxygen as

$$
\begin{bmatrix} \text{rate of} \\ \text{oxygen} \\ \text{ACCUMULATED} \end{bmatrix} = \begin{bmatrix} \text{rate of} \\ \text{oxygen} \\ \text{IN} \end{bmatrix} - \begin{bmatrix} \text{rate of} \\ \text{oxygen} \\ \text{OUT} \end{bmatrix}
$$

$$
+ \begin{bmatrix} \text{rate of} \\ \text{oxygen} \\ \text{PRODUCED} \end{bmatrix} - \begin{bmatrix} \text{rate of} \\ \text{oxygen} \\ \text{CONSUMED} \end{bmatrix}
$$

Because this is not a steady-state situation, the first term is not zero. The "oxygen in" to the black box (the stream) is called *reoxygenation*, and in this case consists of diffusion from the atmosphere. There is no "oxygen out" since the water is less than saturated with oxygen. If the stream is swift and algae do not have time to grow, there is no "oxygen produced." The oxygen used by the microorganisms in respiration is called *deoxygenation* and is the "oxygen consumed" term. The accumulation, as before, is expressed as a differential equation, so that

$$
\begin{bmatrix} \text{rate of O}_2 \\ \text{ACCUMULATED} \end{bmatrix} = \begin{bmatrix} \text{rate O}_2 \text{ IN} \\ \text{(reoxygenation)} \end{bmatrix} - 0 + 0 - \begin{bmatrix} \text{rate O}_2 \text{ CONSUMED} \\ \text{(deoxygenation)} \end{bmatrix}
$$

(7.1)

Both reoxygenation and deoxygenation can be described by first-order reactions. The rate of oxygen use, or deoxygenation, can be expressed as

$$
\text{rate of deoxygenation} = -k_1 C
$$

where k_1 = deoxygenation constant, a function of the type of waste material decomposing, temperature, etc., days^{-1}

C = the amount of oxygen per unit volume necessary for decomposition, measured as mg/L

Later in this text C is defined as the difference between the biochemical oxygen demand (BOD) at any time t and the ultimate BOD demand. For now, consider C as the oxygen necessary for the microorganisms at any time to complete the decomposition. If C is high, then the rate of oxygen use is great.

Water can hold only a limited amount of a gas such as oxygen. The amount of oxygen that can be dissolved in water depends on the water temperature, atmospheric pressure, and the concentration of dissolved solids. The saturation levels of oxygen in deionized water at one atmosphere and at various temperatures is shown in Table 7-2.

The value of k_1, the deoxygenation constant, is measured in the laboratory using analytical techniques, as discussed in the next chapter.

The reoxygenation of water can be expressed as

$$\text{rate of reoxygenation} = k_2 D$$

TABLE 7-2   Oxygen Solubility in Fresh Water

Temperature (°C)	Dissolved Oxygen (mg/L)	Temperature (°C)	Dissolved Oxygen (mg/L)
0	14.60	23	8.56
1	14.19	24	8.40
2	13.81	25	8.24
3	13.44	26	8.09
4	13.09	27	7.95
5	12.75	28	7.81
6	12.43	29	7.67
7	12.12	30	7.54
8	11.83	31	7.41
9	11.55	32	7.28
10	11.27	33	7.16
11	11.01	34	7.05
12	10.76	35	6.93
13	10.52	36	6.82
14	10.29	37	6.71
15	10.07	38	6.61
16	9.85	39	6.51
17	9.65	40	6.41
18	9.45	41	6.31
19	9.26	42	6.22
20	9.07	43	6.13
21	8.90	44	6.04
22	8.72	45	5.95

TABLE 7–3 Empirical Reoxygenation Constants	
	k_2 at 20°C (day^{-1})
Small backwaters	0.1–0.23
Lethargic streams	0.23–0.35
Large streams, low velocity	0.35–0.46
Large streams, normal velocity	0.46–0.69
Swift streams	0.69–1.15
Rapids	1.15

Source: D. J. O'Connor and W. E. Dobbins, *ASCE Trans*, 153 (1958).

where D = deficit in dissolved oxygen, or the difference between saturation (maximum dissolved oxygen the water can hold) and the actual dissolved oxygen (DO), mg/L

k_2 = reoxygenation constant, days^{-1}

The value of k_2 is obtained by conducting a study on a stream using tracers, or by empirical equations that include stream conditions, or by using tables that describe various types of streams.

A popular generalized formula for estimating the reoxygenation constant is[4]

$$k_2 = \frac{3.9 \, v^{1/2} \, ([1.025]^{(T-20)})^{1/2}}{H^{3/2}}$$

where T = temperature of the water, °C

H = average depth of flow, m

v = mean stream velocity, m/s

The temperature term represents the variation of the rate constant with temperature, while the v and H terms describe the kinetic energy in the stream and the ease of O_2 transport due to water depth, respectively. An alternate means of obtaining k_2 values is to use generalized tables, such as Table 7–3.

In a stream loaded with organic material, the simultaneous action of deoxygenation and reoxygenation forms what is called the *dissolved oxygen sag curve*, first described by Streeter and Phelps in 1925.[5]

The shape of the oxygen sag curve, as shown in Figure 7–9, is the result of adding the rate of oxygen use or consumption and the rate of supply or reoxygenation. If the rate of use is great, as in the stretch of stream immediately after the introduction of organic pollution ($t = 0$), the dissolved oxygen level drops because the supply rate cannot keep up with the use of oxygen, creating a deficit. The deficit (D) is defined as the difference between the oxygen concentration in the streamwater (C) and the total amount the water *could* hold, or saturation (S). That is,

$$D = S - C$$

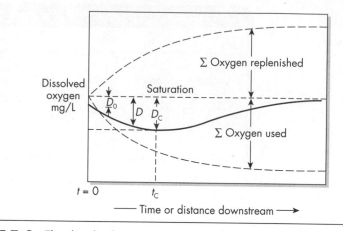

FIGURE 7-9 The dissolved oxygen sag curve in a stream is the difference between the oxygen used and oxygen supplied.

where D = oxygen deficit, mg/L

   S = saturation level of oxygen in the water (the most it can ever hold), mg/L

   C = concentration of dissolved oxygen in the water, mg/L

After the initial high rate of decomposition when the readily degraded material is used by the microorganisms, the rate of oxygen use decreases because only the less readily decomposable materials remain. Since so much oxygen has been used, the deficit, or the difference between oxygen saturation level and actual dissolved oxygen, is great; the supply of oxygen from the atmosphere is high and eventually begins to keep up with oxygen use, and the deficit begins to level off. With time, the dissolved oxygen use decreases (there is little left to decompose) and once again the dissolved oxygen level reaches saturation, producing a dissolved oxygen sag.

This process can be described in terms of the rate of oxygen use and the rate of oxygen resupply, or two concurrent reactions, and expressed as

$$\frac{dD}{dt} = k_1 z - k_2 D$$

where z is the amount of oxygen still required by the microorganisms decomposing the organic material. The rate of change in the deficit (D) depends on the concentration of decomposable organic matter, or the remaining need by the microorganisms for oxygen (z), and the deficit at any time t. As explained more fully in Chapter 8, the demand for oxygen at any time t can be expressed as

$$z = Le^{-k_1 t}$$

where L = ultimate oxygen demand, or the maximum oxygen required, mg/L. As shown on page 123, by substituting this into the above expression and integrating

$$D = \frac{k_1 L_0}{k_2 - k_1} (e^{-k_1 t} - e^{-k_2 t}) + D_0 e^{-k_2 t}$$

where D_0 = the oxygen deficit immediately below the point of pollutant discharge, mg/L

D = oxygen deficit at any time t, mg/L

L_0 = ultimate oxygen demand immediately below the point of pollutant discharge, mg/L

The most serious concern regarding water quality is of course the point in the stream where the deficit is the greatest, or the dissolved oxygen concentration is least. By setting $dD/dt = 0$ (where the dissolved oxygen sag curve flattens out and begins to rise), we can solve for the critical time as

$$t_c = \frac{1}{k_2 - k_1} \ln \left[\frac{k_2}{k_1} \left(1 - \frac{D_0 (k_2 - k_1)}{k_1 L} \right) \right]$$

where t_c = time downstream when the dissolved oxygen is at the lowest concentration

e • x • a • m • p • l • e **7.2**

Problem   A large stream has a reoxygenation constant of 0.4 day^{-1}, a velocity of 0.85 m/sec, and at the point at which an organic pollutant is discharged it is saturated with oxygen at 10 mg/L ($D_0 = 0$). Below the outfall, the ultimate demand for oxygen is found to be 20 mg/L and the deoxygenation constant is 0.2 day^{-1}. What is the dissolved oxygen 48.3 km downstream?

Solution   Velocity = 0.85 m/sec, hence it takes

$$\frac{48.3 \times 10^3}{0.85 \text{ m/sec}} = 56.8 \times 10^3 \text{ sec}$$

or

$$56.8 \times 10^3 \text{ sec} \times \frac{1 \text{ min}}{60 \text{ sec}} \times \frac{1 \text{ hr}}{60 \text{ min}} \times \frac{1 \text{ day}}{24 \text{ hr}} = 0.66 \text{ days}$$

to travel the 48.3 km.

Using the Streeter–Phelps deficit equation,

$$D = \frac{(0.2)(20)}{(0.4) - (0.2)} (e^{-0.2(0.66)} - e^{-0.4(0.66)}) + 0$$

$$D = 2.2 \text{ mg/L}$$

If the saturation dissolved oxygen is 10 mg/L, then the dissolved oxygen at 48.3 km is $10 - 2.2 = 7.8$ mg/L.

◆

An assumption made in the foregoing analysis, that the volume of the polluting stream is very small compared to the stream flow, is not always true. In most cases we cannot assume that the initial dissolved oxygen (DO) deficit in the stream below the outfall is the same as the deficit above the outfall. More accurate would be to calculate the deficit by performing a materials balance on the plug at $t = 0$ and assuming that within this plug the two incoming streams are well mixed. Figure 7-10 shows that the flow of the stream and its dissolved oxygen concentration is Q_s and C_s respectively, and the pollution stream has a flow and dissolved oxygen of Q_p and C_p. Thus a volume flow balance gives

$$[IN] = [OUT]$$

$$Q_s + Q_p = Q_0$$

where Q_s = upstream flow
Q_p = flow from the pollution source
Q_0 = downstream flow

and according to the arguments in Chapter 2, page 50, (mass = volume × concentration); the balance in terms of mass of oxygen can be written as

$$Q_s C_s + Q_p C_p = Q_0 C_0$$

where C_0 = dissolved oxygen concentration in the stream immediately below the confluence of the pollution. Rearranged,

$$C_0 = \frac{Q_s C_s + Q_p C_p}{Q_s + Q_p}$$

Once C_0 is determined (Note that this assumes perfect mixing at the pollution source, not a very realistic assumption, especially if the river is broad

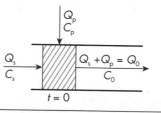

$t = 0$

FIGURE 7–10   The concentration of oxygen in the stream downstream of the source of pollution is a combination of the oxygen concentration in the stream upstream of the discharge (C_S) and the oxygen concentration in the pollutant stream (C_S). Since the deficit is equal to $S - C$, the initial deficit (D_0) must then be a combination of the deficit in the stream (D_s) and the pollutant (D_p).

and lethargic. In this text we ignore such problems.), the water temperature is calculated using equation 6.1 as

$$T_0 = \frac{Q_s T_s + Q_p T_p}{Q_s + Q_p}$$

and the saturation value S_0 is found in Table 7-2. The initial deficit is then calculated as

$$D_0 = S_0 - C_0$$

The dissolved-oxygen sag curve that incorporates the effect of the pollutant stream on the initial deficit D_0 is shown in Figure 7-11. Further, the stream may also have a demand for oxygen at the point at which it reaches the outfall. Assuming once again complete mixing, the oxygen demand below the outfall must be calculated as

$$L_0 = \frac{L_s Q_s + L_p Q_p}{Q_s + Q_p}$$

where L_0 = ultimate oxygen demand immediately below the outfall
$\quad\quad L_s$ = ultimate oxygen demand of the stream immediately above the outfall, mg/L
$\quad\quad L_p$ = ultimate oxygen demand of the pollutant discharge, mg/L

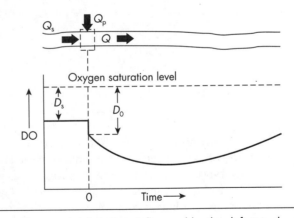

FIGURE 7-11 The initial deficit D_0 is influenced by the deficit in the incoming pollutant stream.

e • x • a • m • p • l • e **7.3**

Problem Using Example 7.2, suppose the waste stream has a dissolved oxygen concentration of 1.5 mg/L, a flow of 0.5 m³/sec, a temperature of 26°C, and an ultimate biochemical oxygen demand of 48 mg/L. The stream water is running at 2.2 m³/sec at a saturated dissolved oxygen concentration, a temperature of 12°C, and an ultimate BOD of 13.6 mg/L. Calculate the dissolved oxygen concentration 48.3 km downstream.

Solution

$$T_0 = \frac{Q_s T_s + Q_p T_p}{Q_s + Q_p} = \frac{2.2(12) + 0.5(26)}{2.2 + 0.5} = 14.6°C$$

From Table 7-2, $S = 10.8$ mg/L at 12°C; and since the stream is saturated, $S = C_s$.

$$C_0 = \frac{Q_s C_s + Q_p C_p}{Q_s + Q_p} = \frac{2.2(10.8) + 0.5(1.5)}{2.2 + 0.5} = 9.1 \text{ mg/L}$$

At $T_0 = 14.6°C$, $S_0 = 10.2$ mg/L from Table 7-2, thus

$$D_0 = S_0 - C_0 = 10.2 - 9.1 = 1.1 \text{ mg/L}$$

The ultimate BOD in the stream immediately below the outfall is

$$L_0 = \frac{L_s Q_s + L_p Q_p}{Q_s + Q_p} = \frac{13.6(2.2) + 48(0.5)}{2.2 + 0.5} = 20 \text{ mg/L}$$

The deficit is then calculated as

$$D = \frac{k_1 L_0}{k_2 - k_1}(e^{-k_1 t} - e^{-k_2 t}) + D_0(e^{-k_2 t})$$

$$D = \frac{0.2(20)}{0.4 - 0.2}(e^{-0.2(0.66)} - e^{0.4(0.66)}) + 1.1(e^{-0.4(0.66)}) = 3.0 \text{ mg/L}$$

and $C = S - D = 10.2 - 3.0 = 7.2$ mg/L.

◆

Obviously, the impact of organics on a stream is not limited to the effect on dissolved oxygen. As the environment changes, the competition for food and survival results in a change in various species of microorganisms in a stream, and the chemical makeup changes as well. Figure 7-12 illustrates the effect of an organic pollutant load on a stream. Note especially the shift in nitrogen species through organic nitrogen to ammonia to nitrite to nitrate. Compare this to the previous discussion of nitrogen in the nutrient cycle (Figure 7-2). The changes in stream quality as the decomposers reduce the oxygen-demanding material, finally achieving a clean stream, is known as *self-purification*. This process is no different from what occurs in a wastewater treatment plant since in both cases energy is intentionally wasted. The organics contain too much energy, and they must be oxidized to more inert materials.

In the next chapter, the means for characterizing water and its pollution are presented, and then the idea of energy wasting in a wastewater treatment plant is discussed in greater detail.

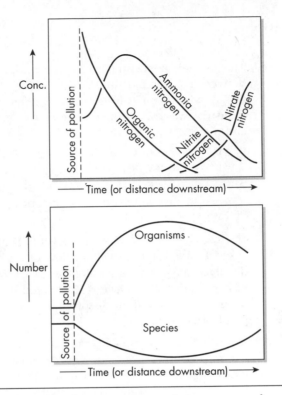

FIGURE 7-12 Nitrogen and aquatic organisms below a source of organic stream pollution.

Abbreviations

C = concentration of oxygen in water, mg/L

C_0 = concentration of oxygen immediately downstream of the introduction of a pollution stream, mg/L

D = deficit of a dissolved gas, mg/L

D_0 = initial oxygen deficit immediately downstream of a discharge, mg/L

DO = dissolved oxygen, mg/L

H = depth of flow, m

k_1 = deoxygenation constant, base e logarithm

k_2 = reoxygenation constant, base e logarithm

L = ultimate oxygen demand, mg/L

L_0 = ultimate oxygen demand immediately downstream of a discharge

Q_p = flow of a pollution stream, m³/sec

Q_s = flow in a stream, m³/sec

t = time

T = temperature, °C

t_c = critical time, point of lowest DO in a stream

v = mean stream velocity, m/sec

z = the amount of oxygen required by microorganisms to decompose organic matter, mg/L

◆ Problems

7-1 The limiting concentration of phosphorus for accelerated eutrophication is between 0.1 and 0.01 mg/L of P. Typical river water might contain 5 mg/L P, 50% of which comes from farm and urban runoff, and 50% from domestic and industrial wastes. Synthetic detergents contribute 50% of the P in municipal and industrial waste.

a. If all phosphorus-based detergents are banned, what level of P would you expect in a typical river?

b. If this river flows into a lake, would you expect the phosphate detergent ban to have much effect on the eutrophication potential of the lakewater? Why or why not?

7-2 A stream has dissolved oxygen level of 9 mg/L, and ultimate oxygen demand (L) of 12 mg/L and an average flow of 0.2 m^3/sec. An industrial waste at zero dissolved oxygen with an ultimate oxygen demand (L) of 20,000 mg/L and a flow rate of 0.006 m^3/sec is discharged into the stream. What is the ultimate oxygen demand and the dissolved oxygen in the stream immediately below the point of discharge?

7-3 Below a discharge from a wastewater treatment plant a 8.6-km long stream has a reoxygenation constant of 0.4 day^{-1}, a velocity 0.15 m/sec, a dissolved oxygen concentration of 6 mg/L, and an ultimate oxygen demand (L_0) of 25 mg/L. The deoxygenation constant is estimated at 0.25 day^{-1}.

a. Will there be fish in this stream?

b. Why should we care if there are fish in the stream? Do the fish deserve moral consideration and protection? What arguments can you muster to support this view? Write a one-page paper supporting the rights of fish to be included in the moral community. (You may wish to review some of the arguments in Chapter 1.)

7-4 A municipal wastewater treatment plant discharges into a stream that at some times of the year has no other flow. The characteristics of the waste are

flow = 0.1 m^3/sec

dissolved oxygen = 6 mg/L

temperature = 18°C

k_1 = 0.23 day^{-1}

ultimate BOD (L) = 280 mg/L

The velocity in the stream is 0.5 m/sec and the reoxygenation constant k_2 is assumed to be 0.45 day^{-1}.

a. Will the stream maintain a minimum 4 mg/L if there is no other flow in the stream?

b. If there is streamwater above the outfall and this water has a temperature of 18°C and no demand for oxygen and is saturated with DO, how great must the upstream flow be to assure a minimum dissolved oxygen of 4 mg/L downstream of the discharge? (*Note*: You can do this on your calculator, but it will be tedious. The easiest way to solve this problem is to either write a simple computer program or to use a spreadsheet.)

7-5　Ellerbe Creek, a large stream at normal velocity, is the recipient for the wastewater from the 10-mgd Durham, North Carolina, Northside Wastewater Treatment Plant. The stream (without the treatment plant discharge) has a mean summertime flow of 0.28 m^3/sec, a temperature of 24°C, and a velocity of 0.25 m/sec. The wastewater characteristics are

$$temperature = 28°C$$
$$ultimate\ BOD\ (L) = 40\ mg/L$$
$$k_1 = 0.23\ day^{-1}$$
$$dissolved\ oxygen = 2\ mg/L$$

The total stream length is 14 miles, at which point it empties into the Neuse River. (*Note*: Use Table 7–3 for the value of k_2.)

a.　The state of North Carolina stream standards require that dissolved oxygen levels be higher than 4 mg/L at all times. Should the state be concerned with this discharge?

b.　Other than legal considerations, why *should* the state be concerned with the oxygen levels in Ellerbe Creek? It isn't much of a creek, actually, and empties out into the Neuse River without being of much use to anyone. And yet the state has set dissolved oxygen levels of 4 mg/L for the 10-year 7-day low flow. Write a letter to the editor of a fictitious local newspaper decrying the spending of tax revenues for the improvements to the Northside Wastewater Treatment Plant just so the dissolved oxygen levels in Ellerbe Creek can be maintained above 4 mg/L.

c.　Pretend you are a fish in Ellerbe Creek. You have read the letter to the editor in part b above, and you are royally ticked off. You take pen in gill and respond. Write a letter to the editor from the standpoint of the fish. The quality of your letter will be judged on the strength of your arguments.(Don't make this a silly letter. Use serious arguments.)

7-6　A large stream with a velocity of 0.85 m/sec, saturated with oxygen, has a reoxygenation constant $k_2 = 0.4$ day^{-1} and a temperature of $T = 12°C$, with an ultimate BOD = 13.6 mg/L and a flow rate $Q = 2.2$ m^3/sec. Into this stream flows a wastewater stream with a flow rate of 0.5 m^3/sec, $T = 26°C$, $L = 220$ mg/L. The deoxygenation constant in the stream downstream from the pollution source is $k_1 = 0.2$ day^{-1} and a dissolved oxygen level of 1.5 mg/L. What is the dissolved oxygen 48.3 km downstream?

7-7　If algae contain P:N:C, in the proportion of 1:16:100, which of the three elements would be limiting algal growth if the concentration in the water is

$$0.20\ mg/L\ P$$
$$0.32\ mg/L\ N$$
$$1.00\ mg/L\ C$$

Show your calculations.

7-8   If offensive football teams are described in terms of the positions as linemen (L), receivers (R), running backs (B), and quarterbacks (Q) in the ratio of L : R : B : Q of 5 : 3 : 2 : 1, and a squad has the following distribution of players: $L = 20$, $R = 16$, $B = 6$, $Q = 12$, how many offensive teams can be created with all the positions filled, and what is the "limiting position"? Show your calculations.

7-9   Suppose you and a nonenvironmental engineering friend are walking by a stream in the woods, and your friend asks, "I wonder if this stream is polluted." How would you answer him or her and what questions would you have to ask him or her before you can answer the question?

7-10   Consider the following poem:

Oh, beautiful for smoggy skies,
Insecticided grain,
For strip-mined mountains' majesty
Above the asphalt plains.
America, America! Man sheds his waste on thee
And hides the pines with billboard signs
From sea to oily sea.[6]

Is this really the state of human impact on our ecosystem? What type of outlook would have caused the author to pen such a poem? Is it really as dismal a situation as that? Write a one-page criticism of this poem, either supporting its basic message or disputing it.

7-11   The Environmental Advisory council of Canada published a booklet entitled "An Environmental Ethic—Its Formulation and Implications,"[7] in which they suggest the following as a concise environmental ethic:

Every person shall strive to protect and enhance the beautiful everywhere his or her impact is felt, and to maintain or increase the functional diversity of the environment in general.

In a one page-essay, critique this formulation of the environmental ethic.

7-12   A discharge from a potential wastewater treatment plant may affect the dissolved oxygen level in a stream. The waste characteristics are expected to be as follows:

flow = 0.56 m³/s

ultimate BOD = 6.5 mg/L

DO = 2.0 mg/L

deoxygenation constant = 0.2 day^{-1}

temperature = 25°C

The water in the stream, upstream from the planned discharge, has the following characteristics:

flow = 1.9 m³/sec

ultimate BOD = 2.0 mg/L

DO = 9.1 mg/L

reoxygenation constant = 0.40 day^{-1}

temperature = 15°C

The state DO standard is 4 mg/L.

a. Will the construction of this plant cause the DO to drop below the state standard?

b. If during a hot summer day, the flow in the stream drops to 0.2 m^3/sec and the temperature in the stream increases to 30°C, will the state standard be met?

c. In the wintertime, the stream is ice covered, so that there cannot be any reaeration ($k_2 = 0$). If the temperature of the water in the stream is 4°C, and all other characteristics are the same as in part a, will the state standard be met?

d. Suppose your calculation showed that in part c the dissolved oxygen dropped to zero during the winter, under the ice. Is this bad? After all, who cares?

7-13 Libby was in good spirits. She loved her job as assistant engineer for the town, and the weather was perfect for working out-of-doors. She had gotten the job of on-site inspector for the new gravity trunk sewer and this not only gave her significant responsibility, but it allowed her to get out of the office. Not bad for a young engineer only a few months out of school.

The town was doing the work in-house, partly because of Bud, an experienced foreman who would see that the job got done right. Bud was wonderful to work with and was full of stories and practical construction know-how. Libby expected to learn a lot from Bud.

This particular morning the job required the cutting and clearing of a strip of woods on the right-of-way. When Libby arrived, the crew was already noisily getting prepared for the morning's work. She decided to walk ahead up the right-of-way to see what the terrain was like.

About 100 yards up the right-of-way, she came upon a huge oak tree, somewhat off the centerline, but still on the right-of-way, and therefore destined for cutting down. It was a magnificent oak, perhaps 300 years old, and had somehow survived the clear-cutting that occurred on this land in the mid-1800s. Hardly any trees here were more than 150 years old, all having fallen to the tobacco farmer's thirst for more land. But there was this magnificent tree. Awesome!

Libby literally dragged Bud up to the tree and exclaimed, "We cannot cut down this tree. We can run the line around it and still stay in the right-of-way."

"Nope. It has to come down," responded Bud. "First, we are running a gravity sewer. You just don't go changing sewer alignment. We'd have to construct additional manholes and redesign the whole line. Most important, you cannot have such a large tree on a sewer line right-of-way. The roots will eventually break into the pipe and cause cracks. In the worst case, the roots will fill up the whole pipe, and this requires a cleaning and possible replacement if the problem is bad enough. We simply cannot allow this tree to remain here."

"But think of this tree as a treasure. It's maybe 300 years old. There probably are no other trees like this in the county," implored Libby.

"A tree is a tree. We're in the business of building a sewer line, and the tree is in the way," insisted Bud.

"Well I think this tree is special, and I insist that we save it. Since I am the engineer in charge," she gulped inwardly, surprised at her own courage, "I say we do not cut down this tree."

She looked around and saw that some of the crew had walked up to them and were standing around, chainsaws in hand, with wry smiles on their faces. Bud was looking very uncomfortable.

"OK," he said, "You're the boss. The tree stays."

That afternoon in her office, Libby reflected on the confrontation and tried to understand her strong feelings for the old tree. What caused her to stand up on her hind legs like that? In order to save a tree? So what if it was special? There were many other trees that were being killed to run the sewer line. What was special about this one? Was it just its age, or was there something more?

The next morning Libby went back to the construction site and was shocked to find that the old tree was gone. She stormed into Bud's construction trailer and almost screamed, "Bud! What happened to the old tree?"

"Don't you get your pretty head upset now. I called the Director of Public Works and described to him what we talked about, and he said to cut down the tree. It was the right thing to do. If you don't agree, you have a lot to learn about construction."

Why did Libby feel so attached to the old tree? Why did she want to save it? Was it for herself, or for other humans, or for the sake of the tree itself? What should Libby do now? Does she have any recourse at all? The tree is already dead, regardless of her future actions.

If you relate to Libby's attitude toward the old tree, write a one-page memorandum from Libby to the Director of Public Works relating the events concerning the tree and expressing her (your) feelings. If, however, you agree with Bud that the tree had to come down, write a one-page memorandum to the Director relating the events as Bud (you) saw them, justifying your (his) actions.

◆ Endnotes

1. K. J. Warren, "The Power and Promise of Ecological Feminism," *Environmental Ethics* 12:2:125 (1990).
2. Story from G. J. C. Smith, et al., *Our Ecological Crisis* Macmillan Publishing Co., New York (1974).
3. This discussion is considerably simplified from what actually occurs in an aquatic ecosystem. For a more accurate representation of such systems, see any modern text on aquatic ecology.
4. D. J. O'Connor and W. E. Dobbins, "The Mechanism of Reaeration of Natural Streams," *Journal of the Sanitary Engineering Division*, ASCE, V82, SA6 (1956).

5. E. B. Phelps, *Stream Sanitation*, John Wiley & Sons, New York (1944).
6. James Coolbaugh, in *Environment* 18 : 6 (1976). Used with permission.
7. "An Environmental Ethic — Its Formulations and Implications," Report No. 2, Norma H. Morse, Ottawa (January 1975).

part
2

APPLICATIONS

When is water dirty? The answer of course depends on what we mean by dirty. For some people, the question is silly, such as the rural judge in a county courthouse who, presiding over a water pollution case, intoned that "Any damn fool knows if water is fit to drink."

But what may be pollution to some people may in fact be an absolutely necessary component in the water to others. For example, trace nutrients are necessary for algal growth (and hence for all aquatic life), while fish require organics as a food source in order to survive. These same constituents, however, may be highly detrimental if the water is to be used for industrial cooling.

In this chapter, various parameters used to measure water quality are discussed first and then the question of what is clean and dirty water is considered further.

8.1 ◆ Measures of Water Quality

Hundreds of parameters can be used to measure water quality. In this discussion we only cover the following:

- *Dissolved oxygen* is a major determinant of water quality in streams, lakes, and other watercourses.

- *Biochemical oxygen demand*, introduced previously (Section 7.2.3) as the "demand for oxygen," is a major parameter indicating the pollutional potential of various discharges to watercourses.

- *Solids* includes suspended solids, which are unsightly in natural waters, and total solids, which include dissolved solids, some of which could be detrimental to aquatic life or to people who drink the water.

- *Nitrogen* is a useful measure of water quality in streams and lakes; ammonia nitrogen is described below as an example of the colorimetric technique for determining the concentration of dissolved chemicals.

- *Bacteriological* measurements are necessary to determine the potential for the presence of infectious agents such as pathogenic bacteria and viruses. These measurements are usually indirect due to the problems of sampling for a literally infinite variety of microorganisms.

8.1.1   Dissolved Oxygen

Dissolved oxygen (DO) is measured with an oxygen probe and meter, shown in Figure 8–1. One of the simplest (and historically oldest) meters operates as a galvanic cell, in which lead and silver electrodes are put in an electrolyte solution with a microammeter between. The reaction at the lead electrode is

$$Pb + 2OH^- \rightarrow PbO + H_2O + 2e^-$$

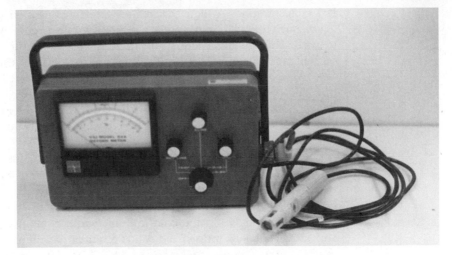

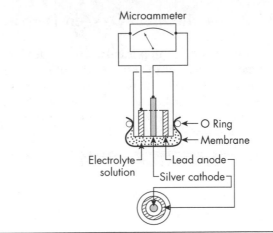

FIGURE 8-1 Dissolved-oxygen meter.

At the lead electrode, electrons are liberated and travel through the micro-ammeter to the silver electrode, where the following reaction takes place:

$$2e^- + \frac{1}{2}O_2 + H_2O \rightarrow 2OH^-$$

The reaction will not go unless free dissolved oxygen is available, and the microammeter will not register any current. The trick is to construct and calibrate a meter in such a manner that the electricity recorded is proportional to the concentration of oxygen in the electrolyte solution.

In the commercial models, the electrodes are insulated from each other with nonconducting plastic and are covered with a permeable membrane with a few drops of an electrolyte between the membrane and electrodes. The amount of oxygen that travels through the membrane is proportional to the

DO concentration. A high DO in the water creates a strong driving force to get through the membrane, while a low DO would force only limited O_2 through to participate in the reaction and thereby create electrical current. Thus the current is proportional to the dissolved oxygen level in the solution.

As noted earlier, the saturation of oxygen in water is a function of temperature and pressure. The saturation levels of O_2 in water also depend on the concentration of dissolved solids, with higher solids reducing oxygen solubility. Table 7–2 lists the saturation level of oxygen in clean water at various temperatures.

The DO meter is calibrated to the maximum (saturated) value in any water by inserting the probe into a sample of the water that has been sufficiently aerated to assure DO saturation. After setting both the zero and saturation values, the meter can be used to read intermediate DO levels in unknown samples. Although most meters automatically compensate for temperature change, a variation in dissolved solids requires recalibration.

8.1.2   Biochemical Oxygen Demand

Perhaps even more important than the determination of dissolved oxygen is the measurement of the rate at which this oxygen is used by microorganisms decomposing organic matter. The demand for oxygen in the decomposition of pure materials can be estimated from stoichiometry, assuming that all the organic material will decompose to CO_2 and water.

e • x • a • m • p • l • e   8.1

Problem   What is the theoretical oxygen demand for a 1.67×10^{-3} molar solution of glucose, $C_6H_{12}O_6$, as it is decomposed to carbon dioxide and water?

Solution   The atomic weights of carbon, hydrogen, and oxygen are 12, 1, and 16 respectively. Balance the equation

$$C_6H_{12}O_6 + O_2 \rightarrow CO_2 + H_2O$$

as

$$C_6H_{12}O_6 + 6O_2 \rightarrow 6CO_2 + 6H_2O$$

That is, for every mole of glucose used, six moles of oxygen are required.

$$\left[\frac{1.67 \times 10^{-3} \text{ g-moles}}{L}\right] \times \left[\frac{6 \text{ moles of } O_2}{\text{mole glucose}}\right] \times \left[\frac{32 \text{ g } O_2}{\text{mole } O_2}\right] \times \left[\frac{1000 \text{ mg}}{g}\right]$$

$$= 321 \frac{\text{mg } O_2}{L}$$

Unfortunately, wastewaters are seldom pure materials, and it is not possible to calculate the demand for oxygen from stoichiometry. A test for the use of oxygen by the microorganisms is used instead.

The rate of oxygen use is commonly referred to as *biochemical oxygen demand* (BOD). BOD is not a measure of some specific pollutant. Rather, it is a measure of the amount of oxygen required by aerobic bacteria and other microorganisms while stabilizing decomposable organic matter. If the microorganisms are brought into contact with a food supply (such as human waste), oxygen is used by the microorganisms during the decomposition.

A low rate of oxygen use indicates either the absence of contamination or that the microorganisms are uninterested in consuming the available organics. A third possibility is that the microorganisms are dead or dying.

The standard BOD test is run in the dark at 20°C for five days. This is defined as five-day BOD, or BOD_5, or the oxygen used in the first five days. The temperature is specified since the rate of oxygen consumption is temperature dependent. The reaction must occur in the dark since algae may be present and, if light is available, may actually produce oxygen in the bottle.

The BOD test is almost universally run using a standard BOD bottle (about 300-mL volume), as shown in Figure 8-2. The bottle is made of special nonreactive glass and has a ground stopper with a lip that is used to create a water seal so no oxygen can get in or out of the bottle.

Although the 5-day BOD is the standard, we can measure a 2-day, 10-day, or any other day BOD. One form of BOD introduced in the previous chapter is the *ultimate BOD*, or the O_2 demand after a very long time, when the microorganisms have oxidized as much of the organics as they can. Ultimate BOD is usually run for 20 days, at which point little additional oxygen depletion will occur.

If the dissolved oxygen is measured every day for five days, a curve such as that in Figure 8-3 on page 209 is usually obtained. Referring to this figure, sample *A* has an initial DO of 8 mg/L, and in five days this drops to 2 mg/L. The BOD therefore is the initial DO minus the final DO, which equals the DO used by the microorganisms, or $8 - 2 = 6$ mg/L. In equation form,

$$BOD = I - F$$

where I = initial DO, mg/L
 F = final DO, mg/L

Referring again to Figure 8-3, sample *B* has an initial DO of 8 mg/L, but the oxygen is used up so fast that it drops to zero in two days. If after five days the DO is measured as zero, all we know is that the BOD of sample *B* is more than $8 - 0 = 8$ mg/L, but we don't know how much more than 8 mg/L, since the organisms might have used more DO if had it been available. For samples with an oxygen demand of greater than about 8 mg/L, it is not possible to measure BOD directly and dilution of the sample is necessary.

Suppose sample *C* shown on Figure 8-3 is really sample *B* diluted with distilled water by 1 : 10. The BOD of sample *B* is therefore 10 times greater than the measured value, or

$$BOD = (8 - 4)10 = 40 \text{ mg/L}$$

FIGURE 8-2   A BOD bottle, made of a special nonreactive glass and supplied with a ground-glass stopper.

With dilution, the BOD equation reads

$$BOD = (I - F)D$$

where D = dilution, represented as a fraction, and defined as

$$D = \frac{\text{total volume of bottle}}{[\text{total volume of bottle}] - [\text{volume of dilution water}]}$$

e · x · a · m · p · l · e   **8.2**

Problem   Three BOD bottles are prepared, with sample and dilution water as follows:

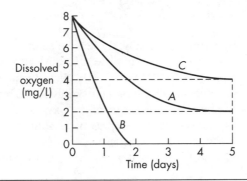

FIGURE 8-3 Decrease in dissolved oxygen in three different BOD bottles: *A* is a valid test for a five-day BOD, *B* is an invalid test since it reaches zero dissolved oxygen before the fifth day, and *C* is the same as *B* but with prior dilution.

Bottle No.	Sample (mL)	Dilution Water (mL)
1	3	297
2	1.5	298.5
3	0.75	299.25

Calculate the dilution (*D*) for each.

Solution   Recall that the volume of a standard BOD bottle is 300 mL. For bottle No. 1, the dilution D is

$$D = \frac{300}{300 - 297} = 100$$

Similarly for the other two bottles, the dilution D is calculated as 200 and 400.

The assumption in the dilution method is that the results from each bottle, when the BOD is calculated, will yield the same BOD value. Sometimes (often) when a series of BOD bottles are run at different dilutions, the calculations do not converge. Consider the following example.

e • x • a • m • p • l • e **8.3**

Problem   A series of BOD tests were run at three different dilutions. The results were as follows:

Bottle No.	Dilution	IDO = I	FDO = F	BOD (mg/L)
1	100	10.0	2.5	750
2	200	10.0	6.0	800
3	400	10.0	7.5	1000

What is the BOD?

Solution   Actually, there is no solution. What this problem represents is what is called a "sliding scale." Ideally, all tests should have resulted in the same calculated BOD, but clearly they do not. So what is the BOD? This example represents a fairly benign sliding scale, and since the middle reading might be the best representative of the oxygen use, one would probably report the BOD as 800 mg/L. A less pragmatic solution would be to clean the bottles again, make sure the dilution water is perfectly clean (so it does not have a BOD of its own), and try it again, hoping for a more consistent result.

◆

The BOD of the sample is seldom known before the BOD test is conducted, and the dilution must be estimated. The test is not very precise if the drop in DO during the five days of incubation is less than about 2 mg/L, or if the DO remaining in the bottle is less than 2 mg/L. Generally a BOD is estimated, and at least two dilutions are used, with one expected to have at least 2 mg/L DO remaining, and one that uses at least 2 mg/L of DO. The dilution is calculated as

$$D = \frac{\text{expected BOD}}{\Delta \text{DO}}$$

e • x • a • m • p • l • e **8.4**

Problem   The five-day BOD of an influent to an industrial wastewater treatment plant is expected to be about 800 mg/L. What dilutions should be used in a five-day BOD test?

Solution   Assume that the saturation is about 10 mg/L. If at least 2 mg/L is to remain in the bottle, the drop in BOD should be $10 - 2 = 8$ mg/L, and the dilution would be

$$D = 800/8 = 100$$

and if at least 2 mg/L of DO is to be used, the dilution of another bottle should be

$$D = 800/2 = 400$$

Two BOD bottles should therefore be run, one with $D = 100$ and one with $D = 400$. Prudence would suggest that a third bottle be run at say $D = 200$.

◆

In the discussion thus far we assume that the waste sample has within it the microorganisms that will decompose the organic material and decrease the DO. It is also possible to measure the BOD of any organic material (e.g., sugar) and thus estimate its influence on a stream that would have plenty of microorganisms that would love to get at the sugar. Pure sugar, however, does not contain the necessary microorganisms for decomposition. *Seeding* is a process in which the microorganisms responsible for the oxygen uptake are added to the BOD bottle with the sample (such as sugar) for the oxygen uptake to occur.

Suppose the water previously described by the A curve in Figure 8–3 is used as seed water since it obviously contains microorganisms (it has a five-day BOD of 6 mg/L). If we now put 100 mL of a sample with an unknown BOD into a bottle, and add 200 mL of seed water, the 300-mL bottle is filled. The dilution (D) is then 3. Assuming that the initial DO of this mixture is 8 mg/L and the final DO is 1 mg/L, the total oxygen used is 7 mg/L. But some of this drop in DO is due to the seed water, since it also has a BOD, and only a portion is due to the decomposition of the unknown material. The DO uptake due to the seed water is

$$\text{BOD}_{seed} = 6(2/3) = 4 \text{ mg/L}$$

since only 2/3 of the bottle is the seed water, which has a BOD of 6 mg/L. The remaining oxygen uptake ($7 - 4 = 3$ mg/L) must be due to the sample, which was diluted $1:3$ (or $D = 3$). Its BOD must then be $3 \times 3 = 9$ mg/L.

If the seeding and dilution methods are combined, the following general formula is used to calculate the BOD:

$$\text{BOD}_t = [(I - F) - (I' - F')(X/Y)]D \tag{8.1}$$

where BOD_t = biochemical oxygen demand, as measured at some time t, mg/L
I = initial DO of bottle with sample and seeded dilution water, mg/L
F = final DO of bottle with sample and seeded dilution water, mg/L
I' = initial DO of bottle with seeded dilution water, mg/L
F' = final DO of bottle with seeded dilution water, mg/L
X = seeded dilution water in sample bottle, mL
Y = seeded dilution water in bottle with only seeded dilution water, mL
D = dilution of sample

e · x · a · m · p · l · e 8.5

Problem Calculate the BOD_5 if the temperature of the sample is 20°C, the initial DO is saturation, and the dilution is $1:30$, with seeded dilution water. The final DO of the seeded dilution water is 8 mg/L and the final DO in the bottle with sample and seeded dilution is 2 mg/L. Recall that the volume of the BOD bottle is 300 mL.

Solution From Table 7–2 at 20°C, saturation is 9.07 mg/L; hence this is the initial DO. Since the BOD bottle contains 300 mL, a $1:30$ solution would have 10 mL of sample and 290 mL of seeded dilution water:

$$\text{BOD}_5 = [(9.07 - 2) - (9.07 - 8)(290/300)]30 = 174 \text{ mg/L}$$

It is important to remember that BOD is a measure of oxygen use, or potential use. An effluent with a high BOD can be harmful to a stream if the oxygen consumption is great enough to eventually cause anaerobic conditions (the dissolved-oxygen sag curve approaches zero dissolved oxygen in the stream). Obviously, a small trickle of wastewater going into a great river probably will have negligible effect, regardless of the mg/L of BOD involved. Similarly, a large flow into a small stream can seriously affect the stream even though the BOD might be low. Accordingly, American engineers often talk of "pounds of BOD," a value calculated by multiplying the concentration by the flow rate with a conversion factor so that

lb BOD/day = [mg/L BOD] × [flow in million gallons per day] × 8.34

Note that this is the same conversion introduced in Chapter 3 for converting volume flows to mass flows, or

(mass flow) = (volume flow) × (concentration)

The BOD of most domestic sewage is about 250 mg/L, although many industrial wastes run as high as 30,000 mg/L. The potential detrimental effect of an untreated dairy waste that might have a BOD of 25,000 mg/L is quite obvious since it represents a 100 times greater effect on the oxygen levels in a stream than raw sewage.

The reactions in a BOD bottle can be described mathematically by first writing a materials balance in terms of the dissolved oxygen, starting as always with

$$\begin{bmatrix} \text{rate of DO} \\ \text{ACCUMULATED} \end{bmatrix} = \begin{bmatrix} \text{rate of DO} \\ \text{IN} \end{bmatrix} - \begin{bmatrix} \text{rate of DO} \\ \text{OUT} \end{bmatrix}$$
$$+ \begin{bmatrix} \text{rate of DO} \\ \text{PRODUCED} \end{bmatrix} - \begin{bmatrix} \text{rate of DO} \\ \text{CONSUMED} \end{bmatrix}$$

Since this is a closed system, and since the test is run in the dark so there is no DO production,

$$\begin{bmatrix} \text{rate of DO} \\ \text{ACCUMULATED} \end{bmatrix} = - \begin{bmatrix} \text{rate of DO} \\ \text{CONSUMED} \end{bmatrix}$$

$$\frac{dz}{dt} = -r$$

where z = dissolved oxygen (necessary for the microorganisms to decompose the organic matter), mg/L

t = time

r = reaction rate

We can assume that this is a first-order reaction (a point of some controversy, incidentally), as initially introduced in Chapter 7:

$$\frac{dz}{dt} = -k_1 z$$

That is, the rate at which the need for oxygen is reduced (dz/dt) is directly proportional to the amount of oxygen necessary for the decomposition to occur (z). Integrated, this expression yields

$$z = z_0 e^{-k_1 t} \tag{8.2}$$

As the microorganisms use oxygen, at any time t the amount of oxygen still to be used is z, and the amount of oxygen already used at any time t is y, or the oxygen already demanded by the organisms. The total amount of oxygen that will ever be used by the microorganisms is the sum of what has been used (y) and what is still to be used (z) or

$$L = z + y$$

where y = DO already used or demanded at any time t (i.e., the BOD), mg/L
  z = DO still required to satisfy the ultimate demand, mg/L
  L = ultimate demand for oxygen, mg/L

Substituting $z = L - y$ into equation 8.2,

$$L - y = z_0 e^{k_1 t}$$

and recognizing from Figure 8–4A that $z_0 = L$,

$$y = L - L(e^{-k_1 t})$$

or

$$y = L(1 - e^{-k_1 t}) \tag{8.3}$$

where y = BOD at any time t in days, mg/L
  L = ultimate BOD, mg/L
  k_1 = deoxygenation constant, day^{-1}

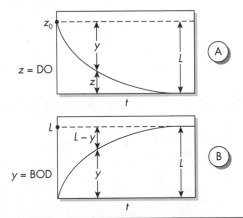

FIGURE 8–4 BOD definitions. Note that L is also used in the dissolved oxygen sag equation.

With reference to Figure 8–4B, note that

$$dy/dt = k_1(L - y)$$

so that as y approaches L, $(L - y) \rightarrow 0$ and $dy/dt \rightarrow 0$. Integrating this expression again produces equation 8.3.

In some applications, such as modeling the DO sag curve in a stream (see the previous chapter), we need to know both k_1 and L. The results of the BOD test, however, produce a curve showing the oxygen used over time. How do we calculate the ultimate oxygen use (L) and the deoxygenation rate (k_1) from such a curve?

There are a number of techniques for calculating k_1 and L, one of the simplest being a method devised by Thomas.[1] Starting with the equation

$$y = L(1 - e^{k_1 t})$$

rearrange it to read

$$\left(\frac{t}{y}\right)^{1/3} = \left(\frac{1}{(k_1 L)^{1/3}}\right) + \left(\frac{k_1^{2/3}}{6L^{1/3}}\right) t$$

This equation is in the form of a straight line

$$x = a + bt$$

where $x = (t/y)^{1/3}$
$a = (k_1 L)^{-1/3}$
$b = (1/6)(k_1^{2/3} L^{-1/3})$

Thus plotting x vs. t, the slope (b) and intercept (a) can be obtained, and

$$k_1 = 6(b/a)$$

$$L = 1/(6ba^2)$$

e • x • a • m • p • l • e **8.6**

Problem   The BOD vs. time data for the first five days of a BOD test are obtained as follows:

Time, t (days)	BOD, y (mg/L)
2	10
4	16
6	20

Calculate the k_1 and L.

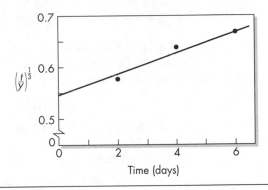

FIGURE 8-5   Plot used for the calculation of k_1 and L.

Solution   The $(t/y)^{1/3}$ values, 0.585, 0.630, and 0.669, are plotted as shown in Figure 8-5. The intercept is $a = 0.545$ and the slope is $b = 0.021$. Thus

$$k_1 = 6(0.021/0.545) = 0.64 \text{ day}^{-1}$$

$$L = 1/[6(0.021)(0.545)^2] = 26.7 \text{ mg/L}$$

But things aren't always that simple. If the BOD of an effluent from a wastewater treatment plant is measured, and if instead of stopping the test after five days, the reaction is allowed to proceed and the DO measured each day, a curve like Figure 8-6 might result. Note that some time after five days the curve takes a sudden jump. This is due to the exertion of oxygen demand by the microorganisms that decompose nitrogenous organics and are converting these to the stable nitrate, NO_3^- (see Chapter 7).

The oxygen-use curve in a BOD test can therefore be divided into two regions, *nitrogenous* and *carbonaceous BOD*. Note also the definition of the ultimate BOD on this curve. If no nitrogenous organics are in the sample, or

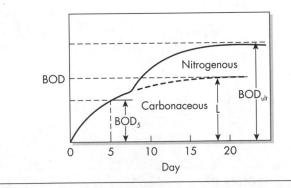

FIGURE 8-6   Idealized BOD curves.

if the action of these microorganisms is suppressed, only the carbonaceous curve results.

For streams and rivers with travel times greater than about five days, the ultimate demand for oxygen must include the nitrogenous demand. In practice, the ultimate BOD is calculated as

$$BOD_{ult} = a(BOD_5) + b(KN)$$

where KN   = Kjeldahl nitrogen (organic plus ammonia mg/L) (see page 219)

    a and b = constants

The state of North Carolina, for example, uses $a = 1.2$ and $b = 4.0$ for calculating the ultimate BOD, which is then substituted for L in the dissolved-oxygen sag equation. This model emphasizes the need for wastewater treatment plants to achieve nitrification (conversion of nitrogen to NO_3^-) or denitrification (conversion to N_2 gas). More on this in Chapter 10.

8.1.3   Solids

The separation of these solids from the water is one of the primary objectives of wastewater treatment. Strictly speaking, in wastewater anything other than water or gas is classified a solid. The usual definition of solids, however, is the residue on evaporation at 103°C (slightly higher than the boiling point of water). These solids are known as *total solids*. The test is conducted by placing a known volume of sample in a large evaporating dish (Figure 8–7) and allowing the water to evaporate. The total solids are then calculated as

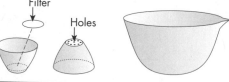

FIGURE 8–7   The Gooch crucible and the evaporating dish used for measuring suspended and total solids, respectively.

$$\text{TS} = \frac{W_{ds} - W_d}{V}$$

where TS = total solids, mg/L
  W_{ds} = weight of dish plus the dry solids after evaporation, mg
  W_d = weight of the clean dish, mg
  V = volume of sample, L

If the volume of the sample is in mL, the most common unit of measure, and the weights are in grams, the equation reads

$$\text{TS} = \frac{W_{ds} - W_d}{V} \times 10^6$$

where TS = total solids, mg/L
  W_{ds} = weight of dish plus the dry solids after evaporation, g
  W_d = weight of the clean dish, g
  V = volume of sample, mL

Total solids can be divided into two fractions: *dissolved solids* and the *suspended solids*. If a teaspoonful of common table salt is placed in a glass of water, the salt dissolves. The water does not look any different, but the salt remains behind if the water is evaporated. A spoonful of sand, however, does not dissolve and remains as sand grains in the water. The salt is an example of dissolved solids while the sand would be measured as a suspended solid.

A *Gooch crucible* is used to separate suspended solids from dissolved solids. As shown in Figure 8-7, the Gooch crucible has holes on the bottom on which a glass fiber filter is placed. The sample is then drawn through the crucible with the aid of a vacuum. The suspended material is retained on the filter while the dissolved fraction passes through. If the initial dry weight of the crucible and filter is known, the subtraction of this from the total weight of crucible, filter, and the dried solids caught on the filter yields the weight of suspended solids, expressed as mg/L. In equation form,

$$\text{SS} = \frac{W_{df} - W_d}{V} \times 10^6$$

where SS = suspended solids, mg/L
  W_{df} = weight of dish plus dry filtered solids, g
  W_d = weight of clean crucible, g
  V = volume of sample, mL

Solids can be classified in another way: those that are volatilized at a high temperature and those that are not. The former are known as *volatile solids*, the latter as *fixed solids*. Although volatile solids are mostly organic, at 600°C, the temperature at which the combustion takes place, some of the inorganics are decomposed and volatilized as well, but this is not considered a serious drawback. In equation form,

$$FS = \frac{W_{du} - W_d}{V} \times 10^6$$

where FS $\ =\ $ fixed solids, mg/L

W_{du} $\ =\ $ weight of dish plus unburned solids, g

The volatile solids can then be calculated as

$$VS = TS - FS$$

where VS = volatile solids, mg/L.

The relationship between the total solids and the total volatile solids can best be illustrated by an example.

e • x • a • m • p • l • e **8.7**

Problem   Given the following data: The weight of a dish (such as that shown in Figure 8-7) = 48.6212 g. A 100-mL sample is placed in the dish and the water is evaporated. The weight of the dish and dry solids = 48.6432 g. The dish is placed in a 600°C furnace for 24 hours and then cooled in a desiccator. The weight of the cooled dish and residue, or unburned solids, = 48.6300 g. Find the total, volatile, and fixed solids.

Solution

$$\text{total solids} = \frac{(\text{dish} + \text{dry solids}) - (\text{dish})}{\text{volume of sample}} \times 10^6$$

$$= \frac{(48.6432) - (48.6212)}{100} \times 10^6$$

$$= 220 \text{ mg/L}$$

$$\text{fixed solids} = \frac{(\text{dish} + \text{unburned solids}) - (\text{dish})}{\text{volume of sample}} \times 10^6$$

$$= \frac{(48.6300) - (48.6212)}{100} \times 10^6$$

$$= 88 \text{ mg/L}$$

$$\text{volatile solids} = TS - FS = 220 - 88 = 132 \text{ mg/L}$$

◆

Sometimes we want to measure the volatile fraction of suspended material, since this is a quick (if gross) measure of the amount of microorganisms present. The *volatile suspended solids* are determined by simply placing the Gooch crucible in a hot oven (600°C), allowing the organic fraction to burn off, and weighing the crucible again. The loss in weight is interpreted as *volatile suspended solids.*

8.1.4 Nitrogen

Recall from Chapter 7 that nitrogen is an important element in biological reactions. Nitrogen can be tied up in high-energy compounds such as amino acids and amines and in this form the nitrogen is known as organic nitrogen. One of the intermediate compounds formed during biological metabolism is ammonia nitrogen. Together with organic nitrogen, ammonia is considered an indicator of recent pollution. These two forms of nitrogen are often combined in one measure, known as *Kjeldahl nitrogen,* named after the scientist who first suggested the analytical procedure.

Aerobic decomposition produces nitrite (NO_2^-) and finally nitrate (NO_3^-) nitrogen forms. A high-nitrate and low-ammonia nitrogen therefore suggests that pollution has occurred, but quite some time ago.

All these forms of nitrogen can be measured analytically by *colorimetric techniques.* The basic idea of colorimetry is that the ion in question combines with some compound and forms a color. The compound is in excess, so that the intensity of the color is proportional to the original concentration of the ion being measured. For example, ammonia can be measured by adding a compound called *Nessler* reagent to the unknown sample. This reagent is a solution of potassium mercuric iodide, K_2HgI_4, and reacts with ammonium ions to form a yellow-brown colloid. Since Nessler reagent is in excess, the amount of colloid formed is proportional to the concentration of ammonia ions in the sample.

The color is measured photometrically. The basic workings of a photometer, illustrated in Figure 8–8, consists of a light source, a filter, the sample, and a photocell. The filter allows only certain wavelengths of light to pass through, thus lessening interferences and increasing the sensitivity of the photocell, which converts light energy to electrical current. An intense color allows only a limited amount of light to pass through, and create little current. However, a sample containing very little of the chemical in question will be clear, allowing almost all the light to pass through, and creating substantial current. If the

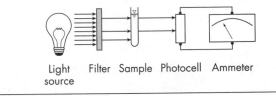

Light   Filter Sample Photocell Ammeter
source

FIGURE 8–8 A photometer used for measuring light penetration through a colored sample, where the intensity of the color is proportional to the chemical constituent being measured.

color intensity (and hence light absorbance) is directly proportional to the concentration of the unknown ion, the color formed is said to obey *Beer's law*.

A photometer can be used to measure ammonia concentration by measuring the absorbance of light by samples containing known ammonia concentration and comparing the absorbance of the unknown sample to these standards.

e • x • a • m • p • l • e   **8.8**

Problem   Several known samples and an unknown sample containing ammonia nitrogen are treated with Nessler reagent and the color measured with a photometer. Find the ammonia concentration of the unknown sample.

Solution

Sample	Absorbance
Standards	
0 mg/L ammonia (distilled water)	0
1 mg/L ammonia	0.06
2 mg/L ammonia	0.12
3 mg/L ammonia	0.18
4 mg/L ammonia	0.24
Unknown sample	0.15

A plot (Figure 8–9) of ammonia concentration of the standards vs. absorbance results in a straight line (Beer's law is adhered to). By entering a point on this chart at 0.15 absorbance (the unknown), the concentration of ammonia in the unknown sample is read as 2.5 mg/L.

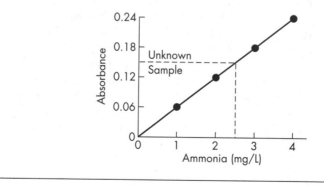

FIGURE 8–9   A typical calibration curve used for measuring ammonia concentration.

8.1.5   Bacteriological Measurements

From the public health standpoint the bacteriological quality of water is as important as the chemical quality. A number of diseases can be transmitted by water, among them typhoid and cholera. However, it's one thing to declare that water must not be contaminated by pathogens (disease-causing organisms) and another to discover the existence or measure the concentration of these organisms in the water.

Seeking the presence of pathogens presents several problems. First, there are many kinds of pathogens. Each pathogen has a specific detection procedure and must be screened individually. Second, the concentration of these organisms can be so small as to make their detection almost impossible. Looking for pathogens in most surface waters is a perfect example of the proverbial needle in a haystack. Yet only one or two organisms in the water might be sufficient to cause an infection if this water is consumed. In the United States, the pathogens of importance today include *Salmonella*, *Shigella*, the hepatitis virus, *Entamoeba histolytica*, *Giardia lamblia*, and *Cryptosporidium*.

Salmonellosis is caused by various species of *Salmonella*, and the symptoms of salmonellosis include acute gastroenteritis, septicemia (blood poisoning), and fever. Gastroenteritis usually consists of severe stomach cramps, diarrhea, and fever and, although it makes for a horrible few days, is seldom fatal. Typhoid fever is caused by the *Salmonella typhi*, and this is a much more serious disease, lasting for weeks; it can be fatal if not treated properly. About 3% of the victims become carriers of *Salmonella typhi*, and although they exhibit no further symptoms, they pass on the bacteria to others mainly through contaminated water.

Shigellosis, also called bacillary dysentery, another gastrointestinal disease, has symptoms similar to salmonellosis. Infectious hepatitis, caused by the hepatitis virus, has been known to be transmitted through poorly treated water supplies. Symptoms include headache, back pains, fever, and eventually jaundiced skin color. While rarely fatal, it can cause severe debilitation. The hepatitis virus can escape dirty sand filters in water plants and can survive for a long time outside the human body.

Amoebic dysentery, or amoebiasis, is also a gastrointestinal disease, resulting in severe cramps and diarrhea. Although its normal habitat is in the large intestine, the amoeba can produce cysts that pass to other persons through contaminated water and cause gastrointestinal infections. The cysts are resistant to disinfection and can survive for many days outside the intestine.

Originally known as "beaver disease" in the north country, giardiasis is caused by *Giardia lamblia*, a flagellated protozoan that usually resides in the small intestine. Out of the intestine, its cysts can inflict severe gastrointestinal problems, often lasting two to three months. Giardiasis was known as beaver disease because beavers can act as hosts, greatly magnifying the concentration of cysts in fresh water. Giardia cysts are not destroyed by usual chlorination levels but are effectively removed by sand filtration. Backpackers should take care with drinking water on the trails by purifying with halogen tablets or small hand-held filters. Giardia has ruined more than one summer for unwary campers.

The latest public health problem has been the incidence of cryptosporidiosis, caused by *Cryptosporidium*, an enteric protozoan. The disease is debilitating, with diarrhea, vomiting, and abdominal pain, and lasts several weeks. There appears to be no treatment for cryptosporidiosis, and it can be fatal. In

Milwaukee in 1993, an outbreak of criptosporidiosis affected more than 30,000 people and resulted in 47 deaths. The usual source seems to be agricultural runoff contaminating water supplies, but the cysts are resistant to the common methods of drinking-water disinfection. Filtration provides the best barrier against water-supply contamination.

Our knowledge of the presence of pathogenic microorganisms in water is actually fairly recent. Diseases such as cholera, typhoid, and dysentery were highly prevalent even in mid-nineteenth century and even though microscopes were able to show living organisms in water, it was not at all obvious that these little critters were able to cause such dreaded diseases. In retrospect, the connection between contaminated water and disease should have been obvious, based on empirical evidence. For example, during the mid-1800's, the Thames River below London was grossly contaminated with human waste, and one Sunday afternoon a large pleasure craft capsized, throwing everyone into the drink. Although nobody drowned, most of the passengers died of cholera a few weeks later!

It took a shrewd public heath physician named John Snow to make the connection between contaminated water and infectious disease. In the mid-nineteenth century, water supply to the citizens of London was delivered by a number of private firms, each pumping water out of the Thames and selling it at pumps in the city. One such company took its supply downstream of waste discharges and provided water by means of a pump on Broad Street. As would be expected, cholera epidemics in London were common, and during one particularly virulent episode in 1854, John Snow noticed that the cases of cholera seemed to be concentrated around the Broad Street pump. He carefully recorded all the cases and marked them on a map, which clearly showed that the center of the epidemic was the pump. He convinced the city fathers to order the pump handle to be removed, and the epidemic subsided. The connection between contaminated water and infectious disease was proven without doubt. John Snow's contribution is immortalized by the naming of the "John Snow Pub" on what is now renamed Broadwick Street in London. A plaque is on the wall of the pub showing the location of the famous pump (Figure 8–10).

Many pathogenic organisms can be carried by water. How then is it possible to measure for bacteriological quality? The answer lies in the concept of indicator organisms. The indicator most often used is a group of microbes called *coliforms*. These critters have five important attributes: They are

- normal inhabitants of the digestive tracts of warm-blooded animals
- plentiful, hence not difficult to find
- easily detected with a simple test
- generally harmless except in unusual circumstances
- hardy, surviving longer than most known pathogens

Because of these five attributes, coliforms have become universal indicator organisms. But the presence of coliforms does not prove the presence of patho-

The Red Granite kerbstone

marks the site of the historic

BROAD STREET PUMP

associated with Dr. John Snow's

discovery in 1854

that Cholera is conveyed by water

FIGURE 8-10 The John Snow Pub and a marker indicating the location of the famous pump.

gens! If a large number of coliforms are present, there is a good chance of recent pollution by wastes from warm-blooded animals, and therefore the water may contain pathogenic organisms, but this is not proof of the presence of such pathogens.

The last point should be emphasized. The presence of coliforms does not mean that there are pathogens in the water. It simply means that there *might* be. A high coliform count is cause for suspicion and although the water may in fact be perfectly safe to drink, it should not be consumed.

The opposite is also true. The absence of coliforms does not prove that there are no pathogens in the water. As pointed out previously, some pathogens are quite resistant to various forms of disinfection and can survive water treatment or wastewater treatment. If the products of wastewater treatment are placed on farmland, some pathogens will surely be present and may represent a public health concern.

There are two principal ways of measuring for coliforms. The simplest is to filter a sample through a sterile filter, thus capturing any coliforms. The filter is then placed in a petri dish containing a sterile agar that soaks into the filter and promotes the growth of coliforms while inhibiting other organisms. After 24 to 48 hours of incubation, the number of shiny dark blue-green dots, indicating coliform colonies, is counted. If it is known how many milliliters of water were poured through the filter, the concentration of coliforms can be expressed as coliforms/100 mL.

The second method of measuring for coliforms is called the most probable number (MPN), a test based on the fact that in a lactose broth coliforms produce gas and make the broth cloudy. The production of gas is detected by placing a small tube upside down inside a larger tube (Figure 8–11) so as not to have air bubbles in the smaller tube. After incubation, if gas is produced, some of it will become trapped in the smaller tube and this along with a cloudy broth will indicate that the tube was inoculated with at least one coliform.

And here is the trouble. Theoretically, one coliform can cause a positive tube just as easily as a million coliforms can. Hence, it is not possible to ascertain the concentration of coliforms in the water sample from just one tube. The solution is to inoculate a series of tubes with various quantities of the sample, the reasoning being that a 10-mL sample would be 10 times more likely to contain a coliform than a 1-mL sample.

For example, using three different inoculation amounts, 10 mL, 1 mL, and 0.1 mL of sample, three tubes are inoculated, and after incubation an array might result as shown below. The plus signs indicate a positive test (cloudy broth with gas formation) and the minus sign represent tubes where no coliforms were found.

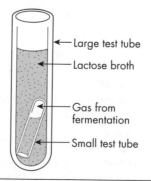

FIGURE 8–11   Illustration of the capture of gas in a tube where lactose is fermented by coliform organisms.

| Amount of Sample, | Tube No. | | |
mL Put in Test Tube	1	2	3
10	+	+	+
1	–	+	–
0.1	–	–	+

Based on these data one would suspect that there is at least 1 coliform per 10 mL, but there still is no firm number. The solution to this dilemma lies in statistics. It can be proven statistically that such an array of positive and negative results will occur most probably if the coliform concentration is 75 coli/100 mL. A higher concentration would most probably result in more positive tubes while a lower concentration would most probably result in more negative tubes. Thus 75 coli/100 mL is the most probable number (MPN).

8.2 ◆ Assessing Water Quality

In the water and wastewater profession, methods for measuring water quality have been standardized in a volume entitled *Standard Methods of the Examination of Water and Wastewater*. The title is commonly shortened to *Standard Methods*.

This volume is continually revised, and the tests are modified and improved as our skills in aquatic chemistry and biology develop. In the most recent volume (18th edition), the total number of standard tests that have undergone rigorous review and parallel testing number more than 500. In other words, there are at least that many standard, quantitative, and useful tests for expressing the quality of water.

To determine which of these tests to use, we need to first define the problem. For example, if the problem is "The water stinks!", we need to decide what measurements will determine just how *much* it stinks (numerically), and next, what the odor-causative agent in the water is. Quantitative measurement of odor is difficult and imprecise simply because it depends on human olfactory senses, and the measurement of the concentration of the causative agent is difficult because it is often not known what to measure. In fortunate circumstances, we can discover the specific culprit, but most often an indirect measure such as volatile solids or ammonia must be used to measure odor.

If, however, the problem is "I drank this water and it made me sick!", we need to know *how* sick and what was the nature of the illness. This information helps determine what may have been in the water that caused the health problem. If the complaint is an upset digestive system (the notorious "green-apple-quick-step"), the suspicion is that microbial contaminants were present. The coliform test can then be used to get an indication of such contamination.

If, instead of digestive upset, the health problem is that people in a community are developing mottled teeth, the immediate suspicion is that the drinking water contains excessive fluoride. The fluoride in this case would be a pollutant, whereas most communities *add* fluoride to drinking water to prevent dental caries in children and teenagers.

A third example of a water quality problem may be that "The fish are all dead!" We then need to find what killed the fish. Some constituent was either present at a high enough concentration to kill the fish, or perhaps another constituent necessary for higher aquatic life was absent. For example, a waste pesticide could have caused the kill, and a chemical screening must be done to identify the pesticide. The fish may also have died due to the lack of oxygen (the most common cause of fish kills), and it is useful to estimate the BOD and measure the DO.

The entire objective of such measurements is to obtain a *quantitative* handle on how clean or dirty a water is. Only then is it possible to proceed to solving pollution problems and avoid the problem of judges and lawyers telling us that "Any damn fool knows if the water is fit to drink."

8.3 ◆ Water Quality Standards

But what use is the quantitative analysis of water quality unless there exist some *standards* that describe the desired quality for various beneficial use of the water? There are, in fact, three main types of water quality standards.

8.3.1 Drinking Water Standards

Based on public health and epidemiological evidence, and tempered by a healthy dose of expediency, the EPA has established national drinking water standards for many physical, chemical, and bacteriological contaminants. Two types of standards are listed: primary and secondary. The primary standards relate to human health, while the secondary standards are for constituents that make water disagreeable to use. Primary standards include physical, chemical, and bacteriological standards.

One example of a physical standard is turbidity, or the interference with the passage of light. A water that has high turbidity is "cloudy," a condition caused by the presence of colloidal solids. Turbidity does not in itself cause a health problem, but the colloidal solids may prove to be convenient vehicles for pathogenic organisms.

The list of chemical standards is quite long and includes the usual inorganics (e.g., lead, arsenic, chromium), volatile organic chemicals (e.g., benzene, carbon tetrachloride), synthetic organic chemicals (e.g., PCBs and many chlorinated pesticides), disinfection by-products (e.g., chloroform and other trihalomethanes), and radioactivity. Bacteriological standards for drinking water are written in terms of the coliform indicators. The normal standard is at present less than about 1 coliform per 100 mL of treated drinking water, although the standard is written so that a drinking water utility cannot obtain more than 5% positive coliform results on all tests run, using 100-mL samples.

The secondary standards include such constituents as chloride, copper, hydrogen sulfide, iron, and manganese. The secondary standard for chloride, for example, is 250 mg/L, a point where water has a distinct salty taste. There is no primary standard for chloride because before the salt can become harmful it will taste so bad that nobody would drink the water. Iron likewise is not a health problem, but high iron concentrations make water appear red (and discolor laundry). Manganese gives water a blue color and similarly can discolor laundry and ceramic surfaces such as bathtubs.

8.3.2   Effluent Standards

The EPA oversees and states operate programs designed to reduce the flow of pollutants into natural watercourses. All discharges are required to obtain a National Pollution Discharge Elimination System (NPDES) permit. Although some detractors have labeled these "permits to continue pollution," the permitting system has nevertheless had a major beneficial effect on the quality of surface waters. Typical effluent standards for a domestic wastewater treatment plant may range from 5 to 20 mg/L BOD, for example. The intent is to tighten these limits as required to enhance water quality.

8.3.3   Surface Water Quality Standards

Tied to the effluent standards are surface water standards, often called "stream standards." All surface waters in the United States are now classified according to a system of standards based on their greatest beneficial use. Each state has its own classification system, and over time these have become quite complex, reflecting the best uses of water for such purposes as trout spawning and shellfish beds. A simple example of a classification system is shown in Table 8–1.

The highest classification (A in Table 8–1) is usually reserved for pristine waters, with the best use as a source of drinking water. The standards are usually quite strict, including a temperature standard. The next highest include waters that have had wastes discharged into them but that nevertheless exhibit high levels of quality. This water can also be best used for domestic purposes, but care has to be taken in its treatment to make sure it adheres to strict drinking water standards. The third category (C in Table 8–1) is for water contact sports such as swimming, but should not be used as a source of drinking water. The last categories are reserved for heavily polluted waters, and in most cases these have ceased to exist in the United States.

The objective is to attempt to establish the highest possible classification for all surface waters and then to use the NPDES permits to turn the screws on polluters and enhance the water quality and increase the classification of

TABLE 8-1 Typical (Simplified) Stream Classification System

Classification	Best Use	DO (mg/L)	Coliforms (number/mL)	Temperature (°C)
A	Drinking water, virgin source, no upstream use permitted	>6	<10/100	<15
B	Drinking water, upstream use permitted	>4	<100/100	<20
C	Water contact sports, fishing	>3	<1000/100	NA
D	Noncontact sports, agriculture	>3	NA	NA
E	Agricultural and industrial use, water transport	>2	NA	NA

the watercourse. Once at a higher classification, no discharge would be allowed that would degrade the water to a lower quality level. The objective is eventually to attain pure water in all surface watercourses. As "pollyannaish" as this may sound, it nevertheless is an honorable goal, and the thousands of engineers and scientists who devote their professional careers toward that end understand the joy of small victories and share in the ultimate dream of pollution-free water.

◆ Abbreviations

a, b = constants
BOD_5 = five-day BOD
BOD_{ult} = ultimate BOD, carbonaceous plus nitrogenous
D = dilution, expressed as a fraction
DO = dissolved oxygen
F = final DO in a BOD test
I = initial DO in a BOD test
KN = Kjeldahl nitrogen
L = ultimate oxygen demand, carbonaceous

TS = total solids
V = volume of sample
W_d = weight of clean dish
W_{ds} = dry weight of dish plus dry solids
X = volume of seeded dilution water
y = oxygen demand at any time t
Y = total volume of the BOD bottle

◆ Problems

8-1 Given the following BOD$_5$ test results:

> initial DO = 8 mg/L
> final DO = 0 mg/L
> dilution = 1 : 10

What can you say about
a. BOD$_5$?
b. BOD ultimate?

8-2 If you have two bottles full of lake water and keep one in the dark and the other in daylight, which one would have a higher DO after a few days? Why?

8-3 The following data are obtained for a wastewater sample:

> total solids = 4000 mg/L
> suspended solids = 5000 mg/L
> volatile suspended solids = 2000 mg/L
> fixed suspended solids = 1000 mg/L

Which of these numbers is questionable (wrong) and why?

8-4 A water has a BOD$_5$ of 10 mg/L. The initial DO in the BOD bottle is 8 mg/L and the dilution is 1 to 10. What is the final DO in the BOD bottle?

8-5 If the BOD$_5$ of a waste is 100 mg/L, draw a curve showing the effect on the BOD$_5$ of adding progressively higher doses of toxic hexavalent chromium.

8-6 Some years ago an industrial plant in New Jersey was having trouble with its downstream neighbors. It seems that the plant was discharging apparently harmless dyes into the water and making the stream turn all sorts of colors. The dye did not seem to harm the aquatic life, and it did not soil boats or docks. It was, in short, an aesthetic nuisance.

The plant wastewater treatment engineer was asked to come up with solutions to the problem. She found that the expansion of the plant, adding activated carbon columns, would cost about $500,000, but that there was a simpler solution. They could build a holding basin and hold the plant effluents in this basin during the day and release it at night, or hold it until they had enough blue and green color to make the resulting effluent appear to be blue-green. The basin would cost only $100,000 to construct. The plant would not be violating any standard or regulation, so the operation would be legal.

In effect, the plant would discharge wastes so as to reduce public complaints, but not actually treat the wastewater to remove the dyes. The total discharge of dye waste would be unchanged.

You are the president of the company and must make a decision either to spend $500,000 and treat the waste or to spend $100,000 and eliminate the complaints from the public.

Write a one-page memo to the engineer advising her how to proceed. Include in the memo your rationale for making the decision.

8-7   Consider the following data from a BOD test:

Day	DO (mg/L)	Day	DO (mg/L)
0	9	5	6
1	9	6	6
2	9	7	4
3	8	8	3
4	7	9	3

What are the

a. BOD_5?

b.   ultimate carbonaceous BOD?

c.   ultimate nitrogenous BOD?

d.   Why do you think there is no oxygen used until the third day?

8-8   A chemical engineer, working for a private corporation, is asked to develop a means for disinfecting their industrial sludge. He decides to use high doses of chlorine to do the job, since this is available at the plant. Laboratory studies show that this method is highly efficient and inexpensive. The sludge disinfection facility is constructed. After years of operation, high concentrations of trihalomethane (e.g., chloroform) are discovered in the plant effluent, and these are traced to the sludge-disinfection unit. Trihalomethanes are carcinogenic, and people downstream have been drinking this water.

It is possible that the company engineer knew about the formation of trihalomethane during the sludge-disinfection process and knew of its health effect, but decided to construct the facility anyway.

It is also possible that the company engineer did not know that the chlorine would cause potential health problems, even though the effect of chlorine and high organic materials such as wastewater sludge have been known for a long time to competent environmental engineers.

In a one-page paper, discuss the engineer's responsibility in both of these cases.

8-9   An industry discharges 10 million gallons a day of a waste that has a BOD_5 of 2000 mg/L. How many pounds of BOD_5 are discharged?

8-10   An industry applies to the state for a discharge permit into a highly polluted stream (DO is zero, it stinks, oil slicks on the surface, black in color, generally nasty). The state denies the permit. The engineer working for the industry is told to write a letter to the state appealing the permit denial based on the premise that the planned discharge is actually *cleaner* than the present streamwater and would actually *dilute* the pollutants in the stream. He is, however, a lousy writer, and asks you to compose a one-page letter for him to send to the state.

a.   Write a letter from the engineer to the state arguing his case.

b.   After you have written the letter arguing for the permit, write a letter back from the state to the industry justifying the state's decision not to allow the discharge.

c.   If the case went to court and a judge had both letters to read as the primary arguments, what would be the outcome? Write an opinion from the judge deciding the case. What elements of environmental ethics might the judge employ to make his or her decision?

8-11   If you dumped half of a gallon of milk into a stream every day, what would be your discharge in lb BOD_5/day? Milk has a five-day BOD of about 20,000 mg/L.

8-12   Given the same standard ammonia samples as in Example 8.8, if your unknown sample measured 20% absorbance, what is the ammonia concentration?

8-13   Suppose you ran a multiple tube coliform test and got the following results: 10-mL samples, all 5 positive; 1-mL samples, all 5 positive; 0.1-samples, all 5 negative. Use the table in *Standard Methods* to estimate the concentration of coliforms.

8-14   If coliform bacteria are to be used as an indicator of viral pollution as well as an indicator of bacterial pollution, what attributes must the coliform organisms have (relative to viruses)?

8-15   Draw a typical DO curve for a BOD run at the following conditions.
a.   stream water, 20°C, dark
b.   unseeded sugar water, 20°C, dark
c.   stream water, 20°C, with light
d.   stream water, 40°C, dark

8-16   Consider the following data for a BOD test:

Day	DO (mg/L)
0	9
1	8
2	7
3	6
4	5
5	4.5
6	4

a.   Calculate BOD_5.
b.   Plot the BOD vs. time.
c.   Suppose you took the sample above, after six days, aerated it, put it into the incubator, and measured the DO every day for five days. Draw this curve on the graph as a dotted line.

8-17   Suppose two water samples have the following forms of nitrogen at day zero:

	Sample A (mg/L)	Sample B (mg/L)
Organic	40	0
Ammonia	20	0
Nitrite	0	0
Nitrate	2	10

For each sample, draw the curves for the four forms of nitrogen as they might exist in a BOD bottle during 10 days of incubation.

8-18   A wastewater sample has a $k_1 = 0.2$ day^{-1} and an ultimate BOD $(L) = 200$ mg/L. What is the final dissolved oxygen in 5 days in a BOD bottle in which the sample is diluted 1 : 20 and where the initial dissolved oxygen is 10.2 mg/L?

8-19   What is the theoretical demand for oxygen if a chemical is identified by the general formula $C_4H_8O_2$?

8-20   A student places two BOD bottles in an incubator, having measured the initial DO of both as 9.0 mg/L. In bottle A, she has 100% sample, and in bottle B she puts 50% sample and 50% unseeded dilution water. The final dissolved oxygen, at the end of five days, is 3 mg/L in bottle A and 4 mg/L in bottle B.

a.   What was the five-day BOD of the sample as measured in each bottle?

b.   What might have happened to make these values different?

c.   Do you think the BOD measure included
   i.   only carbonaceous BOD
   ii.   only nitrogenous BOD
   iii.   both carbonaceous and nitrogenous BOD?

Why do you think so?

8-21   If the BOD_5 of an industrial waste, after pretreatment, is 220 mg/L, and the ultimate BOD is 320 mg/L, what is the deoxygenation constant k_1?

8-22   The ultimate BOD of two wastes is 280 mg/L each. For the first, the deoxygenation constant k_1 is 0.08 day^{-1}, and for the second, k_1 is 0.12 day^{-1}. What is the five-day BOD of each? Show graphically how this can be so.

8-23   You are the chief environmental engineer for a large industry and routinely receive the test results of the wastewater treatment plant effluent quality. One day you are shocked to discover that the level of cadmium is about 1000 times higher than the effluent permit. You call the laboratory technician, and he tells you that he thought that was funny also, so he ran the test several times to be sure.

You have no idea where the cadmium came from or if it will ever show up again. You *are* sure that if you report this peak to the state they may shut down the entire industrial operation since the treated effluent flows into a stream used as a water supply. They will also insist on knowing where the source was so it could not happen again. Such a shutdown would kill the company, already tottering on the verge of bankruptcy. Many people would lose their jobs and the community would suffer.

You have several options, among these are:

1.   Erase the offending data entry and forget the whole thing.

2.   Delay reporting the data to the state and start a massive search for the source, even though you have doubts it will ever be found.

3.   Bring this to the attention of your superiors, hoping that they will make a decision, and you will be off the hook.

4.   Report the data to the state and accept the consequences.

5. Other?

What would you decide, and how would you decide it if

a. You are 24 years old, two years out of school, not married?

b. You are 48 years old, married, two children in college?

How would your decisions differ in these circumstances? Analyze the decisions on the basis of facts, options, people affected by your decision, and final conclusions.

Endnote

1. H. A. Thomas, Jr., "Graphical Determination of BOD Curve Constants," *Water and Sewer Works* 97:123 (1950).

WATER SUPPLY & TREATMENT

As long as population densities are sufficiently low, the ready availability of water for drinking and other uses and the effective disposal of waterborne wastes may not pose a serious problem. For example, in Colonial America, wells and surface streams provided adequate water, and wastes were disposed of into other nearby watercourses without fuss or bother. Even today, much of rural America has no need for water and wastewater systems more sophisticated than a well and a septic tank.

But people are social and commercial animals and, in the process of congregating in cities, have created a problem of adequate water supply and disposal.

In this chapter the availability of water for public use by communities is considered first, followed by a discussion of how this water is treated and then distributed to individual users. In Chapter 10, the collection and treatment of used water, or wastewater, is described.

9.1 ◆ The Hydrologic Cycle and Water Availability

The concept of the *hydrologic cycle*, already presented in Chapter 3, is a useful starting point for the study of water supply. Illustrated in Figure 9–1, this cycle includes the precipitation of water from clouds, percolation into the ground or runoff into surface watercourses, followed by evaporation and transpiration of the water back into the atmosphere.

Precipitation is the term applied to all forms of moisture originating in the atmosphere and falling to the ground. Precipitation is measured with gauges that record in inches of water. The depth of precipitation over a given region is often useful in estimating the availability of water.

Evaporation and *transpiration* are the two processes by which water reenters the atmosphere. Evaporation is loss from free water surfaces, whereas transpiration is loss by plants. The same meteorological factors that influence evaporation are at work in the transpiration process; solar radiation, ambient air temperature, humidity, and wind speed, as well as the amount of soil moisture available to the plants have impact on the rate of transpiration. Because evaporation and transpiration are so difficult to measure separately, they are often combined into a single term, *evapotranspiration*, or the total water loss to the atmosphere by both evaporation and transpiration.

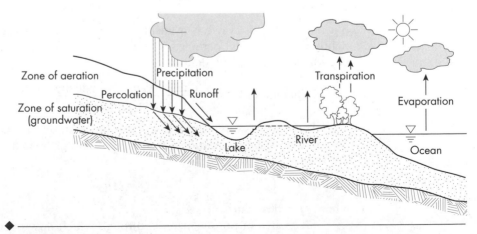

FIGURE 9–1   The hydrologic cycle.

Water on the surface of the earth that is exposed to the atmosphere is called *surface water*. Surface waters include rivers, lakes, oceans, and other open bodies of water. Through the process of *percolation*, some surface water (especially during a precipitation event) percolates into the ground and becomes *groundwater*. Both groundwater and surface water can be used as sources of water for communities.

9.1.1   Groundwater Supplies

Groundwater is both an important direct source of water supply and a significant indirect source of supply since a large portion of the flow to streams is derived from subsurface water. Near the surface of the earth in the *zone of aeration*, soil pore spaces contain both air and water. This zone may have zero thickness in swamplands and be several hundred feet thick in arid regions. Moisture from the zone of aeration cannot be tapped as a water supply source because this water is held on the soil particles by capillary forces and is not readily released.

Below the zone of aeration is the *zone of saturation*, in which the pores are filled with water. Water within the zone of saturation is referred to as *groundwater*. A stratum containing a substantial amount of groundwater is called an *aquifer* and the surface of this saturated layer is known as the *water table*. If the aquifer is underlain by a impervious stratum it is called an *unconfined aquifer*. If the stratum containing water is trapped between two impervious layers, it is known as a *confined aquifer*. Confined aquifers can sometimes be under pressure, just like pipes, and if a well is tapped into a confined aquifer under pressure, an *artesian well* results. Sometimes the pressure is sufficient to allow these artesian wells to flow freely without the necessity of pumping.

The amount of water that can be stored in the aquifer is equal to the volume of the void spaces between the soil grains. The fraction of voids volume to total volume of the soil is termed *porosity*, defined as

$$\text{porosity} = \frac{\text{volume of voids}}{\text{total volume}}$$

But not all of this water is available for extraction and use because it is so tightly tied to the soil particles. The amount of water that can be extracted is known as *specific yield*, defined as

$$\text{specific yield} = \frac{\text{volume of water that will drain freely from a soil}}{\text{total volume of water in the soil}}$$

The flow of water out of a soil can be illustrated using Figure 9-2. The flow rate must be proportional to the area through which flow occurs times the velocity, or

$$Q = Av$$

where Q = flow rate, m³/sec
A = area of porous material through which flow occurs, m²
v = superficial velocity, m/sec

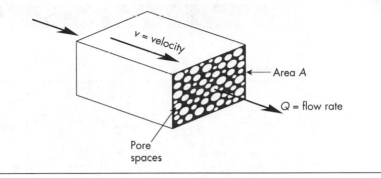

FIGURE 9-2   Flow from a porous medium such as soil.

The superficial velocity is of course not the actual velocity of the water in the soil, since the volume occupied by the soil solid particles greatly reduces the available area for flow. If the area available for flow is a, then

$$Q = Av = av'$$

where v' = the actual velocity of water flowing through the soil
$\quad a$ = the area available for flow

Solving for v',

$$v' = \frac{Av}{a}$$

If a sample of soil is of some length L, then

$$v' = \frac{Av}{a} = \frac{AvL}{aL} = \frac{v}{\text{porosity}}$$

since the total volume of the soil sample is AL and the volume occupied by the water is aL.

Water flowing through the soil at a velocity v' loses energy, just like water flowing through a pipeline or an open channel. This energy loss per distance traveled is defined as

$$\frac{dh}{dL}$$

where h = energy, measured as elevation of the water table in an unconfined aquifer or as pressure in a confined aquifer, m
$\quad L$ = horizontal distance in direction of flow, m

In an unconfined aquifer, the drop in the elevation of the water table with distance is the slope of the water table, dh/dL, in the direction of flow. The elevation of the water surface is the potential energy of the water, and water flows from a higher elevation to a lower elevation, losing energy along the way. Flow through a porous medium such as soil is related to the energy loss using the Darcy equation,

	Porosity (%)	Specific Yield (%)	Coefficient of Permeability (m/sec)
TABLE 9–1 Typical Aquifer Parameters			
Aquifer Material			
Clay	55	3	1×10^{-6}
Loam	35	5	5×10^{-6}
Fine sand	45	10	3×10^{-5}
Medium sand	37	25	1×10^{-4}
Coarse sand	30	25	8×10^{-4}
Sand and gravel	20	16	6×10^{-4}
Gravel	25	22	6×10^{-3}

Source: Adapted from M. Davis and D. Cornwell, *Introduction to Environmental Engineering,* McGraw-Hill, New York (1991).

$$Q = KA \frac{db}{dL}$$

where K = coefficient of permeability, m/sec

A = cross-sectional area, m^2

Table 9–1 shows some typical values of porosity, specific yield and coefficient of permeability.

The Darcy equation makes intuitive sense, in that the flow rate (Q) increases with increasing area (A) through which the flow occurs and with the drop in pressure, db/dL. The greater the driving force (the difference in upstream and downstream pressures), the greater the flow. The fudge factor K is the *coefficient of permeability*, an indirect measure of the ability of a soil sample to transmit water, and varies dramatically for different soils, ranging from about 0.05 m/day for clay, to more than 500 m/day for gravel. The coefficient of permeability is measured commonly in the laboratory using *perme-ameters*, which consist of a soil sample through which is forced a fluid such as water. The flow rate is measured for a given driving force (difference in pressures) through a known area of soil sample, and the permeability calculated.

e • x • a • m • p • l • e **9.1**

Problem   A soil sample is installed in a permeameter as shown in Figure 9–3. The length of the sample is 0.1 m and it has a cross-sectional area of 0.05 m^2. The water pressure on the upflow side is 2.5 m and on the downstream side the water pressure is 0.5 m. A flow rate of 2.0 m^3/day is observed. What is the coefficient of permeability?

Solution   The pressure drop is the difference between the upstream and downstream pressures, or $b = 2.5 - 0.5 = 2.0$ m. Using the Darcy equation, and solving for K,

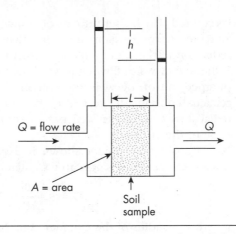

FIGURE 9-3 Permeameter used for measuring coefficient of permeability using the Darcy equation.

$$K = \frac{Q}{A\dfrac{db}{dL}} = \frac{2.0}{0.05 \times \dfrac{2}{0.1}} = 2 \text{ m/day}$$

If a well is sunk into an unconfined aquifer, shown in Figure 9-4, and water is pumped out, the water in the aquifer will begin to flow toward the well. As the water approaches the well, the area through which it flows gets progressively smaller, and therefore a higher superficial (and actual) velocity

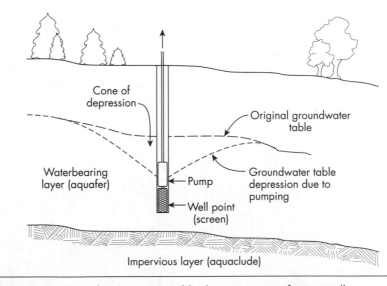

FIGURE 9-4 Drawdown in water table due to pumping from a well.

is required. The higher velocity of course results in an increasing loss of energy, and the energy gradient must increase, forming a *cone of depression*. The reduction in the water table is known in groundwater terms as a *drawdown*. If the rate of water flowing toward the well is equal to the rate of water being pumped out of the well, the condition is at equilibrium, and the drawdown remains constant. If, however, the rate of water pumping is increased, the radial flow toward the well has to compensate, and this results in a deeper cone or drawdown.

Consider a cylinder shown in Figure 9-5 through which water flows toward the center. Using Darcy's equation,

$$Q = KA\frac{db}{dL} = K(2\pi rw)\frac{db}{dr}$$

where r = radius of the cylinder, and $2\pi rw$ is the cross-sectional surface area of the cylinder. If water is pumped out of the center of the cylinder at the same rate as water is moving in through the cylinder surface area, the depth of the cylinder through which water flows into the well, w, can be replaced by the height of the water above the impermeable layer, b. This equation can then be integrated as

$$\int_{r_2}^{r_1} Q\frac{dr}{r} = 2\pi K \int_{b_2}^{b_1} b\, db$$

$$Q \ln\frac{r_1}{r_2} = \pi K(b_1{}^2 - b_2{}^2)$$

or

$$Q = \frac{\pi K(b_1{}^2 - b_2{}^2)}{\ln\dfrac{r_1}{r_2}}$$

Note that the integration is between any two arbitrary values of r and b.

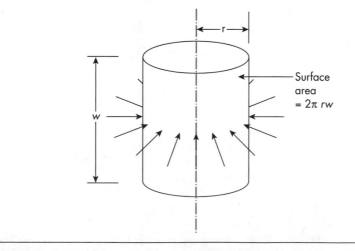

FIGURE 9–5   Cylinder with flow through the surface.

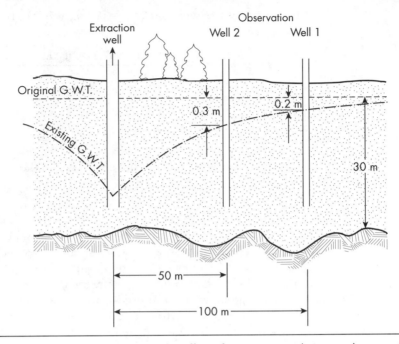

FIGURE 9-6 Multiple wells and the effect of extraction on the groundwater table.

This equation can be used to estimate the pumping rate for a given drawdown any distance away from a well, using the water level measurements in two observation wells in an unconfined aquifer shown in Figure 9-6. Also, knowing the diameter of a well, we can estimate the drawdown at the well, the critical point in the cone of depression. If the drawdown is depressed all the way to the bottom of the aquifer, the well "goes dry"—it cannot pump water at the desired rate. Although the derivation of the foregoing equations is for an unconfined aquifer, the same situation would occur for a confined aquifer, where the pressure would be measured by observation wells.

e • x • a • m • p • l • e **9.2**

Problem A well is 0.2 m in diameter and pumps from an unconfined aquifer 30 m deep at an equilibrium (steady-state) rate of 1000 m³ per day. Two observation wells are located at distances 50 and 100 m, and they have been drawn down by 0.3 and 0.2 m, respectively. What is the coefficient of permeability and estimated drawdown at the well?

Solution

$$K = \frac{Q \ln \dfrac{r_1}{r_2}}{\pi(b_1^2 - b_2^2)} = \frac{1000 \ln(100/50)}{3.14[(29.8)^2 - (29.7)^2]} = 37.1 \ \text{m/day}$$

If the radius of the well is $0.2/2 = 0.1$ m, this can be plugged into the same equation, as

$$Q = \frac{\pi K(b_1^2 - b_2^2)}{\ln \dfrac{r_1}{r_2}} = \frac{3.14 \times 37.1 \times [(29.7)^2 - b_2^2]}{\ln \dfrac{50}{0.1}} = 1000 \text{ m}^3/\text{day}$$

and solving for b_2,

$$b_2 = 28.8 \text{ m}$$

Since the aquifer is 30 m deep, the drawdown at the well is $30 - 28.8 = 1.2$ m.

◆

Multiple wells in an aquifer can interfere with each other and cause excessive drawdown. Consider the situation in Figure 9–7, where an existing well (Well 1) creates a cone of depression. If a second production well (Well 2) is installed, the cones will overlap, causing greater drawdown at each well. If many wells are sunk into an aquifer, the combined effect of the wells could deplete the groundwater resources and all wells would "go dry."

The reverse is also true, of course. Suppose one of the wells is used as an injection well; then the injected water flows from this well into the others, building up the groundwater table and reducing the drawdown. The judicious use of extraction and injection wells is one way that the flow of contaminants from hazardous waste or refuse dumps can be controlled, as discussed further in Chapter 13.

Finally, a lot of assumptions are made in the previous discussion. First, we assume that the aquifer is homogeneous and infinite, that is, it sits on a

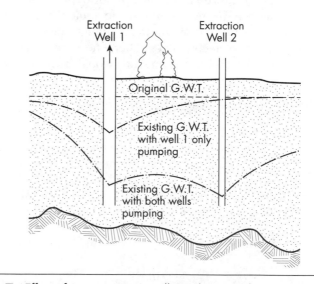

FIGURE 9–7   Effect of two extraction wells on the groundwater table.

level aquaclude and that the permeability of the soil is the same at all places for an infinite distance in all directions, and the moon is made of green cheese. Second, we assume steady state and uniform radial flow, that the well penetrates the entire aquifer and is open for the entire depth of the aquifer, and that the pumping rate is constant. Clearly any of these conditions may cause the analysis to be faulty, and this model of aquifer behavior is only the beginning of the story. Modeling the behavior of groundwater is a complex and sophisticated science.

9.1.2   Surface Water Supplies

Surface water supplies are not as reliable as groundwater sources because quantities often fluctuate widely during the course of a year or even a week, and the quality of surface water is easily degraded by various sources of pollution. The variation in the river or stream flow can be so great that even a small demand cannot be met during dry periods, and storage facilities must be constructed to hold the water during wet periods so it can be saved for the dry ones. The objective is to build these reservoirs sufficiently large to have dependable supplies but not so large as to bankrupt the community.

One method of arriving at the proper reservoir size is by constructing a *mass curve*. In this analysis, the total flow in a stream at the point of a proposed reservoir is summed and plotted against time. On the same curve the water demand is plotted and the difference between the total water flowing in and the water demanded is the quantity that the reservoir must hold if the demand is to be met. The method is illustrated by the following example problem.

e • x • a • m • p • l • e   **9.3**

Problem   A reservoir is needed to provide a constant flow of 15 cubic feet per second (cfs). The monthly stream flow records, in total cubic feet of water for each month, are:

Month	J	F	M	A	M	J	J	A	S	O	N	D
Cubic feet of water $(\times 10^6)$	50	60	70	40	32	20	50	80	10	50	60	80

Calculate the reservoir storage necessary to provide the constant 15 cfs demand.

Solution   The storage requirement is calculated by plotting the cumulative water flows as in Figure 9-8. Note that for January, 50 million cubic feet is plotted, for February 60 is added to that and 110 million cubic feet is plotted and so on. The demand for water is constant at 15 cfs, or 15 (cubic feet/ sec) $\times$ 60 sec/min $\times$ 60 min/hr $\times$ 24 hr/day $\times$ 30 days/month = 38.8 $\times$ 10^6 cubic feet/month. This can be represented as a sloped line in Figure 9-8 and plotted on the curved supply line. The stream flow in May is lower than

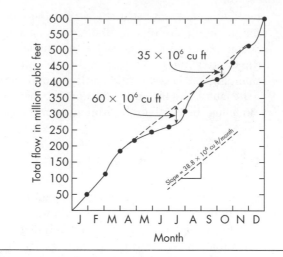

FIGURE 9-8   Mass curve showing required storage volumes.

the demand, the start of a drought lasting into June. The demand slope is greater than the supply, and thus the reservoir has to make up the deficit. In July the rains come and the supply increases until the reservoir can be filled up again, late in August. The reservoir capacity needed to get through that particular drought is 60×10^6 cubic feet. A second drought, starting in September, lasts into November and requires 35×10^6 cubic feet of capacity. If the municipality has a reservoir with at least 60×10^6 cubic feet capacity it can draw water from it throughout the year.

A mass curve such as Figure 9-8 is actually of little use if only limited stream flow data are available. One year's data yield very little information about long-term variations. For example, was the drought in the example the worst drought in 20 years, or was the year shown actually a fairly wet year?

To get around this problem, we need to predict statistically the recurrence of events such as droughts and then design the structures according to a known risk. This procedure is discussed in detail in Chapter 2.

Water supplies are often designed to meet demands 19 out of 20 years. In other words, once in 20 years the drought will be so severe that the reservoir capacity will not be adequate to meet the demand for water. If running out of water once every 20 years is not acceptable, the community can choose to build a bigger reservoir and thus expect to be dry only once every 50 years. The question is one of an increasing investment of capital for a steadily smaller added benefit.

Using a *frequency analysis* of recurring natural events such as droughts, as described in Chapter 2, a "100-year drought" or a "10-year drought" can be calculated. Although this "10-year drought" occurs on the average once every 10 years, there is no guarantee that it would indeed occur once every 10 years.

In fact, it could for example happen 3 years in a row, and then not again for 50 years.

When particularly severe and unanticipated drought occurs, the community can impose sanctions on those who use more water than their share. In recent years, many communities in southern California have imposed severe fines and other penalties for excessive water use.

A far better way of attaining the same end is to have everyone cooperate voluntarily. Such voluntary community cooperation is often difficult to achieve, however, and nowhere is this more evident than in matters of environmental degradation or fair use of resources. The essence of this problem can be illustrated by what has become known as the "prisoner dilemma."

Suppose there are two prisoners, A and B, both of whom are being separately interrogated. Each prisoner is told that if neither of them confess they will both receive only a 2-year sentence. If one prisoner accuses the other, the accuser is set free and the one who was accused receives a 20-year sentence. If, however, they both accuse each other, each one gets a 10-year sentence. Should prisoner A trust prisoner B not to accuse him and do likewise? If prisoner B does not accuse A and A does not accuse B, they both get only 2 years. But all A has to do to go free is to accuse B and hope that B does not accuse him. Prisoner B of course has the same option. Graphically the options look like

		B accuses A	
		NO	YES
A accuses B	NO	2,2	20,0
	YES	0,20	10,10

The same dilemma might occur in the use of a scarce resource such as a water supply. During a severe drought, all industries are asked to cooperate by voluntarily restricting water use. Industry A agrees to do so, and if the other industries cooperate, they will all experience a loss of production, but all of the companies will survive. If Industry A decides not to cooperate, however, and if everyone else does, Industry A wins big because it has full use of the water and its production and profits would increase. By using all the water, Industry A would have deprived others of this resource and the other industries may have been forced to close. If, however, *everyone* behaved like Industry A and decided not to cooperate, then they would all run out of water and all of them would be driven into bankruptcy.

Many theorists hold that human nature prevents us from choosing the cooperation alternative, and that the essence of human nature is competition and not cooperation. In that sense, humans are no different from animals and plants that survive in a stable ecosystem—not by cooperation, but by competition. Within any such system, there is continual competition for resources, and the only sharing that occurs is instinctive sharing such as a mother feeding her young. What might be seen as cooperation in an ecosystem is actually parasitic behavior, the use of another creature for one's own ends. A maple tree, for example, does not exist to have a honeysuckle vine grow on it. If a maple could make a conscious choice, it would no doubt try to prevent the honeysuckle from twisting itself around its trunk and branches and from reducing the amount of sunlight the maple tree receives.

If competition is the essence of human (and other living creature) conduct, can we ever then expect humans to act in a cooperative mode? The answer is yes. Not only are humans competitive, but they can also be caring and sympathetic creatures. Out of altruistic and cooperative motives we help other humans in need. We mount massive food supply efforts for starving people and use international pressures to prevent wars of attrition.

And yet the greatest single threat to the global ecosystem is the cancerous growth of the human population, and this trend has to be reversed if the human species has any hope of long-term survival. If we are truly competitive, then preventing starvation and helping less-advantaged people makes no sense, no more than one loblolly pine seedling helping its neighbor. Both seedlings "know" that only one will survive, and each tries its best to be the taller and stronger of the two.

For years we thought that competition and aggression (wars) would be the undoing of humans on this planet. Ironically, the single most species-destructive trait in *Homo sapiens* may be cooperation.

9.2 ◆ Water Treatment

Many aquifers and isolated surface waters are of high water quality and may be pumped from the supply and transmission network directly to any number of end uses, including human consumption, irrigation, industrial processes, or fire control. However, such clean water sources are the exception to the rule, particularly in regions with dense populations or regions that are heavily agricultural. Here, the water supply must receive varying degrees of treatment prior to distribution.

A typical water treatment plant is diagrammed in Figure 9-9. Such plants are made up of a series of reactors or unit operations, with the water flowing from one to the next, and when stacked in series, achieve a desired end product. Each operation is designed to perform a specific function, and the order of these operations is important. Described in the next section are a number of the most important of these processes.

9.2.1   Coagulation and Flocculation

Raw surface water entering a water treatment plant usually has significant turbidity caused by tiny colloidal clay and silt particles. These particles have a natural electrostatic charge that keeps them continually in motion and prevents them from colliding and sticking together. Chemicals such as alum (aluminum sulfate) are added to the water (Stage 1 in Figure 9-9), first to neutralize the charge on the particles and then to aid in making the tiny particles "sticky" so they can coalesce and form large particles, at Stage 2 in Figure 9-9. The

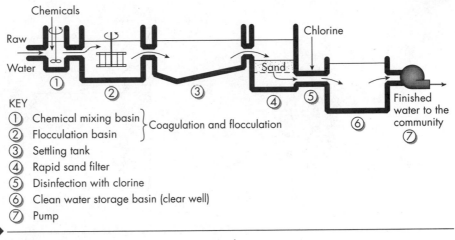

FIGURE 9-9 A typical water treatment plant.

purpose is to clear the water of suspended colloidal solids by building larger particles from the stable colloidal solids so that these larger and heavier particles could be readily settled out of the water.

Coagulation is the chemical alteration of the colloidal particles to make them stick together forming larger particles called *flocs*. When aluminum sulfate, $Al_2(SO_4)_3$, is added to water containing colloidal material, the alum initially dissolves to form aluminum ion, Al^{+++}, and sulfate ion, SO_4^{-2}. But the aluminum ion is unstable and forms various types of charged species of aluminum oxides and hydroxides. The specific forms of these compounds are dependent on the pH of the water, the temperature, and the method of mixing.

Two mechanisms are thought to be important in the process of coagulation. The first is *charge neutralization*, whereby the aluminum ions are used to counter the charges on the colloidal particles, pictured in Figure 9-10. The colloidal particles in natural waters are commonly negatively charged, and when suspended in water, repel each other due to their like charges. This causes the suspension to be stable and prevents the particles from settling out. If aluminum ions are added (regardless of what forms they may assume), some of them have a trivalent positive charge. As these are drawn to the negatively charged particles, they compress the net negative charge on the particles, making them less stable in terms of their charges. Such an increase in colloidal instability makes the particles more likely to collide with each other and form larger particles.

A second mechanism, termed *bridging*, involves the sticking together of the colloidal particles by virtue of the macromolecules formed by the aluminum hydroxides, as illustrated in Figure 9-11. The molecules have positive charge sites, by which the polymers attach themselves to various colloids, and then bridge the gap between adjacent particles, thereby creating larger particles. Because many of the desirable forms of aluminum hydroxides dissolve at low pH,

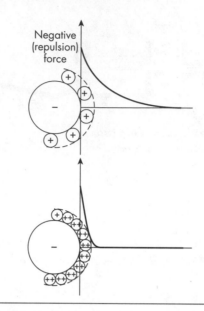

FIGURE 9–10 Effect of multivalent cations on the negative (repulsion) force of colloidal particles, resulting in charge neutralization.

lime [$Ca(OH)_2$] is often added to raise the pH. Some of the calcium precipitates as calcium carbonate [$CaCO_3$], assisting in the settling.

Both of these mechanisms are important in coagulation with alum. The net effect is that the colloidal particles are destabilized and have the propensity to grow into larger particles. The assistance in the growth of these larger particles is a physical process known as *flocculation*.

For particles to come together and stick to each other, through either charge neutralization or bridging, they have to move at different velocities. Consider for a moment the movement of cars on a highway. If all of the cars moved at exactly the same velocity, it would be impossible to have car-to-car collisions. Only if the cars have different velocities (speed and direction), with some cars catching up to others, are accidents possible. The intent of the process of flocculation is to produce differential velocities within the water so that the particles can come into contact. Commonly, this is accomplished in a water treatment plant by simply using a large slow-speed paddle that gently

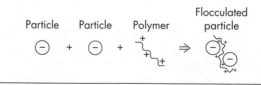

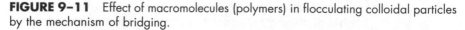

FIGURE 9–11 Effect of macromolecules (polymers) in flocculating colloidal particles by the mechanism of bridging.

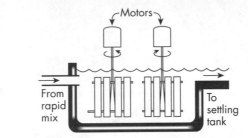

FIGURE 9-12 Typical flocculator used in water treatment.

stirs the chemically treated water and produces the large particles of destabilized colloidal solids and aluminum hydroxides, as depicted as Step 2 in Figure 9-9 and illustrated in Figure 9-12. Once these larger particles are formed, the next step is to remove them using the process of settling.

9.2.2 Settling

When the flocs have been formed they must be separated from the water. This is invariably done in *gravity settling tanks* that simply allow the heavier-than-water particles to settle to the bottom. Settling tanks are designed to approximate a plug-flow reactor. That is, the intent is to minimize all turbulence. The two critical design elements of a settling tank are the entrance and exit configurations since this is where plug flow can be severely compromised. Figure 9-13 shows one type of entrance and exit configuration used for distributing the flow entering and leaving the water treatment settling tank.

The sludge in water treatment plants is comprised of aluminum hydroxides, calcium carbonates, and clays, is not highly biodegradable, and will not decompose at the bottom of the tank. Typically, the sludge is removed every

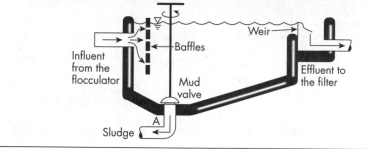

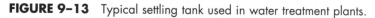

FIGURE 9–13 Typical settling tank used in water treatment plants.

few weeks through a *mud valve* at the bottom and is wasted either into a sewer or into a sludge holding/drying pond.

Settling tanks work because the density of the solids exceeds that of the liquid. The movement of a solid particle through a fluid under the pull of gravity is governed by a number of variables, including

- particle size (volume)
- particle shape
- particle density
- fluid density
- fluid viscosity

The last term may be unfamiliar, but it refers simply to the ability of the fluid to flow. Pancake syrup, for example, has a high viscosity, while water has a relatively low viscosity.

We want to get particles to settle at the highest velocity and this requires large particle volumes, compact shapes, high particle and low fluid densities,

TABLE 9-2 Typical Setting Rates		
Particle Diameter (mm)	Typical Particle	Settling Velocity (m/sec)
1.0	Sand	2×10^{-1}
0.1	Fine sand	1×10^{-2}
0.01	Silt	1×10^{-4}
0.001	Clay	1×10^{-6}

and low viscosities. In practical terms, we cannot control the last three variables, but coagulation and flocculation certainly result in the growth of particles and changes their density and shape.

The reason why coagulation–flocculation is so important in preparing particles for the settling tank is shown by some typical settling rates in Table 9–2. Although the settling velocity of particles in a fluid is also dependent on the particle shape and density, these numbers illustrate that even small changes in particle size for typical flocculated solids in water treatment can dramatically affect the efficiency of removal by settling.

Settling tanks can be analyzed by assuming an "ideal tank" (very much as we analyze "ideal reactors"). Such ideal settling tanks can be visualized hydraulically as perfect plug-flow reactors: A plug of water enters the tank; a column of water moves through the tank without intermixing (Figure 9–14). If a solid particle enters the tank at the top of the column and settles at a velocity of v_0, it should have settled to the bottom as the imaginary column of water exits the tank, having moved through the tank at a horizontal velocity v_h.

Several assumptions are required in the analysis of ideal settling tanks:

- Uniform flow occurs within the settling tank. This is the same as saying that there is ideal plug flow, since uniform flow is defined as a condition where all water flows horizontally at the same velocity.

- All particles settling to the bottom are removed. That is, as the particles drop to the bottom of the column as depicted in Figure 9–14, they are removed from the flow.

- Particles are evenly distributed in the flow as they enter the settling tank.

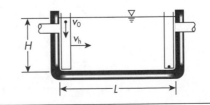

FIGURE 9-14 An ideal settling tank.

- All particles still suspended in the water when the column of water reaches the far side of the tank are not removed and escape the tank.

Consider now a particle entering the settling tank at the water surface. This particle has a settling velocity v_0 and a horizontal velocity v_h such that the resultant vector defines a trajectory as shown in Figure 9–15. In other words, the particle is just barely removed (it hits the bottom at the last instant). Note that if the same particle enters the settling tank at any other height, such as height h, its trajectory always carries it to the bottom. Particles having velocity v_0 are termed *critical particles* in that particles with lower settling velocities are not all removed. For example, the particle having velocity v_s, entering the settling tank at the surface, will not hit the bottom and escape the tank. If this same particle, however, enters at some height h, it should just barely hit the bottom and be removed. Any of these particles that happen to enter the settling tank at height h or lower would thus be removed, and those entering above h would not. Since the particles entering the settling tank are assumed to be equally distributed, the proportion of those particles with a velocity of v_s removed is equal to h/H, where H is the height of the settling tank.

From Chapter 3, the residence time is defined as

$$\bar{t} = \frac{V}{Q}$$

or the volume divided by the flow rate. The volume of rectangular tanks is calculated as $V = HLW$, where W = width of tank. Applying the continuity equation, is $Q = av_h$ where a is the area through which the flow occurs, and v_h is the horizontal velocity through the tank (of both particles and water). In the case of a settling tank acting as a plug-flow reactor, $a = HW$.

$$\bar{t} = \frac{V}{Q} = \frac{HWL}{(HW)v_h} = \frac{L}{v_h}$$

And using similar triangles in Figure 9–15,

$$v_0 = \frac{H}{\bar{t}}$$

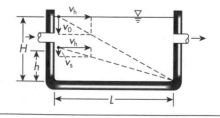

FIGURE 9–15   Particle trajectories in an ideal settling tank.

As noted previously, the residence time $\bar{t}$ is equal to V/Q, where Q is the flow rate and V is the volume of the settling tank, or $V = HWL$, or if the surface area of settling tank is defined as $A = WL$, then $V = AH$. Then

$$v_0 = \frac{H}{\bar{t}} = \frac{H}{AH/Q} = \frac{Q}{A}$$

The equation represents an important design parameter for settling tanks, called the *overflow rate*. Note the units:

$$v_0 = \frac{m}{s} = \frac{Q}{A} = \frac{m^3/\text{sec}}{m^2}$$

The overflow rate has the same units as velocity. Commonly, overflow rate is expressed as "gallons/day-ft^2," but it actually is a velocity term, and is in fact equal to the velocity of the critical particle. When the design of a settling tank is specified by overflow rate, what is really defined then is the critical particle by specifying its velocity.

Note that when any two of the following—overflow rate, residence time, or depth—are defined, the remaining parameter is also fixed.

e • x • a • m • p • l • e 9.4

Problem   A water treatment plant settling tank has an overflow rate of 600 gal/day-ft^2 and a depth of 6 ft. What is its residence time?

Solution

$$v_0 = 600 \text{ gal/day-ft}^2 \times 1/7.48 \text{ ft}^3/\text{gal} = 80.2 \text{ ft/day}$$

$$\bar{t} = H/v_0 = 6/80.2 = 0.0748 \text{ day} = 1.8 \text{ hr}$$

◆

Overflow rate is interesting in that a better understanding of settling can be obtained by looking at individual variables. For example, increasing the flow rate, Q, in a given tank increases the v_0; that is, the critical velocity increases and thus fewer particles are totally removed since fewer particles have a settling velocity greater than v_0.

It would, of course, be advantageous to decrease v_0, so more particles can be removed. This is done by either reducing Q or increasing A. The latter term may be increased by changing the dimensions of the tank so that the depth is shallow, and the length and width are very large. For example, the area can be doubled by taking a 3-m-deep tank, slicing it in half (two 1.5-m slices), and placing them alongside each other. The new shallow tank has the same horizontal velocity because it has the same area through which the flow enters ($a = WH$) but double the surface area ($A = WL$), and hence v_0 is half of the original value.

Why not then make *very* shallow tanks? The problem is first a practical one of hydraulics and the even distribution of flow, and the great expense in concrete and steel. Second, as particles settle they can flocculate, or bump into other slower moving particles and stick together, creating higher settling velocities and enhancing solids removal. As noted previously, as the particles grow in size, they settle faster. Settling tank depth then is an important practical consideration, and typically tanks are built 3 to 4 m deep to take advantage of the natural flocculation that occurs during settling.

e • x • a • m • p • l • e 9.5

Problem   A small water plant has a raw water inflow rate of 0.6 m³/sec. Laboratory studies have shown that the flocculated slurry can be expected to have a uniform particle size (only one size), and it has been found, through experimentation, that all of the particles settle at a rate of $v_s = 0.004$ m/sec. (This is a pretty dumb assumption, of course.) A proposed rectangular settling tank has an effective settling zone of $L = 20$ m, $H = 3$ m, and $W = 6$ m. Can 100% removal be expected?

Solution   Remember that the overflow rate is actually the critical particle settling velocity. What is a critical particle settling velocity for this tank?

$$v_0 = Q/A = 0.6/(20 \times 6) = 0.005 \text{ m/sec}$$

The critical particle settling velocity is greater than the settling velocity of the particle to be settled out, and hence not all the incoming particles will be removed.

The same conclusion can be reached using the particle trajectory. The velocity v through the tank is

$$v = Q/HW = 0.6 /(3 \times 6) = 0.033 \text{ m/sec}$$

Using similar triangles,

$$v_s/v = H/L'$$

where L' is the horizontal distance the particle would need to travel to reach the bottom of the tank,

$$0.004/0.033 = 3/L'$$

$$L' = 25 \text{ m}$$

Hence the particles would need 25 m to be totally removed, but only 20 m are available.

◆

e • x • a • m • p • l • e 9.6

Problem   In Example 9.5, what fraction of the particles will be removed?

Solution   Assume that the particles entering the tank are vertically uniformly distributed. If the length of the tank is 20 m, the settling trajectory from the

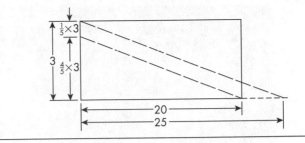

FIGURE 9–16   Ideal settling tank. See Example 9.6.

far bottom corner would intersect the front of the tank at height 4/5(3 m), as shown in Figure 9–16. All those particles entering the tank below this point would be removed, and those above would not. The fraction of particles that will be removed is then 4/5 or 80%.

Alternatively, since the critical settling velocity is 0.005 m/sec and the actual settling velocity is only 0.004 m/sec, the expected effectiveness of the tank is 0.004/0.005 = 0.8 or 80%.

The important similarity between ideal settling tanks and the ideal world is that neither one exists. Yet engineers are continually trying to idealize the world. There is nothing wrong with this, of course, since simplification is a necessary step in solving engineering problems, as demonstrated in Chapter 1. The danger of idealizing is that we can forget all the assumptions used in the process. A settling tank for example *never* has uniform flow. Wind, density, and temperature currents, as well as inadequate baffling at the tank entrance can all cause nonuniform flow. So don't expect a settling tank to behave ideally but design them with large safety factors.

If the (overdesigned) settling tank works well, the water leaving a settling tank is essentially clear. It is not yet acceptable for domestic consumption, however, and one more polishing step is necessary, usually using a rapid sand filter.

9.2.3   Filtration

As noted in the discussion of groundwater quality, the movement of water through soil removes many of the contaminants in water. Environmental engineers have learned to apply this natural process to water treatment and supply systems and developed what is now known as the *rapid sand filter*. The operation of a rapid sand filter involves two phases: filtration and backwashing.

A slightly simplified version of the rapid sand filter is illustrated in a

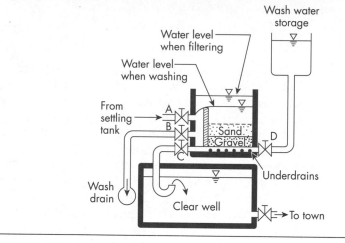

FIGURE 9-17   A rapid sand filter used in water treatment.

cutaway drawing in Figure 9–17. Water from the settling basins enters the filter and seeps through the sand and gravel bed, through a false floor and out into a clear well that stores the finished water. During filtration valves A and C are open. The design parameter for filters is known as "filter loading" and is usually about 4 gal/min-ft^2.

The suspended solids that escape the flocculation and settling steps are caught on the filter sand particles and eventually the rapid sand filter becomes clogged and must be cleaned. This cleaning is performed hydraulically using a process called *backwashing*. The operator first shuts off the flow of water to the filter (closing valves A and C), then opens valves D and B, allowing wash water (clean water stored in an elevated tank or pumped from the clear well) to enter below the filter bed. This rush of water forces the sand and gravel bed to expand and jolts individual sand particles into motion, rubbing them against their neighbors. The suspended solids trapped within the filter are released and escape with the wash water. After a few minutes, the wash water is shut off and filtration is resumed.

9.2.4 Disinfection

Following filtration, and before storage in the *clear well*, the water is *disinfected* to destroy whatever pathogenic organisms might remain. Commonly disinfection is accomplished with chlorine, purchased as a liquid under pressure and released into the water as chlorine gas using a chlorine feeder system. The dissolved chlorine oxidizes organic material, including pathogenic organisms. The presence of a residual of active chlorine in the water is an indication that no further organics remain to be oxidized and that the water can be assumed to be free of disease-causing organisms. Water pumped into distribution systems usually contains a residual of chlorine to guard against any contamination in the distribution system. This is why water from drinking fountains or faucets often has a slight taste of chlorine.

9.3 ◆ Distribution of Water

From the clear well in the water treatment plant the finished water, which is now both safe and pleasing to drink, is pumped into the distribution system. Such systems are under pressure, so that any tap into a pipe, whether it be a fire hydrant or domestic service, will yield water.

Because the demand for finished water varies with day of the week and hour of the day, storage facilities must be used in the distribution system. Most communities have an elevated storage tank that is filled during periods of low water demand and supply water to the distribution system during periods of high demand. Figure 9–18 shows how such a storage tank can assist in providing water during peak demand periods and emergencies.

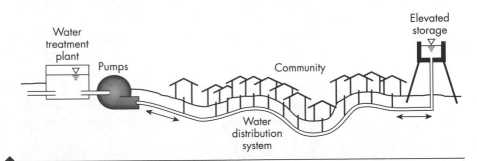

FIGURE 9–18 The water transmission mains connect the water treatment plant and the elevated storage reservoir. During periods of high water demand by the community, the water flows from both the water plant and the elevated storage to meet the demand. During the low demand periods, the pumps fill up the elevated storage tanks.

The calculation of the required elevated storage capacity requires both a frequency analysis and a materials balance, as illustrated in the next example.

e • x • a • m • p • l • e **9.7**

Problem   It has been determined that a community requires a maximum flow of 10 million gallons per day (MGD) of water during 10 hours in a peak day, beginning at 8 A.M. and ending at 6 P.M. During the remaining 14 hours it needs a flow of 2 MGD. During the entire 24 hours, the water treatment plant is able to provide a constant flow of 6 MGD, which is pumped into the distribution system. How large must the elevated storage tank be to meet this peak demand?

Solution   Assume the tank is full at 8 A.M. and run a materials balance on the community over the next 10 hours:

$$\begin{bmatrix} \text{Rate of water} \\ \text{ACCUMULATED} \end{bmatrix} = \begin{bmatrix} \text{Rate of} \\ \text{water IN} \end{bmatrix} - \begin{bmatrix} \text{Rate of} \\ \text{water OUT} \end{bmatrix}$$

$$+ \begin{bmatrix} \text{Rate of water} \\ \text{PRODUCED} \end{bmatrix} - \begin{bmatrix} \text{Rate of water} \\ \text{CONSUMED} \end{bmatrix}$$

The flow to the community comes from the tower (Q_1) and the plant (6 MGD).

$$0 = [Q_1 + 6] - [10] + 0 - 0$$

Solving,

$$Q_1 = 4 \text{ MGD}$$

This flow must be provided from the tank over a 10-hour period. The required volume of the tank is

$$4 \text{ MGD} \times 1 \text{ day}/24 \text{ hr} \times 10 \text{ hr} = 1.67 \text{ mil gal}$$

So if the tank holds 1.67 million gallons, the community can get the water it needs.

But can the tank be full at 8 A.M.? A materials (water) balance on the community between 6 P.M. and 8 A.M. yields

$$0 = [6] - [2 + Q_2] + 0 - 0$$

The flow coming into the community is 6 MGD, while the flow out is 2 MGD (the water used) plus Q_2, which is water needed to fill up the tank.

$$Q_2 = 4 \text{ MGD}$$

This is spread over 14 hours, so that 4 MGD × 1 day/24 hr × 14 hr = 2.3 mil gal. So there is no problem filling the tank.

◆

Abstract

◆ # Abbreviations

a = actual area of porous spaces through which flow occurs, m^2	$\bar{t}$ = retention time, sec
a = area of incoming flow = WH	v = horizontal velocity in an ideal settling tank, m/sec
A = area of soil through which flow occurs, m^2	v = superficial velocity through soil, m/day
A = surface area or settling tank = WL = m^2	v' = actual velocity within the soil pores, m/day
b = depth of water in an aquifer above the impermeable layer, m	v_0 = critical settling velocity in an ideal settling tank, m/sec
b = height at which a particle enters the settling tank, m	v_h = horizontal velocity through an ideal settling tank, m/sec
b = height or depth, m	v_s = settling velocity of any particle, m/sec
H = height of the settling tank, m	
L = length of sample or aquifer, m	V = volume, m^3
L = length of the settling tank, m	w = depth of a cylinder through which flow occurs, m
Q = flow rate, m^3/day	
Q = flow rate, m^3/sec	W = width of the settling tank, m

◆ # Problems

9–1   Suppose you are asked to recommend a series of laboratory tests to be run on a small drinking water treatment plant in a developing country. The plant consists of alum flocculation, settling, rapid sand filtration, and chlorination. What tests would you suggest they run, and at what frequency? Justify your answers, considering cost, human health, and environmental protection.

9–2   One day Farmer Brown drilled a well 200 ft deep into an aquifer that is 1000 ft deep and has a groundwater table 30 ft below the ground surface. On the same day, a neighbor of Farmer Brown, Farmer Jones, drilled a well only 50 ft deep. They both pumped the same quantity of water on that first day.

a.   Explain, using sketches, why it was worth the extra money for Farmer Brown to drill a deep well even though they both obtained the same yield from their wells the first day.

b.   Does Farmer Brown *own* the water he pumps from the ground? Is such ownership of natural resources possible? What would happen to a resource such as groundwater in a capitalistic market economy in the absence of governmental controls?

c.   Suppose the government believes that it owns the water, and then sells it to Farmer Brown. Would that necessarily result in a higher level of conservation of natural resources, or would it encourage the rapid depletion of resources? (Consider the recent experience in Eastern Europe.)

d.   What would the "deep ecologist" say about such governmental or private ownership of natural resources such as groundwater?

e.   If you were Farmer Jones, and discovered what Farmer Brown had done, what would you do, and why?

9-3   During the years following the Civil War, the New Orleans Water Company installed sand filters to treat Mississippi River water and sell it to the folks in New Orleans. The filters resembled the rapid sand filters in use today, and water from the river was pumped directly into the filters. Unfortunately, after the plant was built, the facility failed to produce the expected quantity of water and the company went bankrupt. Why do you think this happened? What would you have done as company engineer to save the operation?

9-4   A typical colloidal clay particle suspended in water has a diameter of 1.0 μm. If coagulation and flocculation with other particles manages to increase its size 100 times its initial diameter (at the same shape and density), how much shorter will be the settling time in 10 feet of water, such as a settling tank?

9-5   It is necessary to maintain a constant flow of 15 million gallons per month to a power plant cooling system. The runoff records for a stream are as follows:

Month	Total Flow during that Month (million gallons)
1	940
2	122
3	45
4	5
5	5
6	2
7	0
8	2
9	16
10	7
11	72
12	92
13	21
14	55
15	33

If a reservoir were to be constructed, what would be the storage requirement?

9-6   A storage tank at an oil refinery receives a constant flow into the tank at 0.1 m^3/sec. It is used to distribute the oil for processing only during the 8-hour working day. What must be the flow out of the tank, and how big must the tank be?

9-7   An unconfined aquifer is 10 m thick and is being pumped so that one observation well placed at a distance of 76 m shows a drawdown of 0.5 m. On the opposite side of the extraction well is another observation well, 100 m from the extraction well, and this well shows a drawdown of 0.3 m. Assume the coefficient of permeability is 50 m/day.

      a.   What is the discharge of the extraction well?

      b.   Suppose the well at 100 m from the extraction well is now pumped. Show with a sketch what this will do to the drawdown.

      c.   Suppose the aquifer sits on an aquaclude that has a slope of 1/100. Show with a sketch how this would change the drawdown.

9-8   A settling tank in a water treatment plant has an inflow of 2 m³/min and a solids concentration of 2100 mg/L. The effluent from this settling tank goes to sand filters. The concentration of sludge coming out of the bottom of the settling tank (the underflow) is 18,000 mg/L and the flow to the filters is 1.8 m³/min.

      a.   What is the underflow flow rate?

      b.   What is the solids concentration in the effluent (flow to the filters)?

      c.   How large must the sand filters be (in m²) if the filter loading is 4 gal/min-ft²?

9-9   A settling tank is 20 m long, 10 m deep, and 10 m wide. The flow rate to the tank is 10 m³/min. The particles to be removed all have a settling velocity of 0.1 m/min.

      a.   What is the hydraulic residence time?

      b.   Will all the particles be removed?

9-10   The settling basins for a 50-MGD wastewater treatment plant are operated in parallel with flow split evenly to 10 settling tanks, each of which is 3 meters deep and 25 meters wide, with a length of 32 meters.

      a.   What is the expected theoretical percent removal for particles of 0.1-mm diameter that settle at 1×10^{-2} m/sec?

      b.   What theoretical percent removal is expected for particles of 0.01-mm diameter that settle at 1×10^{-4} m/sec?

9-11   A 0.1-m diameter well fully penetrates an unconfined aquifer 20 m deep. The permeability is 2×10^{-3} m/sec. How much can it pump for the drawdown at the well to reach 20 m and the well to start sucking air?

9-12   The settling velocity of a particle is 0.002 m/sec and the overflow rate of a settling tank is 0.008 m/sec.

      a.   What percentage of the particles does the settling tank capture?

      b.   If the particles are flocculated so that their settling rate becomes 0.05 m/sec, what fraction of the particles is captured?

      c.   If the particles are not changed, and another settling tank is constructed to run in parallel with the original settling tank, will all the particles be captured?

9-13   A water treatment plant is being designed for a flow of 1.6 m³/sec.

      a.   How many rapid sand filters, using only sand as the medium, are needed for this plant if each filter is 10 × 20 m? What assumption do you have to make to solve this problem?

      b.   How can you reduce the number of filters?

9-14   Engineer Jaan, fresh out of school, is busily at work at a local consulting firm. As one of his first jobs, he is asked to oversee the operation of a hazardous waste cleanup job at a Superfund site. The remediation plan is to drill a series of interception wells and capture the contaminated groundwater. The depth of the wells is based on a series of borings, and it is intended that the wells reach to bedrock, an impermeable layer.

One day, Jaan is on the job and the foremen calls him over to the well being drilled.

"Strange. We were supposed to hit bedrock at 230 feet, and we are already at 270 and haven't hit anything yet. You want me to keep going?" the foreman asks Jaan.

Not knowing exactly how to respond, Jaan calls the office and talks to his immediate superior, engineer Robert.

"What do you mean we haven't hit rock?" inquires Robert. "We have all the borings to show it's at 230 feet."

"Well, the foremen says he hasn't hit anything yet," replies Jaan. "What do you want me to do?"

"We are working on a contract, and we were expecting to hit rock here. Maybe the drill has bent, or we are in some kind of seam in the rock. Whatever, we can't afford to keep on drilling. Let's just stop here, show on the drilling log that we hit rock at 270 feet. Nobody will ever know."

"We can't do that! Suppose there *is* a seam down there? The hazardous waste could seep out of the containment."

"So what. It'll be years, maybe decades before anyone will know. And besides, it probably will be so diluted by the groundwater that nobody will even be able to detect it. Just say on the drilling log that you hit rock, and move to the next site."

This is a case of falsification of data, but it could be totally harmless, and nobody will ever know. Should Jaan obey the direct order from Robert, or should he take some other action? What would happen if he simply told the foremen to keep drilling, and never told Robert that this had happened? He could mark the drilling log with the false information, and Robert would never know. Discuss the wisdom of such a course of action.

chapter 10

WASTEWATER TREATMENT

Water has many uses, including drinking, commercial navigation, recreation, fish propagation, and waste disposal (!). It is easy to forget that a major use of water is simply as a vehicle for transporting wastes. In isolated areas where water is scarce, waste disposal becomes a luxury use, and other methods of waste carriage are employed, such as pneumatic pipes or containers. But in most of the Western world, a beneficial use of water for waste transport is almost universal, and this of course results in large quantities of contaminated water.

10.1 ◆ Wastewater Transport

Wastewater is discharged from homes, commercial establishments, and industrial plants by means of *sanitary sewers*, usually large pipes flowing partially full (not under pressure). Sewers flow by gravity drainage downhill, and the system of sewers has to be so designed that the *collecting sewers*, which collect the wastewater from homes and industries, all converge to a central point where the waste flows by *trunk sewers* to the wastewater treatment plant. Sometimes it is impossible or impractical to install all gravity sewers and the waste has to be pumped by pumping stations through *force mains*, or pressurized pipes.

Design and operation of sewers is complicated by the inflow of stormwater, which is supposed to flow off in separate *storm sewers* but often seeps into the wastewater sewers through loose manhole covers and broken lines. Such an additional flow to the wastewater sewers is called *inflow*. Further, sewers often have to be installed below the groundwater table and any breaks or cracks in the sewer (such as from the roots of trees seeking water during dry spells) can result in water seeping into the sewers. This additional flow is known as *infiltration*. Local communities often spend considerable time and expense in rehabilitating sewerage systems to prevent such inflow and infiltration because every gallon that enters the sewerage system has to be treated at the wastewater treatment plant.

The wastewater, diluted by stormwater inflow or groundwater infiltration, flows downhill and eventually to the edge of the community that the sewerage system serves. In years past, this wastewater simply entered a convenient natural watercourse and was forgotten by the community. Increased population and the awareness of public health problems created by the raw sewage makes such a discharge untenable and illegal, and wastewater treatment is required.

Although water can be polluted by many materials, the most common contaminants found in domestic wastewater that cause damage to natural watercourses or create human health problems are

- organic materials, as measured by the demand for oxygen (BOD)
- nitrogen (N)
- phosphorus (P)
- suspended solids (SS)
- pathogenic organisms (as estimated by coliforms)

Treatment plants used for the treatment of municipal wastewater must respond to these contaminants and treat the wastewater to remove the objectionable characteristics. Such wastewater treatment plants can vary considerably in design, but often take a general form as shown in Figure 10-1.

The typical wastewater treatment plant is divided into five main areas:

- Preliminary treatment—removal of large solids to prevent damage to the remainder of the unit operations.

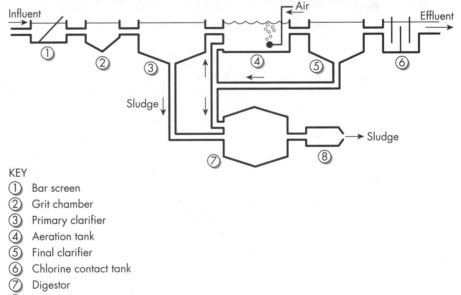

KEY
1. Bar screen
2. Grit chamber
3. Primary clarifier
4. Aeration tank
5. Final clarifier
6. Chlorine contact tank
7. Digestor
8. Dewatering

FIGURE 10–1 A typical wastewater treatment plant.

- Primary treatment—removal of solids by settling. Primary treatment systems are usually physical processes, as opposed to biological or chemical.

- Secondary treatment—removal of the demand for oxygen. These processes are commonly biological in nature.

- Tertiary treatment—a name applied to any number of polishing or clean-up processes, one of which is the removal of nutrients such as phosphorus. These processes can be physical (e.g., filters), biological (e.g., oxidation ponds), or chemical (e.g., precipitation of phosphorus).

- Solids treatment and disposal—the collection, stabilization and subsequent disposal of the solids removed by other processes.

10.2 ◆ Primary Treatment

10.2.1 Preliminary Treatment

The most objectionable aspect of discharging raw sewage into watercourses is the presence of floating material. It is only logical, therefore, that *screens* were the first form of wastewater treatment used by communities, and even today, screens are used as the first step in treatment plants. Typical screens, shown in Figure 10–2 consist of a series of steel bars, which might be about 2.5 cm (1 in.) apart. The purpose of a screen in modern treatment plants is the removal of materials that might damage equipment or hinder further treatment. In some older treatment plants screens are cleaned by hand, but mechanical cleaning equipment such as shown in Figure 10–2 is used in almost all new plants. The cleaning rakes are automatically activated when the screens become sufficiently clogged to raise the water level in front of the bars.

In many plants the next treatment step is a *comminutor*, a circular grinder designed to grind the solids coming through the screen into pieces about 0.3 cm (1/8 in.) or smaller. Many designs are in use; one common design is shown in Figure 10–3.

The third treatment step involves the removal of grit or sand. This is necessary because grit can wear out and damage such equipment as pumps and flow meters. The most common *grit chamber* is simply a wide place in the channel where the flow is slowed down sufficiently to allow the heavy grit to settle out. Sand is about 2.5 times as heavy as most organic solids and thus settles much faster than the light solids. The objective of a grit chamber is to remove sand and grit without removing the organic material. The latter must be further treated in the plant, but the sand can be dumped as fill without undue odor or other problems. One way of assuring that the light biological solids do not settle out is to aerate the grit chamber, allowing the sand and

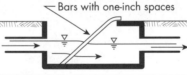

FIGURE 10-2 A typical bar screen.

other heavy particles to sink, but keeping everything else afloat. The aeration has the additional advantage of driving some oxygen into the sewage, which may have become devoid of oxygen in the sewerage system. Figure 10-4 shows such a grit chamber with a chain collection mechanism and a screw conveyor for getting the grit out of the chamber and into a trash can.

10.2.2 Settling or Clarification

Following the grit chamber most wastewater treatment plants have a *settling tank* to settle out as much of the solid matter as possible. These tanks in principle are no different from the settling tanks introduced in the previous chapter. Settling tanks are designed to operate as plug-flow reactors, and turbu-

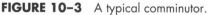

FIGURE 10-3   A typical comminutor.

lence is therefore kept to a minimum. Solids settle to the bottom and are removed while the clarified liquid escapes over a *V-notch weir*, a notched steel plate over which the water flows, promoting equal distribution of the liquid discharge all the way around a tank. Settling tanks can be circular (Figure 10-5) or rectangular (Figure 10-6).

Settling tanks are also known as *sedimentation tanks* and often as *clarifiers*. Settling tanks that follow preliminary treatment such as screening and grit removal are known as *primary clarifiers*. The solids that drop to the bottom of a primary clarifier are removed as *raw sludge*, a name that doesn't do justice to the undesirable nature of this stuff.

FIGURE 10–4 A grit chamber used in wastewater treatment.

Raw sludge is generally odoriferous, can contain pathogenic organisms, and is full of water, three characteristics that make its disposal difficult. Raw primary sludge must be both stabilized to reduce its possible public health impact and retard further decomposition, and dewatered for ease of disposal. In addition to the solids from the primary clarifier, solids from other processes must similarly be treated and disposed of. The treatment and disposal of waste-

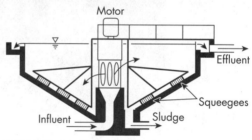

FIGURE 10-5   Circular settling tank (primary clarifier) used in wastewater treatment.

water solids (sludge) is an important part of wastewater treatment and is discussed further in a subsequent section.

Primary treatment, in addition to removing about 60% of the solids, also removes about 30% of the demand for oxygen and perhaps 20% of the phosphorus (both as a consequence of the removal of raw sludge). If this removal is adequate and the dilution in the watercourse is such that the adverse effects are acceptable, then a primary treatment plant is sufficient wastewater treatment. Governmental regulations, however, have forced all primary plants to add secondary treatment, whether needed or not.

Sometimes when primary treatment is judged to be inadequate, higher rates of solids, BOD, and phosphorus removal can be enhanced by the addition of chemicals such as aluminum sulfate (alum) or calcium hydroxide (lime) to the primary clarifier influent. With such addition, the effluent BOD can be reduced to about 50 mg/L, and this BOD level may meet required effluent

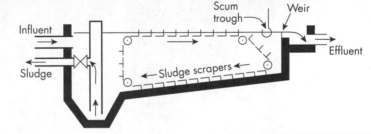

FIGURE 10-6 Rectangular settling tank (primary clarifier) used in wastewater treatment.

standards. Chemical addition to primary treatment is especially attractive for large coastal cities that can achieve high dilution in the dispersion of the plant effluent.

In a more typical wastewater treatment plant, primary treatment without chemical addition is followed by secondary treatment, designed specifically to remove the demand for oxygen.

10.3 ◆ Secondary Treatment

The water leaving the primary clarifier has lost much of the suspended organic matter but still contains a high demand for oxygen due to the dissolved organics: that is, it is composed of high-energy molecules that will decompose by microbial action, creating a biochemical oxygen demand (BOD). This demand for oxygen must be reduced (energy wasted) if the discharge is not to create unacceptable conditions in the watercourse. The objective of secondary treatment is to remove BOD while, by contrast, the objective of primary treatment is to remove solids. Except in rare circumstances, almost all secondary treatment methods use microbial action to reduce the energy level (BOD) of the waste. The basic differences among all these alternatives are how the waste is brought into contact with the microorganisms.

10.3.1   Fixed Film Reactors

Although there are many ways the microorganisms can be put to work, the first really successful modern method of secondary treatment was the *trickling filter*. The trickling filter, shown in Figure 10-7, consists of a filter bed of fist-size rocks over which the waste is trickled. An active biological growth forms

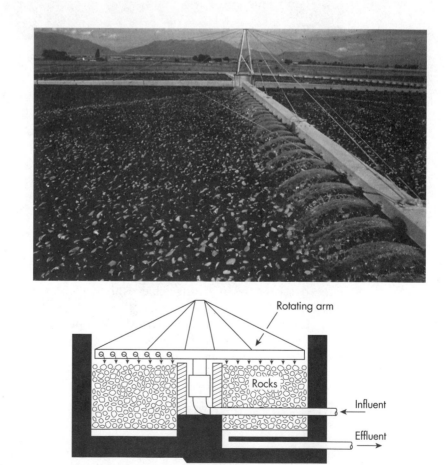

FIGURE 10-7   A trickling filter.

on the rocks, and the organisms obtain their food from the waste stream dripping over the bed of rocks. Air is either forced through the rocks or, more commonly, air circulation is obtained automatically by a temperature difference between the air in the bed and ambient temperature. In the older filters the waste is sprayed onto the rocks from fixed nozzles, while the newer designs use a rotating arm that moves under its own power, like a lawn sprinkler, distributing the waste evenly over the entire bed. Often the flow is recirculated, obtaining a higher degree of treatment. The name "trickling filter" is obviously a misnomer since no filtration takes place.

A modern variation of the trickling filter is filling the tank up with rubberized corrugated blocks that provide great surface area while taking up very little volume (as do the rocks). Often old rock–media trickling filters are upgraded by building up the sides of the trickling filter to a height of perhaps 8 meters and filling it with corrugated blocks. Such devices are able to achieve more than 90% BOD removal and represent a fairly inexpensive upgrade for older treatment plants.

Another modern modification of the trickling filter is the *rotating biological contactor*, or rotating disk, pictured in Figure 10-8. The microbial growth occurs on rotating disks, which are slowly dipped into the wastewater. When the disks are brought out into the open air, the microbes are able to obtain the necessary oxygen to keep the growth aerobic.

10.3.2 Suspended Growth Reactors

Around the turn of the nineteenth century, when trickling filtration was already firmly established, some researchers began musing about the wasted space in a filter taken up by the rocks. Could the microorganisms not be allowed to float free and could they not be fed oxygen by bubbling in air? Although this concept was quite attractive, it was not until 1914 that the first workable pilot plant was constructed. It took some time before this process became established as what we now call the *activated sludge system*.

The key to the activated sludge system is the reuse of microorganisms. The system, illustrated in Figure 10-9, consists of a tank full of waste liquid (from the primary clarifier) and a mass of microorganisms. Air is bubbled into this tank (called the *aeration tank*) to provide the necessary oxygen for the survival of the aerobic organisms. The microorganisms come into contact with the dissolved organics and rapidly adsorb these organics on their surface. In time, the microorganisms use the energy and carbon by decomposing this material to CO_2, H_2O, some stable compounds, and in the process produce more microorganisms. The production of new organisms is relatively slow, and most of the aeration tank volume is used for this purpose.

Once most of the food has been used up, the microorganisms are separated from the liquid in a settling tank, sometimes called a *secondary* or *final clarifier*. The liquid escapes over a V-notch weir and can be discharged into the receiving watercourse (the recipient).

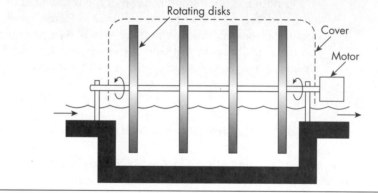

FIGURE 10-8   Rotating disk fixed film biological reactor.

The separated microorganisms exist on the bottom of the final clarifier without additional food and become hungry, waiting for more dissolved organic matter for food. These microorganisms are "activated"; hence the term *activated sludge*.

When these settled and hungry microorganisms are pumped to the head of the aeration tank, they find more food (organics in the effluent from the primary clarifier) and the process starts all over again. The sludge pumped from the bottom of the final clarifier to the aeration tank is known as *return activated sludge*.

The activated sludge process is a continuous operation, with continuous sludge pumping and clean water discharge. Unfortunately, one of the end products of this process is excess microorganisms. If none of the microorganisms is removed, their concentration eventually increases to the point where the system is clogged with solids. Some of the microorganisms must therefore

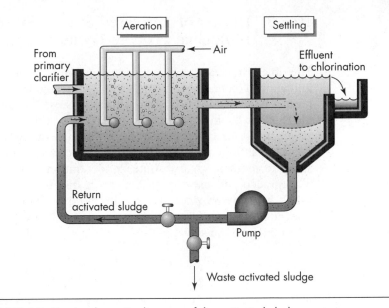

FIGURE 10–9 A schematic diagram of the activated sludge system.

be wasted and this *waste activated sludge* must be processed and disposed of. Its disposal is one of the most difficult aspects of wastewater treatment.

Activated sludge systems are designed on the basis of loading, or the amount of organic matter (food) added relative to the microorganisms available. This ratio is known as the food-to-microorganisms ratio (F/M) and is a major design parameter. Unfortunately neither F nor M is easy to measure accurately, and engineers have approximated these by BOD and suspended solids in the aeration tank, respectively. That is, the food is indirectly measured as the BOD, and the microorganisms are assumed to the suspended solids.

The combination of the liquid and microorganisms undergoing aeration is known (for some unknown reason) as *mixed liquor*, and the suspended solids are called *mixed liquor suspended solids* (MLSS). The ratio of incoming BOD to MLSS, the F/M ratio, is also known as the *loading* on the system and is calculated as pounds of BOD/day per pound of MLSS in the aeration tank.

If this ratio is low (little food for a lot of microorganisms) and the aeration period (retention time in the aeration tank) is long, the microorganisms make maximum use of available food, resulting in a high degree of treatment. Such systems are known as *extended aeration* and are widely used for isolated sources (e.g., motels, small developments). An added advantage of extended aeration is that the ecology within the aeration tank is quite diverse and little excess biomass is created, resulting in little or no waste activated sludge to be disposed of—a significant saving in operating costs and headaches. At the other extreme is the *high-rate* system where the aeration periods are very short (thus saving money by building smaller tanks) and the treatment efficiency is lower.

10.3.3   Design of Activated Sludge Systems Using Biological Process Dynamics

The objective of an activated sludge system is to degrade the organics in the influent and oxidize them to CO_2 and H_2O, recognizing that some of this energy must also be used to build new microorganisms. These influent organics provide the food for the microorganisms and in biological process dynamics are known as *substrate*. As noted previously, substrate is usually measured indirectly as the BOD, although methods such as organic carbon, for example, may be more accurate measures of substrate concentration. The substrate (S) or food is biodegraded and used by the microorganisms (X), expressed as suspended solids, at a rate dS/dt. As the food is used up, new organisms are produced. The rate of new cell mass (microorganisms) production as a result of the destruction of the substrate is

$$\frac{dX}{dt} = Y\frac{dS}{dt} \qquad (10.1)$$

where Y = the yield, or mass of microorganisms produced per mass of substrate used, commonly expressed as kg SS produced per kg BOD used, or

$$Y = \frac{dX}{dS} \qquad (10.2)$$

The expression for substrate utilization commonly employed

$$\frac{S}{t} = \frac{X}{Y}(\mu) = \frac{X}{Y}\left(\frac{\hat{\mu}S}{K_S + S}\right) \qquad (10.3)$$

is the Monod model. (The argument for the model validity is beyond the scope of this text. Suffice it to say that the model is empirical, but reasonable. If you are interested in the development of this model, see any number of modern textbooks on wastewater processing.)

$$\frac{dS}{dt} = \frac{dX}{dt}\left(\frac{1}{Y}\right) \qquad \mu = \frac{\hat{\mu}S}{K_S + S} \qquad (10.4)$$

where μ = the specific growth rate, in terms of days^{-1}, and equal to $(dX/dt)(1/X)$
  $\hat{\mu}$ = maximum growth rate constant, days^{-1}
  K_S = saturation constant, mg/L

This expression is an empirical model, based on experimental work with pure cultures. The two constants, $\hat{\mu}$ and K_S, must be evaluated for each substrate and microorganism culture, but remain constant for a given system as S and X are varied. They are a function of the substrate and microbial mass, not of the reactor.

The application of biological process dynamics to the activated sludge process is best illustrated by considering a system shown in Figure 10-9. This is a simple continuous biological reactor, of volume V and a flow rate of Q.

The reactor is completely mixed. Recall that this means that the influent is dispersed within the tank immediately upon introduction; thus there are no concentration gradients in the tank and the quality of the effluent is exactly the same as that of the tank contents.

In such a continuous reactor, it is possible to develop two types of materials balances: in terms of solids (microorganisms) and BOD (substrate). There are also two retention times: liquid and solids. The liquid, or hydraulic, retention time is introduced earlier and is expressed as

$$\bar{t} = \frac{V}{Q}$$

where $\bar{t}$ = hydraulic retention time, min
$\quad V$ = volume, m³
$\quad Q$ = flow rate to the reactor, m³/min

Recall that $\bar{t}$ can also be defined as the average time the liquid remains in the reactor.

The *solids retention time* is analogous to the hydraulic retention time and represents the average time *solids* stay in the reactor. The solids retention time is also known as *sludge age*, or the average time microorganisms stay in the system. Sludge age or solids retention time is known also by a third name, the *mean cell residence time*, a term used in biotechnology research. All three names represent the same parameter, which can be calculated as

$$\theta_c = \frac{\text{mass of solids (microorganisms) in the system}}{\text{mass of solids wasted/time}} = \text{time}$$

where θ_c = sludge age = mean cell residence time = solids retention time.

Consider first a simple completely mixed-flow reactor pictured in Figure 10–10. This is not an activated sludge system at all, since there is no solids recycle, but serves to introduce the notation and terminology. The amount of solids in this reactor is expressed as VX (volume × solids concentration) and the solids wasted is equal to QX (flow rate × solids concentration). Thus the mean cell residence time is

$$\theta_C = \frac{VX}{QX} = \frac{V}{Q} \tag{10.5}$$

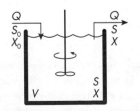

FIGURE 10–10 A suspended growth reactor with no recycle.

The amount of solids wasted must also be equal to the rate at which they are produced, or dX/dt. Substituting for equation 10.1,

$$\theta_C = \frac{XV}{Y(dS/dt)V} = \frac{X}{Y(dS/dt)} \tag{10.6}$$

Note that the concentration terms have to be multiplied by volume to obtain mass.

Remember that equations 10.5 and 10.6 are for a reactor *without recycle* (Figure 10-10). Note that θ_C in this case is also equal to the hydraulic retention time. The units of θ_C commonly are expressed in days.

For the system in Figure 10-10, we now assume that there are no microorganisms in the influent ($X_0 = 0$) and that the reactor has attained steady state (the growth rate of microorganisms is balanced by the loss rate of microorganisms in the effluent).

A mass balance in terms of the microorganisms, using the same equation as before, is:

$$\begin{bmatrix} \text{rate of} \\ \text{ACCUMULATION} \end{bmatrix} = \begin{bmatrix} \text{rate} \\ \text{IN} \end{bmatrix} - \begin{bmatrix} \text{rate} \\ \text{OUT} \end{bmatrix}$$

$$+ \begin{bmatrix} \text{rate of} \\ \text{microorganism} \\ \text{GROWTH} \end{bmatrix} - \begin{bmatrix} \text{rate of} \\ \text{microorganism} \\ \text{DEATH} \end{bmatrix}$$

If the growth and death rates are combined as *net* growth, this equation reads

$$V\frac{dX}{dt} = QX_0 - QX + Y(dS/dt)V \tag{10.7}$$

The last term comes from equation 10.1. In a steady-state system, $(dX/dt)V = 0$, and since there are no cells in the inflow, $QX_0 = 0$. Substituting the Monod substrate utilization model (equation 10.3),

$$\frac{dS}{dt} = \frac{X}{Y}\left(\frac{\hat{\mu}S}{K_S + S}\right)$$

into the previous relationship and introducing the mean cell residence time, θ_C.

$$\frac{1}{\theta_C} = \frac{\hat{\mu}S}{K_S + S} \tag{10.8}$$

and

$$S = \frac{K_S}{\hat{\mu}\theta_C - 1} \tag{10.9}$$

This is an important expression since we can infer that the substrate concentration, S, is a function of the kinetic constants (which are beyond our control for a given substrate) and the mean cell residence time. The value of S, which in real life would be the effluent BOD, is influenced then by the mean cell residence time (or the sludge age as previously defined). Since there is no recycle, the only way that the sludge age can be increased is by increasing the concentration of microorganisms in the reactor. The more critters, the better the treatment.

---

e • x • a • m • p • l • e **10.1**

Problem A biological reactor such as the one pictured in Figure 10-10 must be operated so that an influent BOD of 600 mg/L is reduced to 10 mg/L. The kinetic constants are $K_S = 500$ mg/L and $\hat{\mu} = 4$ days^{-1}. If the flow is 3 m³/day, how large should the reactor be?

Solution Using equation 10.9;

$$S = \frac{K_S}{\hat{\mu}\theta_C - 1}$$

$$\theta_C = \frac{K_S + S}{S\mu} = \frac{500 + 10}{10(4)} = 12.75 \text{ days}$$

From equation 10.5

$$\theta_C = \frac{V}{Q}$$

$$V = \theta_C Q = (12.75)(3) = 38.25 \text{ m}^3$$

Remember that the substrate concentration S in the effluent is exactly the same as S in the reactor if the reactor is assumed to be perfectly mixed.

---

e • x • a • m • p • l • e **10.2**

Problem Given the conditions in Example 10.1, suppose the only reactor available has a volume of 24 m³. What would be the percent reduction in substrate (substrate removal efficiency)?

Solution

$$\theta_C = \frac{V}{Q} = \frac{24}{3} = 8 \text{ days}$$

$$S = \frac{K_S}{\mu\theta_C - 1} = \frac{500}{4(8) - 1} = 16.1 \text{ mg/L}$$

Using equation 3.4

$$\text{recovery} = (600 - 16.1)/600 \times 100 = 97.3\%$$

The system pictured in Figure 10–10 is not very efficient, however, since long hydraulic residence times are necessary to prevent the microorganisms from being flushed out of the tank. Their growth rate has to be faster than the rate of being flushed out, or the system will fail. The success of the activated sludge system for wastewater treatment is based on a significant modification: the recycle of the microorganisms. Such a system is shown in Figure 10–11.

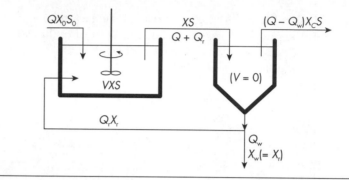

FIGURE 10–11   A suspended growth reactor with recycle (the activated sludge system).

Some simplifying assumptions are necessary before this model can be used. First, again assume the $X_0 = 0$, and further that the microorganism separator (the final settling tank or clarifier) is a perfect device, so that there are no microorganisms in the effluent ($X_c = 0$). Also once again assume steady-state conditions and perfect mixing. The excess microorganisms (waste activated sludge) are removed from the system at flow rate Q_w and a solids concentration X_r, which is the settler underflow concentration and the concentration of the solids being recycled to the aeration tank. And last, assume that there is no substrate removal in the settling tank and that the settling tank has no volume, so that all the microorganisms in the system are in the reactor (aeration tank). The volume of the aeration tank is thus the only active volume, and the settling (microorganism separation) is assumed to take place magically in a zero-volume tank, an obviously incorrect assumption. Note that the mean cell residence time in this case is

$$\theta_C = \frac{\text{microorganisms in the system}}{\text{microorganisms wasted/time}}$$

$$= \frac{XV}{Q_w X_r + (Q - Q_w)X_C} \tag{10.10}$$

and since $X_C = 0$

$$\theta_C = \frac{XV}{Q_w X_r} \tag{10.11}$$

The removal of substrate is often expressed in terms of a *substrate removal velocity* (q) and defined as

$$q = \frac{\text{mass substrate removed/time}}{\text{mass microorganisms under aeration}}$$

Using the previous notation,

$$q = \frac{\left(\dfrac{(S_0 - S)}{\bar{t}}\right) V}{XV}$$

$$q = \frac{S_0 - S}{X\bar{t}} \qquad (10.12)$$

The substrate removal velocity is a rational measure of the substrate removal activity, or the mass of BOD removed in a given time per mass of microorganisms doing the work. This is also called, in some texts, the *process loading factor*, with the clear implication that it is a useful operation and design tool since, as you recall, the basic design parameter for an activated sludge system is its loading.

The substrate removal velocity can be derived by conducting a mass balance in terms of substrate on a continuous system with microorganism recycle (Figure 10-11).

$$\left[\begin{array}{c} \text{rate of} \\ \text{ACCUMULATION} \end{array}\right] = [\text{rate IN}] - [\text{rate OUT}]$$

$$+ \left[\begin{array}{c} \text{rate of} \\ \text{substrate} \\ \text{PRODUCTION} \end{array}\right] - \left[\begin{array}{c} \text{rate of} \\ \text{substrate} \\ \text{CONSUMPTION} \end{array}\right]$$

The production term is of course zero, and the equation reads

$$\frac{dS}{dt} V = QS_0 - QS + 0 - qSV$$

The rate of substrate utilization is

$$qXV = \left[\begin{array}{c} \text{mass substrate} \\ \text{removed/time} \\ \hline \text{mass microorganisms} \end{array}\right] \times \left[\begin{array}{c} \text{mass} \\ \text{microorganisms} \\ \hline \text{reactor volume} \end{array}\right] \times \left[\begin{array}{c} \text{reactor} \\ \text{volume} \end{array}\right]$$

Assuming steady-state conditions, that is, $(dS/dt)V = 0$, and solving for q,

$$q = \frac{S_0 - S}{X\bar{t}}$$

The Monod growth rate (μ) is defined previously (equation 10.3) by

$$\frac{dS}{dt} = \frac{\hat{\mu}SX}{Y(K_s + S)} = \frac{X}{Y}(\mu)$$

$$\mu = \frac{dX}{dt}\frac{1}{X} = \frac{\text{mass microorganisms produced}}{(\text{time}) \times (\text{mass microorganisms})}$$

and the yield (Y) is (equation 10.2)

$$Y = \frac{dX}{dS} = \frac{\text{mass microorganisms produced}}{\text{mass substrate removed}}$$

Using now a mass balance in terms of the microorganisms:

$$\begin{bmatrix} \text{rate} \\ \text{ACCUMULATED} \end{bmatrix} = \begin{bmatrix} \text{rate} \\ \text{IN} \end{bmatrix} - \begin{bmatrix} \text{rate} \\ \text{OUT} \end{bmatrix}$$

$$+ \begin{bmatrix} \text{rate} \\ \text{microorganisms} \\ \text{PRODUCED} \end{bmatrix} - \begin{bmatrix} \text{rate} \\ \text{microorganisms} \\ \text{CONSUMED} \end{bmatrix}$$

Again, the last two terms are combined, and the equation reads

$$\frac{dX}{dt} V = QX_0 - Q_w X_r - (Q - Q_w)X_C + \mu XV$$

Assuming again steady state ($[dX/dt] = 0$) and $X_0 = X_C = 0$,

$$\mu = \frac{X_r Q_w}{XV}$$

Note that this is the reciprocal of previously defined sludge age, or mean cell residence time, so that

$$\theta_C = \frac{XV}{X_r Q_w} = \frac{1}{\mu} \tag{10.13}$$

The substrate removal velocity can then be expressed as

$$q = \frac{\mu}{Y} = \begin{bmatrix} \dfrac{\text{mass microorganisms}}{\text{produced/time}} \\ \dfrac{}{\text{mass microorganisms}} \\ \text{in the reactor} \end{bmatrix} \times \begin{bmatrix} \dfrac{\text{mass substrate}}{\text{removed}} \\ \dfrac{}{\text{mass microorganisms}} \\ \text{produced} \end{bmatrix} \tag{10.14}$$

and substituting

$$q = \frac{\hat{\mu} S}{Y(K_s + S)} \tag{10.15}$$

The substrate removal velocity is also defined as

$$q = \frac{(S_0 - S)}{X\bar{t}}$$

Equating these two expressions and solving for ($S_0 - S$), the substrate removal (reduction in BOD) is

$$S_0 - S = \frac{\hat{\mu} SX\bar{t}}{Y(K_s + S)} \tag{10.16}$$

Since the substrate removal velocity q is defined as

$$q = \frac{\mu}{Y}$$

the mean cell residence time (sludge age) is

$$\theta_c = \frac{1}{qY} \tag{10.17}$$

and the concentration of microorganisms in the reactor (mixed liquor suspended solids) is

$$X = \frac{S_0 - S}{\bar{t}q} \tag{10.18}$$

e • x • a • m • p • l • e   **10.3**

Problem   An activated sludge system operates at a flow rate (Q) of 400 m³/day, with an incoming BOD (S_0) of 300 mg/L. Through pilot plant work, the kinetic constants for this system are determined to be $Y = 0.5$ kg SS/kg BOD, $K_s = 200$ mg/L, $\hat{\mu} = 2$ day^{-1}. A solids concentration of 4000 mg/L in the aeration tank is considered appropriate. A treatment system must be designed that will produce an effluent BOD of 30 mg/L (90% removal). Determine:

   a. the volume of the aeration tank

   b. the sludge age

   c. the quantity of sludge wasted daily

Solution   The mixed liquor suspended solids concentration is usually limited by the ability to keep an aeration tank mixed and to transfer sufficient oxygen to the microorganisms. A reasonable value for the solids under aeration would be $X = 4000$ mg/L as stated in the problem. The hydraulic retention time is then obtained from equation 10.16.

$$S_0 - S = \frac{\hat{\mu} S X \bar{t}}{Y(K_s + S)}$$

Rearranged,

$$\bar{t} = \frac{Y(S_0 - S)(K_s + S)}{\hat{\mu} S X} = \frac{0.5(300 - 30)(200 + 30)}{2(30)(4000)} = 0.129 \text{ days} = 3.1 \text{ hr}$$

The volume of the tank is then $V = \bar{t}Q = 400(0.129) = 51.6$ m³. The sludge age is (equation 10.17)

$$\theta_c = \frac{1}{qY}$$

where, according to equation 10.12,

$$q = \frac{S_0 - S}{X\bar{t}} = \frac{300 - 30}{(4000)(0.129)} = 0.523 \frac{\text{kg BOD removed/day}}{\text{kg SS in the reactor}}$$

or equivalently, using equation 10.15,

$$q = \frac{\hat{\mu}S}{Y(K_s + S)} = \frac{2(30)}{0.5(200 + 30)} = 0.522 \text{ day}^{-1}$$

and

$$\theta_C = \frac{1}{qY} = \frac{1}{0.522(0.5)} = 3.8 \text{ days}$$

Using equation 10.13,

$$\theta_C = \frac{XV}{X_r Q_w}$$

$$X_r Q_w = \frac{XV}{\theta_C} = \frac{4000(51.6)(10^3 \text{ L/m}^3)(1/10^6 \text{ kg/mg})}{3.8} = 54.3 \text{ kg/day}$$

◆

e · x · a · m · p · l · e   10.4

Problem   Using the same data as in the previous example, what mixed liquor solids concentration is necessary to attain a 95% BOD removal (i.e., $S = 15$ mg/L)?

Solution   The substrate removal velocity is

$$q = \frac{\hat{\mu}S}{Y(K_s + S)} = \frac{2(15)}{0.5(200 + 15)} = 0.28 \text{ day}^{-1}$$

and

$$X = \frac{S_0 - S}{\bar{t}q} = \frac{300 - 15}{0.129(0.28)} = 7890 \text{ mg/L}$$

The sludge age would be

$$\theta_C = \frac{1}{qY} = \frac{1}{0.28(0.5)} = 7.2 \text{ days}$$

◆

Note that more microorganisms are required in the aeration tank if higher removal efficiencies are to be attained. This, however, depends on the efficiency of settling in the final clarifier. If the sludge does not settle well, the return sludge solids concentration is low and there is no way to increase the solids concentration in the aeration tank.

Results from settling the sludge in a liter cylinder can be used to estimate the return sludge pumping rates. After 30 minutes of settling, the solids in the cylinder are at a SS concentration that would be equal to the expected return sludge solids, or

$$X_r = \frac{H}{h}(X)$$

where X_r — the expected return suspended solids concentration, mg/L
$\quad X$ = mixed liquor suspended solids, mg/L
$\quad H$ = height of cylinder, m
$\quad h$ = height of settled sludge, m

The mixed liquor solids concentration is of course a combination of the return solid diluted by the influent, or

$$X = \frac{Q_r X_r + Q X_0}{Q_r + Q}$$

Again assume no solids in the influent ($X_0 = 0$),

$$X = \frac{Q_r X_r}{Q_r + Q}$$

The success or failure of an activated sludge system often depends on the performance of the final clarifier. If this settling tank is not able to achieve the required return sludge solids, the MLSS will drop, and of course the treatment efficiency will be reduced.

The microorganisms in the system are sometimes very difficult to settle out, and the sludge is said to be a *bulking sludge*. Often this condition is characterized by a biomass comprised almost totally of filamentous organisms, which form a kind of lattice structure with the filaments and refuse to settle. Treatment plant operators must keep a close watch on settling characteristics because a trend toward poor settling can be the forerunner of a badly upset (and hence ineffective) plant. When the sludge does not settle, the return activated sludge becomes thin (low suspended solids concentration) and thus the concentration of microorganisms in the aeration tank drops. This results in a higher *F/M* ratio (same food input, but fewer microorganisms) and a reduced BOD removal efficiency.

The two principal means of introducing sufficient oxygen into the aeration tank are by bubbling compressed air through porous diffusers (Figure 10-12) and by beating air in mechanically (Figure 10-13). In both cases, the intent is to transfer one gas (oxygen) from the air into the liquid, and simultaneously transfer another gas (carbon dioxide) out of the liquid. These processes are commonly called *gas transfer*.

Air header

Diffuser

FIGURE 10-12   Diffused aeration used in the activated sludge system.

10.3.4   Gas Transfer

Gas transfer means simply the process of allowing any gas to dissolve in a fluid, or the opposite of that, promoting the release of a dissolved gas from a fluid. One of the critical aspects of the activated sludge system is the supply of oxygen to a suspension of microorganisms in a wastewater treatment plant.

Figure 10-14 shows a system where air is forced through a tube and a porous diffuser, creating very small bubbles that rise through clean water.

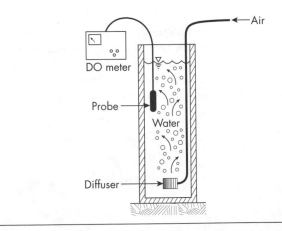

FIGURE 10-13 Mechanical aeration used in the activated sludge system.

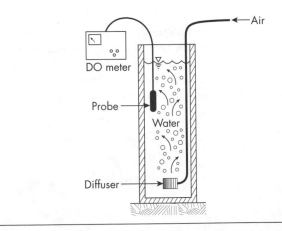

FIGURE 10-14 Gas transfer experiment where air is bubbled into the tank through a diffuser and the dissolved oxygen is measured using a probe and dissolved oxygen meter.

Assume that the water in this system does not contain microorganisms, and nothing uses up the dissolved oxygen. Of interest here is only how the aeration system performs. What happens to the oxygen once it is dissolved in the water is not important at this time.

The transfer of oxygen takes place through the bubble gas–liquid interface, as shown in Figure 10–15. If the gas inside the bubble is air, and an oxygen deficit exists in the water, the oxygen transfers from the bubble into the water. In some cases, depending on the concentration of gases already dissolved in the water, transfer of gases such as CO_2 from solution and into the bubble can also occur. Before discussing gas transfer further, however, it is necessary to briefly review some concepts of gas solubility.

Most gases are only slightly soluble in water; among these are hydrogen, oxygen, and nitrogen. Other gases are very soluble, including sulfur dioxide (SO_2) chlorine (Cl_2), and carbon dioxide (CO_2). Most of these readily soluble gases dissolve and then ionize in the water. For example, when CO_2 dissolves in water, the following reactions occur:

$$CO_2 \text{ (gas)} \leftrightarrows CO_2 \text{ (dissolved)} + H_2O \leftrightarrows H_2CO_3$$

$$H_2CO_3 \leftrightarrows H^+ + HCO_3^- \leftrightarrows H^+ + CO_3^{-2}$$

All these reactions are in equilibrium, so that the more CO_2 is bubbled in, the more CO_2 is dissolved, and eventually the more carbonate ion (CO_3^{-2}) is produced. (The equations are driven to the right.) At any given pH and temperature, the quantity of a given gas dissolved in a liquid is governed by *Henry's law*, which states that

$$S = KP$$

where S = solubility of a gas (the maximum amount that can be dissolved), mg gas/L water

P = partial pressure of the gas, as measured in pounds per square inch (psi), kilo pascals (kPa), atmospheres, or other pressure terms

K = solubility constant

Note that this is different from the solubility *coefficient*. The solubility constant defines the equilibrium condition for various species at a constant pressure.

Henry's law states that the solubility is a direct function of the partial pressure of the gas being considered. In other words, if the partial pressure P is doubled, the solubility of the gas S is likewise doubled, and so on.

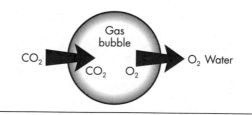

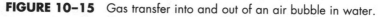

FIGURE 10–15   Gas transfer into and out of an air bubble in water.

Solubility is influenced by many variables such as the presence of impurities and the temperature. The effect of temperature on the solubility of oxygen in water is tabulated in Table 7-2 (page 186).

The partial pressure is defined as the pressure of the gases over the fluid interface exerted by the gas of interest. For example, at atmospheric pressure (1 atmosphere = 101 KPa), a gas that is 60% O_2 and 40% N_2 has a partial pressure of oxygen $0.60 \times 101 = 60.6$ KPa, and a partial pressure of nitrogen of $0.4 \times 101 = 40.4$ KPa. The total pressure is always the sum of the partial pressures of the individual gases. This is known as *Dalton's law*.

e • x • a • m • p • l • e   **10.5**

Problem   At one atmosphere, the solubility of pure oxygen in water is 46 mg/L if the water contains no dissolved solids. What would be the quantity of dissolved oxygen in pure water if the gas above the water is oxygen (Figure 10-16A)? What would be solubility if the oxygen is replaced by nitrogen (Figure 10-16B) and then by air (Figure 10-16C)? Assume that in all three cases the total pressure of the gases above the water is one atmosphere and the temperature is 20°C.

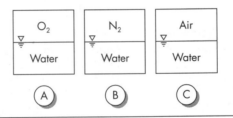

FIGURE 10-16   Three gases at atmospheric pressure above water. See Example 10.5.

Solution   Henry's law reads

$$S = KP$$

$$46 = K \times 1 \quad \text{or} \quad K = 46 \text{ mg/L-atm}$$

The water in Figure 10-16A therefore would contain 46 mg//L of dissolved oxygen since the entire pressure is due to oxygen. When the oxygen is replaced by nitrogen, the partial pressure of oxygen is zero, and $S = 0$. In the third case, since air is 20% oxygen,

$$S = KP = (46)(1/5) = 9.2 \text{ mg/L}$$

Referring to Table 7-2, at 20°C, the solubility of oxygen from air into pure water is about 9.1 mg/L. Close enough.

Henry's law defines the solubility of gas at equilibrium. That is, it is assumed that the gas in contact with the water has had sufficient time to come to a dissolved-gas concentration which, over time, will not change. Consider next the changes in dissolved-gas concentrations with time.

Using oxygen in air as an example of a gas and water as the liquid, the dissolved-oxygen concentration at any time may be visualized as in Figure 10–17. Above the water surface the air is well mixed so that there are no concentration gradients in the air. If the system is allowed to come to equilibrium, the concentration of dissolved oxygen in the water will eventually attain saturation, S. Before equilibrium occurs, however, at some time t, the concentration of dissolved oxygen in the water is C, some value less than S. The difference between the saturation value S and the concentration C is the deficit D, so that $D = (S - C)$. As time passes, the value of C increases until it meets S, producing a saturated solution and reducing the deficit D to zero. Below the air–water interface, it is visualized that there exists a diffusion layer through which the oxygen has to pass. The concentration of the gas decreases with depth until the concentration C is reached, and assume that the water is well mixed so that except for the thin layer at the air interface, the concentration of dissolved oxygen in the water is everywhere C mg/L. At the interface, the concentration increases at a rate of

$$\frac{dC}{dx}$$

where $x =$ thickness of diffusion layer. If this slope is large (dC is big compared to dx), the rate at which oxygen is driven into the water is large. Conversely, when dC/dx is small, C approaches S and the rate of change is small. This idea can be expressed as

$$\frac{dC}{dx} \propto (S - C)$$

Note that as $(S - C)$ approaches zero, $dC/dx \rightarrow 0$, and when $(S - C)$ is large, dC/dx is large. It can also be argued that the rate of change in concentration with time must be large when the slope is large (i.e., dC/dx is large), and

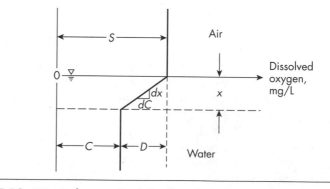

FIGURE 10–17   Definition sketch for describing gas transfer.

conversely, as dC/dx approaches zero, the rate of change should also approach zero. The proportionality is

$$\frac{dC}{dt} \propto (S - C)$$

The proportionality constant is symbolically written as K_La and given the name *gas transfer coefficient*, and the equation is written as

$$\frac{dC}{dt} = K_La(S - C)$$

where K_La = gas transfer coefficient.

Note that the rate at which the oxygen is driven into the water is high when the driving force, $(S - C)$, is high, and conversely, as C approaches S, the rate decreases. This driving force can be expressed equally well by using the deficit, or the difference between the concentration and saturation (how much oxygen could still be driven into the water). Written in terms of the deficit,

$$\frac{dD}{dt} = -K_LaD$$

Since the deficit is *decreasing* with time as aeration occurs, K_La is negative.

This equation can be integrated to yield:

$$\ln\frac{D}{D_0} = -K_Lat$$

where D_0 is the initial deficit.

K_Las are measured using aeration tests as illustrated in Figure 10–14. The water is first stripped of oxygen, usually by chemical means, with C approaching zero. The air is then turned on and the dissolved oxygen concentration measured with time using a dissolved oxygen meter. If, for example, different types of diffusers are to be tested, the K_La is calculated for all types, using identical test conditions. The higher K_La implies that the diffuser is more effective in driving oxygen into the water, and presumably the least expensive to operate in a wastewater treatment plant.

K_La is a function of, among other factors, type of aerator, temperature, size of bubbles, volume of water, path taken by the bubbles, and presence of surface active agents. The explanation of how K_La is influenced by these variables is beyond the scope of this text.

e • x • a • m • p • l • e **10.6**

Problem   Two diffusers are to be tested for their oxygen-transfer capability. Tests were conducted at 20°C using a system as shown in Figure 10–18, with the following results:

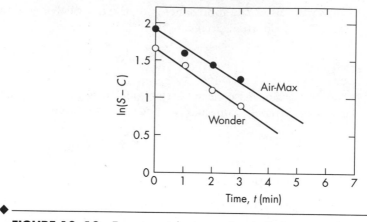

FIGURE 10–18   Experimental gas transfer data. See Example 10.6.

	Dissolved Oxygen, C (mg/L)	
Time (min)	Air-Max Diffuser	Wonder Diffuser
0	2.0	3.5
1	4.0	4.8
2	4.8	6.0
3	5.7	6.7

Note that the test does not need to start at $C = 0$ at $t = 0$.

Solution   With $S = 9.2$ mg/L (saturation at 20°C),

	Air-Max Diffuser	Wonder Diffuser
t	$(S - C)$	$(S - C)$
0	7.2	5.7
1	5.2	4.4
2	4.4	3.2
3	3.5	2.5

These numbers are now plotted by first calculating $\ln(S - C)$ and plotting against the time t (or using semi-log paper, plotting $(S - C)$ vs. t). The slope of such a plot is the proportionality factor, or in this case, the gas transfer coefficient $K_L a$. Calculating the slopes, the $K_L a$ for the Air-Max diffuser is found to be 2.37 min^{-1}, while the Wonder diffuser has a $K_L a$ of 2.69 min^{-1}. The latter seems to be the better diffuser based on oxygen-transfer capability.

In Chapter 7, the transfer of oxygen to streamwater is described by the equation

$$\frac{dD}{dt} = -k_2 D$$

where k_2 is called the *reaeration constant* and is identical to $K_L a$ as defined previously. The reaeration constant is used to describe how oxygen is transferred from the atmosphere into a stream to provide sufficient dissolved oxygen for the aerobic aquatic microorganisms. In this chapter, the same mechanism is used to describe how oxygen is driven into the liquid in an activated sludge aeration tank.

10.3.5 Solids Separation

An activated sludge system does not work if the sludge in the final clarifier does not settle. Treatment plant operators must keep a close watch on the sludge settling characteristics because a trend toward poor settling can be the forerunner of a badly upset (and hence ineffective) plant. The settlability of activated sludge is most often described by the sludge volume index (SVI), which is determined by measuring the milliliters of volume occupied by a sludge after settling for 30 minutes in a 1-L cylinder, and calculated as

$$SVI = \frac{(\text{volume of sludge after 30 min settling, mL}) \times 1000}{\text{mg/L suspended solids}}$$

e • x • a • m • p • l • e **10.7**

Problem   A sample of mixed liquor is found to have SS = 4000 mg/L. After settling for 30 minutes in a 1-L cylinder the sludge occupied 400 mL. Calculate the SVI.

Solution

$$SVI = (400 \times 1000)/4000 = 100$$

◆

SVI values less than 100 are usually considered acceptable, while sludges with SVI greater than 200 are badly bulking sludges and will be difficult to settle in the final clarifier. By convention, SVI has no units.

The causes of poor settling (high SVI) are not always known, and hence the solutions are elusive. Wrong or variable F/M ratios, fluctuations in temperature, high concentrations of heavy metals, and deficiencies in nutrients in the incoming wastewater have all been blamed for bulking. Cures include reducing the F/M ratio, changing the dissolved oxygen level in the aeration tank, and dosing with hydrogen peroxide (H_2O_2) to kill the filamentous microorganisms.

When the sludge does not settle, the return activated sludge becomes thin (low suspended-solids concentration) and the concentration of microorganisms in the aeration tank drops. This results in a higher *F/M* ratio (same food input, but fewer microorganisms) and a reduced BOD removal efficiency.

Effluent from secondary treatment often has a BOD of about 15 mg/L and a suspended solids concentration of about 20 mg/L. This is most often quite adequate for disposal into watercourses, since the BOD of natural streamwater can vary considerably, from about 2 mg/L to far greater than 15 mg/L. The effluent from wastewater treatment plants is often diluting the stream.

Before discharge, however, modern wastewater treatment plants are required to disinfect the effluents to reduce the possibility of disease transmission. Often chlorine is used for the disinfection since it is fairly inexpensive. Chlorination occurs in simple holding basins designed to act as plug-flow reactors (Figure 10–19). The chlorine is injected at the beginning of the tank, and it is assumed that all of the flow is in contact with the chlorine for 30 minutes.

Usually an excess of chlorine is used and the residual chlorine is destroyed by a *dechlorination* process such as bubbling in sulfur dioxide. The chlorine is reduced to chloride while the SO_2 oxidizes to sulfate. Discharging excess chlorine to the receiving watercourse can be toxic to many aquatic organisms.

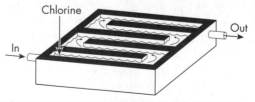

FIGURE 10–19   Plug-flow reactor for chlorination.

But dechlorination does not solve another problem, the production of chlorinated compounds such as chloroform and other trihalomethanes, many of them carcinogens, and as a result the chlorination of effluents has been discouraged. Other disinfection methods such as ultraviolet light are finding wide use in newer wastewater treatment plants.

Actually, there is no epidemiological evidence that undisinfected treatment plant effluents cause any public health problems. Regulatory agencies, however, are unwilling to eliminate what might be another level of public health protection.

10.4 ◆ Tertiary Treatment

Sometimes secondary treatment (even with chlorination) is inadequate to protect the watercourse from harm due to a wastewater discharge. Nutrients such as nitrogen and phosphorus may cause a problem if the effluent is discharged into a still body of water. If the water downstream is used for recreational purposes a high degree of treatment is necessary, especially in solids and pathogen removal. When such situations occur, the effluents from secondary treatment are treated further to achieve whatever quality is required, or the primary + secondary plant is upgraded to produce a higher quality effluent. Such processes are collectively called tertiary treatment.

Rapid sand filters similar to those in drinking-water treatment plants can be used to remove residual suspended solids and to polish the water.

Oxidation ponds are commonly used for BOD removal. The oxidation or polishing pond is essentially a hole in the ground, a large pond used to confine the plant effluent before it is discharged into the natural watercourse. Such ponds are designed to be aerobic, and since light penetration for algal growth is important, a large surface area is needed. The reactions occurring within an oxidation pond are depicted in Figure 10-20. Oxidation ponds are sometimes used as the only treatment step if the waste flow is small and pond area is large.

Activated carbon adsorption is another method of BOD removal, and this process has the added advantage that inorganics as well as organics are removed. The mechanism of adsorption on activated carbon is both chemical and physical, with tiny crevices catching and holding colloidal and smaller particles. An activated carbon column is a completely enclosed tube with dirty water pumped up from the bottom and the clear water exiting at the top. As the carbon becomes saturated with various materials, these must be removed from the column and the carbon regenerated, or cleaned. Removal is often continuous, with clean carbon being added at the top of the column. The cleaning or regeneration is usually done by heating the carbon in the absence of oxygen, driving off the organic matter. A slight loss in efficiency is noted with regeneration, and some virgin carbon must always be added to ensure effective performance.

Nitrogen removal is accomplished by first treating the waste thoroughly enough in secondary treatment to oxidize all the nitrogen to nitrate. This usually

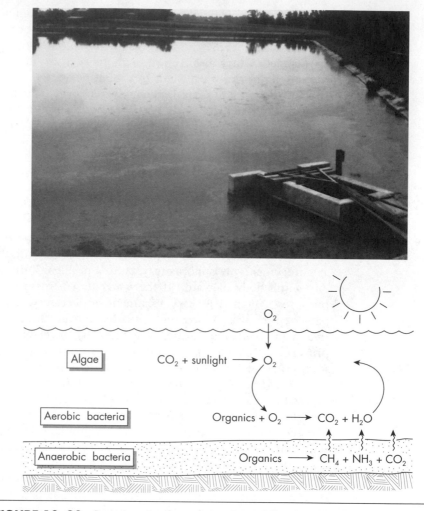

FIGURE 10–20 Reactions in an oxidation (or stabilization) pond.

involves longer detention times in secondary treatment, during which bacteria such as *Nitrobacter* and *Nitrosomonas* convert ammonia nitrogen to NO_3^-, a process called *nitrification*. These reactions are

$$2NH_4^+ + 3O_2 \xrightarrow{\quad Nitrosomonas \quad} 2NO_2^- + 2H_2O + 4H^+$$

$$2NO_2^- + O_2 \xrightarrow{\quad Nitrobacter \quad} 2NO_3^-$$

Both these reactions are slow and require sufficient oxygen and long retention times in the aeration tank. The rate of microorganism growth is also low, resulting in low net sludge production, making washout a constant danger.

Once the ammonia has been converted to nitrate, it can be reduced by a broad range of facultative and anaerobic bacteria such as *Pseudomonas*. This reduction, called *denitrification*, requires a source of carbon, and methanol

(CH_3OH) is often used for that purpose. The sludge containing NO_3^- is placed in an *anoxic* condition where the microorganisms use the oxygen from the nitrate ion and nitrogen becomes the hydrogen acceptor, as discussed in Chapter 7. Using the methanol as the source of carbon, the facultative microorganisms convert the nitrate to nitrogen gas, N_2, which then bubbles out of the sludge into the atmosphere. Sometimes anoxic conditions are not desirable, such as in a primary clarifier. When the sludge in the primary clarifier is not pumped out and all oxygen is depleted in the bottom sludge, the tank begins to bubble, carrying with it some of the solids from the sludge zone.

Modern secondary treatment plants using activated sludge technology often have a series of anoxic zones where denitrification can occur. In such cases there is no need for a carbon source since the remaining dissolved organics provide the energy.

Phosphorus removal is accomplished by either chemical or biological means. In wastewater, phosphorus exists as orthophosphate (PO_4^{---}), polyphosphate (P_2O_7), and organically bound phosphorus. Polyphosphate and organic phosphate may be as much as 70% of the incoming phosphorus load. In the metabolic process, microorganisms use the poly- and organophosphates and produce the oxidized form of phosphorus, orthophosphate. Phosphate removal is either a chemical precipitation process or a biological process designed as part of the activated sludge system.

Chemical phosphorus removal requires that the phosphorus be fully oxidized to the orthophosphate, and hence the most effective chemical removal occurs at the end of the secondary biological treatment system. The most popular chemicals used for phosphorus removal are lime, $Ca(OH)_2$, and alum, $Al_2(SO_4)_3$. The calcium ion, in the presence of high pH, will combine with phosphate to form a white, insoluble precipitate called calcium hydroxyapatite, which is settled out and removed. Insoluble calcium carbonate is also formed and removed and can be recycled by burning in a furnace.

The aluminum ion from alum precipitates out as poorly soluble aluminum phosphate,

$$Al^{+++} + PO_4^{---} \rightarrow AlPO_4 \downarrow$$

and also forms aluminum hydroxides

$$Al^{+++} + 3OH^- \rightarrow Al(OH)_3 \downarrow$$

which are sticky flocs and help to settle out the phosphates. The most common point of alum dosing is in the final clarifier.

The amount of alum required to achieve a given level of phosphorus removal depends on the amount of phosphorus in the water, as well as on other constituents. The sludge produced can be calculated using stoichiometric relationships.

Biological methods of phosphorus removal seem to be becoming increasingly popular, especially since the process does not produce more solids for disposal. Most of the biological phosphorus removal systems rely on the fact that microorganisms can be stressed by cutting off their supply of oxygen (anoxic condition) and fooling them into thinking that all is lost and they will surely die! If this anoxic condition is then followed by a sudden reintroduction of oxygen, the cells will start to store the phosphorus in their cellular material and do so at levels far exceeding their normal requirement. This *luxury uptake* of phosphorus is followed by the removal of the cells from the liquid stream, thereby removing much of the phosphorus. Several proprietary processes that use microorganisms to store excess phosphorus can produce effluents that challenge the chemical precipitation techniques.

Fortunately, the use of anoxic zones in an aeration tank will effectively remove both nitrogen and phosphorus, without the use of chemicals, and it is easy to see why these systems are gaining wide popularity.

The primary-secondary-tertiary wastewater treatment process described is not only complex, but it is also expensive, and alternative wastewater management strategies have been sought. One such alternative is to spray secondary effluent on land and allow the soil microorganisms to degrade the remaining organics. Such systems, known as *land treatment*, have been employed for many years in Europe, but only recently have been used in North America. They appear to represent a reasonable alternative to complex and expensive systems, especially for smaller communities where land is plentiful.

Probably the most promising land treatment method is irrigation. But again the amount of land area required is substantial, and disease transmission is possible since the waste carries pathogenic organisms. Commonly, from 1000 to 2000 hectares of land are required for every 1 m³/sec of wastewater flow, depending on the crop and soil. Nutrients such as N and P remaining in the secondary effluent are of course beneficial to the crops.

10.5 ◆ Sludge Treatment and Disposal

The slurries produced as underflows from the settling tanks, from the primary treatment as well as secondary treatment, must be treated and eventually disposed of. Generally speaking, two types of sludges are produced in the wastewater treatment plants described above—*raw primary sludge* and *biological* or *secondary sludge*. The raw primary sludge comes from the bottom of the primary clarifier, and the biological sludge is either solids that have grown on the fixed film reactor surfaces and sloughed off the rocks or other surfaces, or waste activated sludge grown in the activated sludge system.

The quantity of sludges produced in a treatment plant can be analyzed using the mass-flow technique. Figure 10–21 is a schematic representation of a typical municipal wastewater treatment plant with activated sludge as the secondary treatment step. The symbols in this figure are as follows:

S_0 = influent BOD, lb/day (kg/hr)

X_0 = influent suspended solids, lb/day (kg/hr)

b = fraction of BOD not removed in the primary clarifier

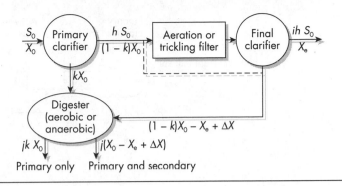

FIGURE 10–21 Typical wastewater treatment plant used for calculating sludge quantities.

i = fraction of BOD not removed in activated sludge system

X_e = plant effluent BOD, lb/day (kg/hr)

k = fraction of influent solids removed in primary clarifier

ΔX = net solids produced by biological action, lb/day (kg/hr)

Y = yield, or the mass of biological solids produced in the aeration tank per mass of BOD destroyed, or

$$\frac{\Delta X}{\Delta S}$$

where
$$\Delta S = hS_0 - ihS_0$$

e • x • a • m • p • l • e **10.8**

Problem   A wastewater enters the 6-mgd treatment plant with a BOD of 200 mg/L and suspended solids of 180 mg/L. The primary clarifier is expected to be 60% effective in removing the solids, while it also removes 30% of the BOD. The activated sludge tank removes 95% of the BOD that it receives, produces an effluent with a suspended solids concentration of 20 mg/L, and the yield of biological sludge is expected to be 0.5 lb solids produced per lb of BOD destroyed. The plant is shown schematically in Figure 10–22A. Find the quantity of both raw primary sludge and waste activated sludge produced in this plant.

Solution   The raw primary sludge from the primary clarifier is simply the fraction of solids removed, $k = 0.60$, times the influent solids. The influent solids flow is $X_0 = 180 \times 6 \times 8.34 = 9007$ lb/day, and the production of raw primary sludge is $0.6 \times 9007 = 5404$ lb/day. Recall that the conversion factor 8.34 converts mg/L × mgd to lb/day.

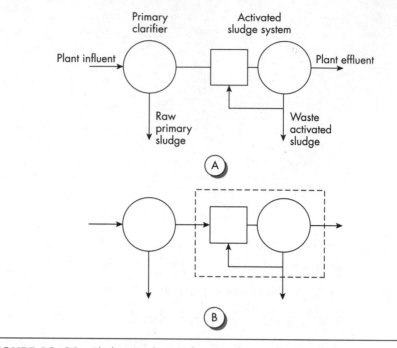

FIGURE 10–22 Sludge production from primary and secondary treatment.

Drawing a dotted line around the activated sludge system and setting up the solids mass balance for it, (Figure 10–22B)

$$
\begin{bmatrix} \text{rate of} \\ \text{solids} \\ \text{ACCUMULATED} \end{bmatrix} = \begin{bmatrix} \text{rate of} \\ \text{solids} \\ \text{IN} \end{bmatrix} - \begin{bmatrix} \text{rate of} \\ \text{solids} \\ \text{OUT} \end{bmatrix}
$$

$$
+ \begin{bmatrix} \text{rate of} \\ \text{solids} \\ \text{PRODUCED} \end{bmatrix} - \begin{bmatrix} \text{rate of} \\ \text{solids} \\ \text{CONSUMED} \end{bmatrix}
$$

Assuming steady state and no consumption of solids,

$$0 = [\text{rate IN}] - [\text{rate OUT}] + [\text{rate PRODUCED}] - 0$$

The solids into the activated sludge system are from the solids not captured in the primary clarifier, or $(1 - k)X_0 = 3603$ lb/day. The solids out of the system are of two kinds, the effluent solids and waste activated sludge. The effluent solids are $X_e = 20 \times 6 \times 8.34 = 1001$ lb/day. The waste activated sludge is the unknown.

The biological sludge is produced as the BOD is used. The amount of BOD entering the activated sludge system is $bS_0 = 0.7 \times (200 \times 6 \times 8.34) = 7006$ lb/day. The activated sludge system is 95% effective in removing this BOD, or $i = 0.95$, so the amount of BOD destroyed within the system is $i \times bS_0 = 0.95 \times 7006 = 6655$ lb/day. Since the yield is assumed to be 0.5 lb solids produced per lb of BOD destroyed, then the biological solids produced must be 0.5 × 6655 = 3328 lb solids per day.

Using the mass balance,

$$0 = [(1 - k)X_0] - [X_e + X_w] + [Y(bS_0)t] - 0$$

$$0 = 3603 - [1001 + X_w] + 3328 - 0$$

or

$$X_w = 5930 \text{ lb/day}$$

or about 3 tons of dry solids per day!

◆

A great deal of money could be saved, and troubles averted, if sludge could be disposed of without further treatment, just as it is drawn off the main process train. Unfortunately, the sludges produced in wastewater treatment have three characteristics that make such simple disposal unlikely: they are aesthetically displeasing, they are potentially harmful, and they contain too much water.

The first two problems are often solved by stabilization, and the third problem requires sludge dewatering. The next three sections cover the topics of stabilization, dewatering, and finally considerations of ultimate disposal.

10.5.1 Sludge Stabilization

The objective of sludge stabilization is to reduce the problems associated with two of the detrimental characteristics listed above, sludge odor and putrescence, and the presence of pathogenic organisms.

There are three primary means of sludge stabilization

- lime
- aerobic digestion
- anaerobic digestion

Lime stabilization is achieved by adding lime (either as hydrated lime, $Ca(OH)_2$ or as quicklime, CaO) to the sludge, raising the pH to about 11 or above. This significantly reduces the odor and helps in the destruction of pathogens. The major disadvantage of lime stabilization is that it is temporary. With time (days) the pH drops and the sludge once again becomes putrescent.

Aerobic digestion is merely a logical extension of the activated sludge system. Waste activated sludge is placed in dedicated aeration tanks for a very long time, and the concentrated solids allowed to progress well into the endogenous respiration phase, in which food is obtained only by the destruction of other viable organisms. This results in a net reduction in total and volatile solids. Aerobically digested sludges are, however, more difficult to dewater than anaerobic sludges.

Anaerobic digestion is the third commonly employed method of sludge stabilization. The biochemistry of anaerobic decomposition of organics is illustrated in Figure 10–23. Note that this is a staged process, with the solution of organics by extracellular enzymes being followed by the production of organic acids by a large and hearty group of anaerobic microorganisms known, appropriately enough, as the *acid formers*. The organic acids are in turn degraded further by a group of strict anaerobes called *methane formers*. These microorganisms are the "prima donnas" of wastewater treatment, getting upset at the least change in their environment. The success of anaerobic treatment boils down to the creation of a suitable condition for the methane formers. Since they are strict anaerobes, they are unable to function in the presence of oxygen and are very sensitive to environmental conditions such as temperature, pH, and the presence of toxins. If a digester goes "sour," the methane formers have been inhibited in some way. The acid formers, however, keep chugging away, making more organic acids. This has the effect of further lowering the pH and making conditions even worse for the methane formers. A sick digester is therefore difficult to cure without massive doses of lime or other antacids.

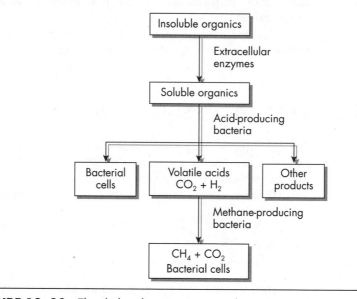

FIGURE 10–23   The sludge digestion process dynamics.

Most treatment plants have two kinds of anaerobic digesters: primary and secondary (Figure 10–24). The primary digester is covered, heated, and mixed to increase the reaction rate. The temperature of the sludge is usually about 35°C (95°F). Secondary digesters are not mixed or heated and are used for storage of gas and for concentrating the sludge by settling. As the solids settle, the liquid supernatant is pumped back to the main plant for further treatment. The cover of the secondary digester often floats up and down, depending on the amount of gas stored. The gas is high enough in methane to be used as a fuel and is in fact usually used to heat the primary digester.

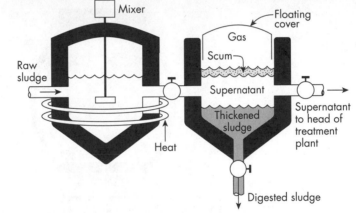

FIGURE 10–24 Two-stage anaerobic digestion; primary and secondary anaerobic digesters.

Sludge digesters such as those pictured in Figure 10–24 have been notoriously difficult to operate. When the methane-forming organisms are adversely affected by such environmental factors as temperature change, overfeeding, or inadequate mixing, they often cease to perform. The acid formers are not so affected, and continue to make organic acids, much to the detriment of the methane formers, which are then even less likely to start working again. A "stuck" digester is very difficult to restart.

Often the reason for such problems is the difficulty of mixing the sludge in the digester. No good mixing techniques have been developed for the dumpy digesters usually used in American wastewater treatment plants. In Europe, egg-shaped digesters (Figure 10–25) have found favor mainly because of the ease of mixing. The digester gas is pumped into the bottom and an effective

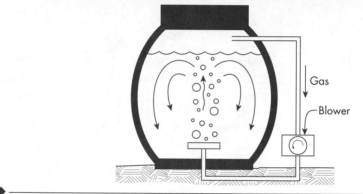

◆───────────────────────────────

FIGURE 10–25 Egg-shaped anaerobic digesters.

circulation pattern is set up. Egg-shaped digesters are only now being introduced into the United States.

   Anaerobic digesters run at elevated temperatures should achieve substantial pathogen reduction, but the process is not perfect, and many pathogenic organisms survive. An anaerobic digester cannot, therefore, be considered a method of sterilization.

10.5.2   Sludge Dewatering

In most wastewater plants, dewatering is the final method of volume reduction prior to ultimate disposal. In the United States three dewatering techniques are widely used at present: sand beds, belt filters, and centrifuges. Each of these is discussed as follows.

Sand beds have been in use for a great many years and are still the most cost-effective means of dewatering when land is available and labor costs are not exorbitant. As shown in Figure 10-26, sand beds consist of tile drains in gravel, covered by about 0.25 m (10 in.) of sand. The sludge to be dewatered is poured on the beds at about 15 cm (6 in.) deep. Two mechanisms combine to separate the water from the solids: seepage and evaporation. Seepage into the sand and through the tile drains, although important in the total volume of water extracted, lasts for only a few days. As drainage into the sand ceases, evaporation takes over and this process is actually responsible for the conversion of liquid sludge to solid. In some northern areas sand beds are enclosed under

FIGURE 10-26   Sand drying bed for sludge dewatering.

greenhouses to promote evaporation as well as prevent rain from falling into the beds.

For mixed digested sludge, the usual design is to allow for three months' drying time. Some engineers suggest that this period be extended to allow a sand bed to rest for a month after the sludge has been removed, which seems to be an effective means of increasing the drainage efficiency once the sand beds are again flooded.

Raw sludge will not drain well on sand beds and will usually have an obnoxious odor, and therefore it is seldom dried on sand beds. Raw secondary sludges have a habit of either seeping through the sand or clogging the pores so quickly that no effective drainage takes place. Aerobically digested sludge can be dried on sand, but usually with difficulty.

If dewatering by sand beds is considered impractical, mechanical dewatering techniques must be employed. One mechanical method of dewatering is the *belt filter*, which operates as both a pressure filter and a gravity filter (Figure 10–27). As the sludge is introduced onto a moving belt, the free water

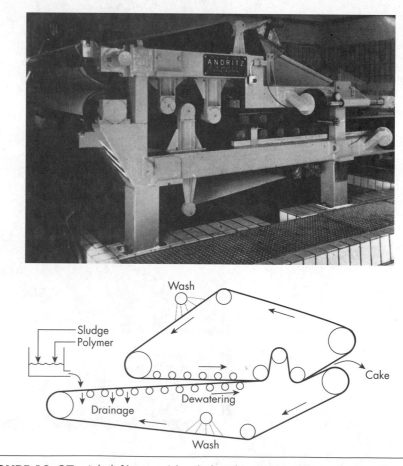

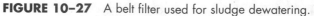

FIGURE 10–27　A belt filter used for sludge dewatering.

surrounding the sludge solid particles drips through the belt. The solids are retained on the surface of the belt. The belt then moves into a dewatering zone, where the sludge is squeezed between two belts, forcing the *filtrate* from the sludge solids. The dewatered solids, called the *cake*, are then discharged when the belts separate.

The *centrifuge* most widely used in wastewater sludge dewatering is the solid bowl decanter, which consists of a bullet-shaped body rotating on its long axis. The sludge is placed into the bowl, and the solids settle out under a centrifugal force about 500 to 1000 times gravity and are scraped out of the bowl by a screw conveyor (Figure 10–28). The solids coming out of a centrifuge are known as *cake*, as in filtration, but the decanted liquid is known as *centrate*.

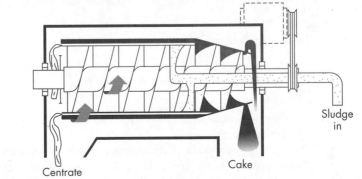

FIGURE 10–28 A solid bowl decanter centrifuge used for sludge dewatering.

The objective of a dewatering process such as centrifugation is twofold: to produce a solids cake of high solids concentration and to make sure all of the solids end up in the cake. Unfortunately the centrifuge is not an ideal device and some of the solids still end up in the centrate, and some of the liquid in the cake. The centrifuge is a two-material black box, as introduced in Chapter 3.

Assuming steady-state operation, and recognizing that no liquid or solids are produced or consumed in the centrifuge, the mass balance equations are

$$[\text{rate of material IN}] = [\text{rate of material OUT}]$$

The volume balance in terms of the sludge flowing in and out is

$$Q_0 = Q_k + Q_c$$

and the mass balance in terms of the sludge solids is

$$Q_0 C_0 = Q_k C_k + Q_c C_c$$

where Q_0 = flow of sludge as the feed, volume/unit time
Q_k = flow of sludge as the cake, volume/unit time
Q_c = flow of sludge as the centrate, volume/unit time

and subscripts 0, k, and c refer to feed, cake, and centrate solids concentrations, respectively.

The cake solids is then simply C_k. The recovery of solids, as defined previously, is

$$\text{solids recovery} = \frac{\text{mass of dry solids as cake}}{\text{mass of dry feed solids}} \times 100$$

$$= \frac{C_k Q_k}{C_0 Q_0} \times 100$$

From these materials balances, by substitution,

$$Q_k = \frac{Q_0(C_0 - C_c)}{C_k - C_c}$$

and a further substitution yields

$$\text{solids recovery} = \frac{C_k(C_0 - C_c)}{C_0(C_k - C_c)} \times 100$$

This expression allows for the calculation of solids recovery using only the concentration terms.

The centrifuge is an interesting solids separation device because it can operate at almost any cake solids and solids recovery, depending on the condition of the sludge and the flow rate. For any given sludge, a centrifuge will have an *operating curve* as shown in Figure 10-29. As the flow rate

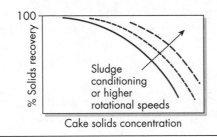

FIGURE 10-29 Centrifuge operating curve.

is increased, the solids recovery begins to deteriorate because the residence time in the machine is shorter, but at the same time the cake solids improve because the machine discriminates among the solids and spews out only those solids that are easy to remove, while also producing a high solids concentration in the cake. At lower feed flows, high residence times allow for complete settling and solids removal, but the cake becomes wetter because all the soft and small particles that carry a lot of water with them are also removed.

This seems like a difficult problem. It appears that one can only move up or down on the operating curve, trading off solids recovery for cake solids. There are, however, two ways of moving the entire curve to the right, that is, obtaining a better operating curve so that both solids recovery and cake solids are improved. The first method is to increase the centrifugal force imposed on the solids.

The creation of the centrifugal force is best explained by recalling first Newton's law

$$F = ma$$

where F = force, N
$\quad m$ = mass, kg
$\quad a$ = acceleration, m/sec^2

If the acceleration is due to gravity,

$$F = mg$$

where g = acceleration due to gravity, 9.8 m/sec^2. When a mass is spun around, it requires a force toward the center of rotation to keep it from flying off into space. This force is calculated as

$$F_c = m(w^2r)$$

where F_c = centrifugal force, N
$\quad m$ = mass, kg
$\quad w$ = rotational speed, radians/sec
$\quad r$ = radius of rotation, m

The term w^2r is called centrifugal acceleration, and has units of m/sec². The number of *gravities*, G, produced by a centrifuge is

$$G = \frac{w^2r}{g}$$

where G = number of "gravities" produced by centrifugal acceleration.
$\quad\quad w$ = rotational speed, radians/sec (recall that each rotation equals 2π radians)
$\quad\quad r$ = inside radius of the centrifuge bowl

Increasing the speed of the bowl therefore increases the centrifugal force and moves the operating curve to the right.

The second method of simultaneously improving cake solids and solids recovery is to condition the sludge with chemicals prior to dewatering. This is commonly done using organic macromolecules called *polyelectrolytes*. These large molecules, with molecular weights of several million, have charge sites on them that appear to attach to sludge particles and to bridge the smaller particles so that larger flocs are formed. These larger flocs are able to expel water and compact the small solids, yielding both a cleaner centrate and a more compact cake, in effect moving the operating curve to the right.

The performance of centrifuges is measured by sampling the feed and the centrate and cake coming out of an operating machine. The hardware is such that it is very difficult to measure the flow rates of the centrate and cake, and only the solids concentrations can be sampled for these flows. Mass balances can be used to calculate recovery and cake solids (purity).

e • x • a • m • p • l • e   **10.9**

Problem   A wastewater sludge centrifuge operates with a feed of 10 gpm and a feed solids concentration of 1.2% solids. The cake solids concentration is 22% solids and the centrate solids is 500 mg/L. What is the recovery of solids?

Solution   The solids concentrations can first be converted to similar units. Recall that if the density of the solids is almost one (a good assumption for wastewater solids), 1% solids = 10,000 mg/L.

	Solids Concentration mg/L
Feed solids	12,000
Cake solids	220,000
Centrate solids	500

$$\text{solids recovery} = \frac{220,000(12,000 - 500)}{12,000(220,000 - 500)} \times 100 = 96\%$$

10.5.3 Ultimate Disposal

The options for ultimate disposal of sludge are limited to air, water, and land. Strict controls on air pollution usually make incineration expensive, although this may be the only option for some communities. Disposal of sludges in deep water (such as oceans) is being phased out due to adverse or unknown detrimental effects on aquatic ecology. Land disposal can be achieved by either dumping into a landfill, or spreading the sludge out over land and allowing natural biodegradation to assimilate the sludge back into the environment. Finally, sludge can be disposed of by either giving it away or, better yet, selling it as a valuable fertilizer and soil conditioner.

Strictly speaking, incineration is *not* a method of disposal but rather a sludge treatment step in which the organics are converted to H_2O, CO_2, and many other partially oxidized compounds, and the inorganics drop out as a nonputrescent residue. Two types of incinerators have found use in sludge treatment: *multiple hearth* and *fluid bed*. The multiple hearth incinerator, as the name implies, has several hearths stacked vertically, with rabble arms pushing the sludge progressively downward through the hottest layers, and finally into the ash pit (Figure 10–30). The fluid bed incinerator has a boiling cauldron of hot sand into which the sludge is pumped. The sand provides mixing and acts as a thermal flywheel for process stability.

The second method of disposal—land spreading—depends on the ability of land to absorb sludge and to assimilate it into the soil matrix. This assimilative capacity depends on such variables as soil type, vegetation, rainfall, slope, and the like. In addition, the important variable of the sludge itself influences the capacity of a soil to assimilate sludge. Generally, sandy soils with lush vegetation, low rainfall, and gentle slopes have proven most successful. Mixed digested sludges have been spread from tank trucks, and activated sludges have been sprayed from both fixed and moving nozzles. The application rate has been variable but 100 dry tons/acre/yr is not an unreasonable estimate. Most unsuccessful land application systems can be traced to overloading the soil. Given enough time (and absence of toxic materials) most soils will assimilate sprayed liquid sludge.

Transporting liquid sludge is often expensive, however, and volume reduction by dewatering is necessary. The solid sludge can then be deposited on land and either worked into the soil; alternatively, trenches can be dug with a backhoe and the sludge deposited and covered. The sludge seems to assimilate rapidly, with no undue leaching of nitrates or toxins.

The toxicity of sludge can be interpreted in several ways: toxicity to vegetation, toxicity to the animals (including people) who eat the vegetation, and the contamination of groundwater supplies. Most domestic sludges do not contain sufficient toxins such as heavy metals to cause harm to vegetation. The total body burden of heavy metals is of some concern, however. The most effective means of controlling heavy metals seems to be to prevent these metals from entering the sewerage system in the first place by reducing the contribution of metals from industrial waste. Strictly enforced sewer ordinances are necessary and can be cost effective.

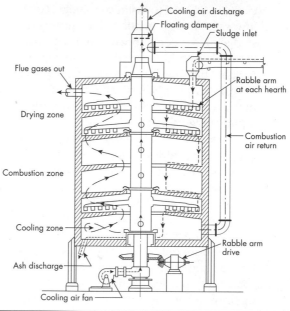

FIGURE 10–30 A multiple-hearth incinerator.

10.6 ◆ Selection of Treatment Strategies

Who is to choose which of the treatment strategies will be used for given wastewater and aquatic conditions? The simple (but wrong) answer is "decision makers," these ambiguous, anonymous bureaucrats who decide how to spend our money. In fact, decisions on which treatment strategy to adopt are made by engineers who define what is to be achieved and then propose the treatment to accomplish the objective.

This process is not unlike the way architects work. A building is commissioned, and the architect first spends considerable time with the client trying to understand just what the client intends to do with the building. But then the architect designs the building based on the architect's own tastes and ideas. In a similar manner, the design engineers establish what the objectives of treatment are and then design a facility to meet these objectives.

In this role engineers have considerable latitude and hence responsibility. They are asked by society to design something that not only works, but that works at the least possible cost, does not prove to be nuisance to its neighbors, and looks nice as well. In this role, engineers become the repositories of the public trust.

Because of this public and environmental responsibility, engineering is a profession, and as such, all engineers are expected to adhere to high professional standards. Historically, and especially overseas, engineering has been an honored and highly respected profession. Russian dissident and Nobel Prize winning author Aleksandr I. Solzhenitsyn describes the Russian engineers of his day in this manner:

> An engineer? I had grown up among engineers, and I could remember the engineers . . . their open, shining intellects, their free and gentle humor, their agility and breadth of thought, the ease with which they shifted from one engineering field to another, and for that matter, from technology to social concerns and art.[1]

A lot to live up to, isn't it?

If engineers are indeed a special breed, do they also have social responsibilities that exceed their roles as private citizens? Is an engineer simply a robot that does certain chores and after 5 o'clock becomes a person with no more social responsibilities than any other person? Or is the engineer's special training, experience, and status in society such that he or she cannot help being a special person who others will look up to and whose opinions will be respected?

If it is the latter, the role of the environmental engineer takes on special significance. Not only is the environmental engineer responsible for performing a job, but he or she also has another "client," the environment. Explaining the conflicts that arise in working with environmental concerns can be a major role of the environmental engineer, and his or her responsibilities can far exceed those of an ordinary citizen.

Sometimes engineers are placed in what might appear to be anti-environmental positions. A classical case would be engineers involved in the construction of a massive secondary treatment plant for a large coastal city where the discharge is to the deep waters of the bay. Even though the efficiency of treatment provided by a primary treatment facility is quite adequate to meet all environmental concerns about public health and water quality, the political situation might demand the expenditure of public funds for the construction of secondary treatment facilities. These billion (yes, billions) of dollars could be well spent elsewhere, with no detrimental effect to the aquatic environment. The new secondary plant, however, is a political necessity. Are there times when the engineer ought to speak out in favor of *less* environmental control?

◆ Abbreviations

BOD	= biochemical oxygen demand		S	= substrate (BOD) concentration
C	= concentration		SS	= suspended solids
D	= deficit = $S - C$		SVI	= sludge volume index
k_1	= deaeration constant		X	= microorganism (SS) concentration
k_2	= reaeration constant			
K	= solubility constant		Y	= yield constant
$K_l a$	= gas transfer coefficient, base e		μ	= growth rate
			$\hat{\mu}$	= maximum growth rate constant
K_S	= saturation constant			
P	= partial pressure of a gas		θ_C	= sludge age or mean cell residence time
q	= substrate removal velocity			
S	= solubility of a gas in a liquid			

◆ Problems

10–1   The following data were reported on the operation of a wastewater treatment plant:

Constituent	Influent (mg/L)	Effluent (mg/L)
BOD_5	200	20
SS	220	15
P	10	0.5

a.   What percent of removal was experienced for each constituent?

b. What kind of treatment plant would produce such an effluent? Draw a block diagram showing one configuration of the treatment steps that would result in such a plant performance.

10-2 Describe the condition of a primary clarifier one day after the raw sludge pumps broke down. What do you think would happen?

10-3 One operational problem with trickling filters is *ponding*, the excessive growth of slime on the rocks and subsequent clogging of the spaces so that the water no longer flows through the filter. Suggest some cures for the ponding problem.

10-4 One problem with sanitary sewers is illegal connections from roof drains. Suppose a family of four, living in a home with a roof area of 70 × 40 ft, connects the roof drain to the sewer. A typical rain is 1 in./hr.

a. What percentage increase will there be in the flow from their house over the dry weather flow (assumed at 50 gal/capita-day)?

b. What is so wrong with connecting the roof drains to the sanitary sewers? How would you explain this to someone who has just been fined for such illegal connections? (Don't just say "It's the law." Explain *why* there is such a law.) Use ethical reasoning to fashion your argument.

10-5 Draw block diagrams of the unit operations necessary to treat the following wastes to effluent levels of BOD_5 = 20 mg/L, SS = 20 mg/L, P = 1 mg/L.

Waste	BOD_5 (mg/L)	SS (mg/L)	P (mg/L)
A. Domestic	200	200	10
B. Chemical industry	40,000	0	0
C. Pickle cannery	0	300	1
D. Fertilizer mfg.	300	300	200

10-6 The success of an activated sludge system depends on the settling of the solids in the final settling tank. Suppose the sludge in a system starts to bulk (not settle very well), and the suspended solids concentration of the return activated sludge drops from 10,000 mg/L to 4000 mg/L.

a. What will this do to the mixed liquor suspended solids?

b. What will this in turn do to the BOD removal? Why? (Do not answer quantitatively.)

10-7 The MLSS in an aeration tank is 4000 mg/L. The flow from the primary settling tank is 0.2 m³/sec with a SS of 50 mg/L and the return sludge flow is 0.1 m³/sec with a SS of 6000 mg/L. Do these two sources of solids make up the 4000 mg/L SS in the aeration tank? If not, how is the 4000 mg/L level attained? Where do the solids come from?

10-8 A 1-L cylinder is used to measure the settleability of 0.5% suspended solids sludge. After 30 min, the settled sludge solids occupy 600 mL. Calculate the SVI.

10-9   What measures of stability would you need if a sludge from a wastewater treatment plant was to be

    a.   placed on the White House lawn?

    b.   dumped into a trout stream?

    c.   sprayed on the playground?

    d.   spread on a vegetable garden?

10-10   A sludge is thickened from 2000 mg/L to 17,000 mg/L. What is the reduction in volume, in percent?

10-11   Sludge age (sometimes called mean cell residence time) is defined as the mass of sludge in the aeration tank divided by the mass of sludge wasted per day. Calculate the sludge age if the aeration tank has a volume of 180 m³ and the suspended solids concentration is 2000 mg/L. The return sludge solids concentration is 12,000 mg/L, and the flow rate of waste activated sludge is 0.5 m³/min.

10-12   The block diagram in Figure 10-31 shows a secondary wastewater treatment plant.

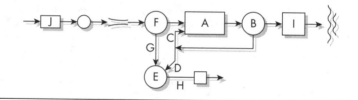

FIGURE 10-31   Schematic of a secondary wastewater treatment plant. See Problem 10-12.

    a.   Identify the various unit operations and flows, and state their purpose or function. Why are they there, or what do they do?

    b.   Suppose you are a senior engineer in charge of wastewater treatment for a metropolitan region. You have retained a consulting engineering firm to design the plant shown. Four firms were considered for the job, and this firm was selected, over your objection, by the metropolitan authority board. There are indications of significant political contributions from the firm to the members of the board. While this plant will probably work, it clearly will not be the caliber of plant you desired, and you will probably spend a lot of (public) money upgrading it. What responsibilities do you, a professional engineer, have in this matter? Analyze the problem, identify the people involved, state the options you have, and come to a conclusion. All things considered, what ought you to do?

10-13   Suppose a law requires that all wastewater discharges to a watercourse meet the following standards:

BOD: less than 20 mg/L

Suspended solids: less than 20 mg/L

Phosphorus (total): less than 0.5 mg/L

Design treatment plants (block diagrams) for the following wastes:

Waste 1	BOD = 250 mg/L
	SS = 250 mg/L
	P = 10 mg/L
Waste 2	BOD = 750 mg/L
	SS = 30 mg/L
	P = 0.6 mg/L
Waste 3	BOD = 30 mg/L
	SS = 450 mg/L
	P = 20 mg/L

10-14 Draw a block diagram for a treatment plant, showing the necessary treatment steps for achieving the desired effluent.

	Influent	Effluent
BOD	1000 mg/L	10 mg/L
Suspended solids	10 mg/L	10 mg/L
Phosphorus	50 mg/L	5 mg/L

Be sure to use only those treatment steps necessary to achieve the desired effluent. Extraneous steps will be considered wrong.

10-15 A 36-inch diameter centrifuge for dewatering sludge is rotated at 1000 rotations per minute. How many gravities does this machine produce? (Remember that each rotation represents 2π radians.)

10-16 A gas transfer experiment results in the following data:

Time (min)	Dissolved Oxygen (mg/L)
0	2.2
5	4.2
10	5.0
15	5.5

The water temperature is 15°C. What is the gas transfer coefficient $K_l a$?

10-17 Derive the equation for calculating centrifuge solids recovery from concentration terms. That is, recovery is to be calculated only as a function of the solids concentrations of the three streams, feed, centrate, and cake. Show each step of the derivation, including the materials balances.

10-18 Counterattacks by people who do not agree with the environmental movement have been an interesting phenomenon in the environmental ethics field. These people do not understand nor appreciate the values that most people hold about environmental concerns, and try to discredit the environmentalists. Often this is too much like shooting fish in a barrel. It's too easy. The environmental movement is full of dire warnings, disaster predictions, and apocalypses that never happened. As an example of this genre of writing, read the first chapter in one of the classics in the field, *The Population Bomb* by Paul Erlich. Note the date it was written and be prepared to discuss the effect this book probably had on public opinion.

10-19   A community with a wastewater flow of 10 mgd is required to meet effluent standards of 30 mg/L for both BOD_5 and SS. Pilot plant results with the influent of $BOD_5 = 250$ mg/L estimate the kinetic constants at $K_S = 100$ mg/L, $\hat{\mu} = 0.25$ day^{-1}, and $Y = 0.5$. The MLSS is to be 2000 mg/L. What are the hydraulic retention time, the sludge age, and the required tank volume?

10-20   A wastewater from a peach-packaging plant was tested in a pilot activated sludge plant, and the kinetic constants were found to be $\hat{\mu} = 3$ day^{-1}, $Y = 0.6$, $K_S = 450$ mg/L. The influent BOD is 1200 mg/L and a flow rate of 19,000 m^3/day is expected. The aerators to be used will limit the suspended solids in the aeration tank to 4500 mg/L. The available aeration volume is 5100 m^3.

a.   What efficiency of BOD removal can be expected?

b.   Suppose we found that the flow rate is actually much higher, say 35,000 m^3/day, and the flow is more dilute, $S_0 = 600$ mg/L. What removal efficiency might we expect now?

c.   At this flow rate and S_0, suppose we could not maintain 4500 mg/L solids in the aeration tank (why?). If the solids were only 2000 mg/L, and if 90% BOD removal was required, how much extra aeration tank volume would be needed?

10-21   An activated sludge system has a flow of 4000 m^3/day with a BOD_5 of 300 mg/L. From pilot plant work, the kinetics constants are $Y = 0.5$, $\hat{\mu} = 3$ day^{-1}, $K_S = 200$ mg/L. We need to design an aeration system that will remove 90% of the BOD_5. Assume mixed liquor solids of 4000 mg/L. Specifically, we need to know

a.   the volume of the aeration tank

b.   the sludge age

c.   amount of waste activated sludge

10-22   A centrifuge manufacturer is trying to sell your city a new centrifuge that is supposed to dewater your sludge to 35% solids. You know, however, based on a few telephone calls to your friends in other cities, that their machines will most likely achieve only about 25% solids in the cake.

a.   How much extra volume of sludge cake will you have to be able to handle and dispose of 35% vs. 25%?

b.   You know that none of the competitors can provide a centrifuge that works better, and you like this machine and convince the city to buy it. A few weeks before the purchase is final, you receive a set of drinking glasses with the centrifuge company logo on them, along with a thank you note from the salesman. What do you do? Why? Use ethical reasoning to develop your answer.

10-23   In Westminster Abbey in London are honored many immortal historical characters of English history, each of whom made a significant contribution to Western civilization. Numbered among these, perhaps inadvertently, is one Thomas Crapper, the inventor of the "pull-and-let-go" flush toilet. A manhole cover, shown in Figure 10–32, is Crapper's memorial. American soldiers during the First World War apparently were so enamored and impressed with Crapper's invention that after the war they brought both the idea and the name back to the United States. In England, if you ask someone for directions "to the crapper" they will not understand you. You have to ask for the "loo." So in his own country, poor Thomas is without honor. But in the colonies, his fame and glory live on.

For this problem, investigate the flushing mechanism of an ordinary toilet and explain its operation using pictures and a verbal description.

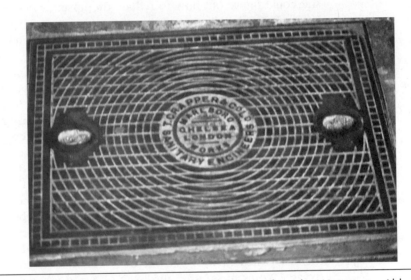

FIGURE 10–32 Thomas Crapper's manhole cover from the Westminster Abbey.

10–24 Your city is contemplating placing its wastewater sludge on an abandoned strip mine to try to reclaim the land for productive use, and you, as the city consulting engineer, are asked for an opinion. You know that the city has been having difficulties with its anaerobic digesters, and although it can probably meet the standards on pathogens most of the time, it probably will not be able to do so consistently. Reclaiming the strip mine is a positive step, as is the low cost of disposal. The city engineer tells you that if they cannot use the strip mine disposal method, they must buy a very expensive incinerator to meet the new EPA regulation. The upgrade of the anaerobic digesters will also be prohibitively expensive for this small town.

What do you recommend? Analyze this problem in terms of the affected parties, possible options, and final recommendations for action.

10-25   The state had promised to come down hard on Domestic Imports, Inc., if it violates its NPDES permit one more time, and Sue is on the spot. It is her treatment works, and she is expected to operate it. Her requests for improvements and expansion have been turned down and she has been told that the treatment works has sufficient capacity to treat the waste. This is true, if the average flow is used in the calculation. Unfortunately, the manufacturing operation is such that slug flows could come unexpectedly, and the biological system simply cannot adapt fast enough. The effluent BOD can shoot sky high for a few days, and then settled back again to below NPDES standards. If the state shows up on one of the days during the upset, she is in big trouble.

One day she is having lunch with a friend of hers, Emmett, who works in the quality control lab. Sue complains to Emmett about her problem. If only she could figure out how to reduce the upsets when the slug loads come in. She has already talked to the plant manager to get him to assure her that the slug loads would not occur, or to build an equalization basin, but both requests were denied.

"The slug load is not your problem, Sue," Emmett suggested. "It's the high BOD that results from the slug load upsetting your plant."

"OK, wiseguy. You are right. But that doesn't help me."

"Well, maybe I can. Your problem is that you are running high BOD and you need to reduce it. Suppose I told you that you could reduce the BOD by simply bleeding a chemical additive into the line, and that you can add as much of this chemical as you need to reduce your BOD. Would you buy it?" asked Emmett.

"Just for fun, suppose I would. What is this chemical?"

"I have been playing around with a family of chemicals that slows down microbial metabolism, but does not kill the microbes. If you add this stuff to your line following the final clarifier, your effluent BOD will be reduced because the metabolic activity of the microbes will be reduced, hence they will use less oxygen. You can set up a small can of it and whenever you see one of those slugs coming down from manufacturing you start to bleed in this chemical. Your BOD will stay within the effluent limits, and in a few days when things have calmed down, you turn it off. Even if the state comes to visit, there is no way they can detect it. You are doing nothing illegal. You are simply slowing down the metabolic activity."

"But is this stuff toxic?" asks Sue.

"No, not at all. It shows no detrimental effect in bioassay tests. You want to try it out?"

"Wait a minute. This is complicated. If we use this magic bullet of yours, our effluent BOD will be depressed, we pass the state inspection, but we haven't treated the wastewater. The oxygen demand will still occur in the stream."

"Yes, but many miles and many days downstream. They will never be able to associate your discharge with a fish kill, if in fact this would occur. What do you say? Want to give it a try?"

Assuming that it is highly unlikely that Sue will ever get caught adding this chemical to the effluent, why should she not do it? What values is she struggling with? Who are the parties involved and what stakes do they have in her decision? What are all her options, and which one do you recommend?

◆ Endnote

1. Aleksandr I. Solzhenitsyn, *The Gulag Archipelago*, Harper-Row, New York (1974).

chapter 11

AIR QUALITY

The air we breathe, like the water we drink, is necessary to life. As with water, we want to be assured that the air will not cause us harm. We expect to breathe "clean air."

But what exactly is clean air? This question is just as difficult to answer as defining what is clean water. Recall from Chapter 8 that many parameters are needed to describe the quality of water, and that only with the selective and judicious use of these parameters is it possible to describe what is meant by water quality. Recall also that water quality is a relative term and that it is unrealistic to ask for all water to be pure H_2O, and that in many cases such as in streams and lakes pure water would actually be unacceptable.

An analogous situation exists with air quality. Pure air is a mixture of gases, which contains

- 78.0% nitrogen
- 20.1% oxygen
- 0.9% argon
- 0.03% carbon dioxide
- 0.002% neon
- 0.0005% helium

and so on. But such air is not found in nature and is of interest only as a reference, such as pure H_2O would be.

If this is pure air, then it may be useful to define as pollutants those materials (gases, liquid, or solid) that, when added to pure air at sufficiently high concentrations, will cause adverse effects. For example, sulfur compounds emitted into the atmosphere reduce the pH of rain and result in acidic rivers and lakes, causing widespread damage. This is clearly unacceptable, and the sulfur compounds can be without much argument classified as air pollutants. Yet the problem is not so easily settled, since some sulfur may be emitted from natural sources such as volcanos and hot springs. Sulfur cannot therefore be classified as a pollutant without specifying its sources.

Further, the pollutants emitted into the atmosphere must travel through the atmosphere to reach people, animals, plants, or things to have an effect. Whereas in water pollution this carriage of pollutants is by water currents, in air pollution wind is the means for transport of pollutants.

In this chapter some basic meteorology is discussed first to illustrate how the transport and dispersion of pollutants takes place. Then the methods of air-quality measurement are discussed, followed by a discussion of the sources and effects of some major air pollutants. Finally, some air pollution law is introduced to show how the government can influence the attainment of quality air.

11.1 ◆ Meteorology and Air Movement

The earth's atmosphere can be divided into easily recognizable strata, depending on the temperature profile. Figure 11-1 shows a typical temperature profile for four major layers. The troposphere, where most of our weather occurs, ranges from about 5 km at the poles to about 18 km at the equator. The

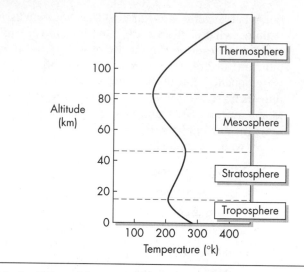

FIGURE 11-1   The earth's atmosphere.

temperature here decreases with altitude. More than 80% of the air is within this well-mixed layer. On top of the troposphere is a layer of air where the temperature profile is inverted, and in this stratosphere little mixing takes place. Pollutants that migrate up to the stratosphere can stay there for many years. The stratosphere has a high ozone concentration, and the ozone adsorbs the sun's shortwave ultraviolet radiation. Above the stratosphere are two more layers, the mesosphere and the thermosphere, which contain only about 0.1% of the air.

Other than the problems of global warming and stratospheric ozone depletion, air pollution problems occur in the troposphere. Pollutants in the troposphere, whether produced naturally (such as terpenes in pine forests) or emitted from human activities (such as smoke from power plants) are moved by air currents, which we commonly call wind. Meteorologists identify many different kinds of winds, ranging from global wind patterns caused by the differential warming and cooling of the earth as it rotates under the sun, to local winds caused by differential temperatures between land and water masses. A sea breeze, for example, is a wind caused by the progressive warming of the land during a sunny day. The temperature of a large water body such as an ocean or large lake does not change as rapidly during the day, and the air over the warm land mass rises, creating a low-pressure area toward which air coming horizontally over the cooler large water body flows.

Wind not only moves the pollutants horizontally, but it also causes the pollutants to disperse, reducing the pollutant concentration with distance away from the source. The amount of dispersion is directly related to the stability of the air, or how much vertical air movement is taking place. The stability of the atmosphere is best explained using an ideal parcel of air.

As an imaginary parcel of air rises in the earth's atmosphere, it experiences lower and lower pressure from surrounding air molecules and thus expands. This expansion lowers the temperature of the air parcel. Ideally a rising parcel of air cools at about 1°C/100 m or 5.4°F/1000 ft (or warms at 1°C/100 m if it is coming down). This warming or cooling is termed the *dry adiabatic lapse rate*. (Recall from Chapter 6 that adiabatic is a term denoting no heat transfer such as between the air parcel and the surrounding air.) The adiabatic lapse rate is independent of prevailing atmospheric temperatures. The 1°C/100 m *always* holds (for dry air), regardless of the actual temperature at various elevations. When there is moisture in the air the lapse rate becomes the *wet adiabatic lapse rate* because the evaporation and condensation of water influences the temperature of the air parcel. This is an unnecessary complication for the purpose of the following argument and a moisture-free atmosphere is assumed.

The actual temperature–elevation measurements are called *prevailing lapse rates* and can be classified as shown in Figure 11–2. A *superadiabatic lapse rate*, also called a *strong lapse rate*, occurs when the atmospheric temperature drops more than 1°C/100 m. A *subadiabatic lapse rate*, also called a *weak lapse rate*, is characterized by drop of less than 1°C/100 m. A special case of the weak lapse rate is the *inversion*, a condition that has warmer air above colder air.

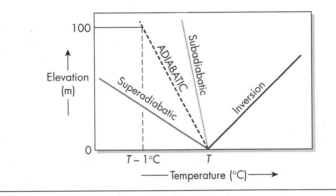

FIGURE 11–2   Prevailing and dry adiabatic lapse rates.

During a superadiabatic lapse rate the atmospheric conditions are unstable; a subadiabatic and especially an inversion characterizes a stable atmosphere. This can be demonstrated by depicting a parcel of air at 500 m (see Figure 11–3). If the temperature at 500 m is 20°C, during a superadiabatic condition the temperature at ground level might be 30°C and at 1000 m it might be 10°C. Note that this represents a change of more than 1°C/100 m.

If the parcel of air at 500 m and 20°C is moved upward to 1000 m, what would be its temperature? Remember that assuming adiabatic condition, the parcel would cool 1°C/100 m. The temperature of the parcel at 1000 m is thus 5°C less than 20°C or 15°C. The *prevailing* temperature (the air surrounding the parcel), however, is 10°C, and the air parcel finds itself surrounded by cooler air. Will it rise or fall? Obviously, it will rise, since warm air rises. We

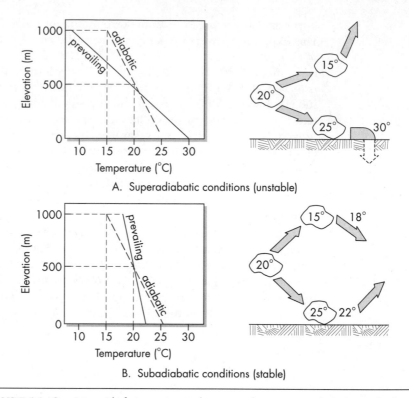

FIGURE 11-3   A parcel of air moving in the atmosphere; superadiabatic and subadiabatic prevailing lapse rates.

then conclude that once a parcel of air under superadiabatic conditions is displaced upward, it keeps on going.

Similarly, if a parcel of air under superadiabatic conditions is displaced downward, say to ground level, the air parcel is 20°C + (500 m × [1/100 m]) = 25°C. It finds the air around it a warm 30°C and thus the cooler air parcel would just as soon keep going down if it could. Superadiabatic conditions also promote the downward air movement, adding to the instability. Superadiabatic atmospheres are characterized by a great deal of vertical air movement and turbulence, since any upward or downward movement tends to continue and not be dampened out.

The subadiabatic prevailing lapse rate is by contrast a very stable system. Consider again, as in Figure 11-3, a parcel of air at 500 m and at 20°C. A typical subadiabatic system has a ground level temperature of 21°C and 19°C at 1000 m. If the parcel is displaced to 1000 m, it will cool by 5°C to 15°C. But, finding the air around it a warmer 10°C, it will fall right back to its point of origin. Similarly, if the air parcel were brought to ground level, it would be at 25°C, and finding itself surrounded by 21°C air, it would rise back to 500 m. Thus the subadiabatic system would tend to dampen out vertical movement and is characterized by a limited vertical mixing.

The movement of plumes from smokestacks is governed by the lapse rate into which they are emitted, as illustrated by Example 11.1.

e • x • a • m • p • l • e 11.1

Problem   A stack 100 m tall emits a plume at 20°C. The prevailing lapse rates are shown in Figure 11–4. How high will the plume rise (assuming perfect adiabatic conditions)?

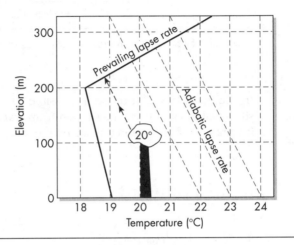

FIGURE 11-4   A typical prevailing lapse rate.

Solution   Note that the prevailing lapse rate is subadiabatic to 200 m and an inversion exists above 200 m. The smoke at 20°C finds itself surrounded by colder (18.5°C) air, and rises. As it rises, it cools, so that at 200 m it is 19°C. At about 220 m, the surrounding air is at the same temperature as the smoke (about 18.7°C) and the smoke ceases to rise.

An inversion is an extreme subadiabatic condition, and the vertical air movement within an inversion is almost nil. Some inversions, called *subsidence inversions*, are due to the movement of a large warm air mass over cooler air. Such inversions, typical in Los Angeles, last for several days and are responsible for the serious air pollution episodes. A more common type of inversion is the *radiation inversion*, caused by the thermal radiation of heat to the atmosphere from the earth. During the night, as the earth cools, the air close to the ground loses heat, thus causing an inversion (Figure 11-5). The pollution emitted during the night is caught under this lid and does not escape until the earth warms sufficiently to break the inversion.

In addition to inversions, serious air pollution episodes are almost always accompanied by fogs. These tiny droplets of water are detrimental in two ways.

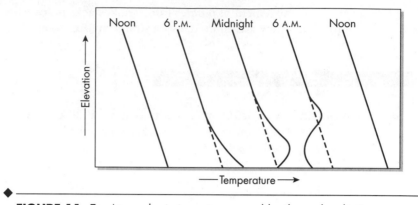

FIGURE 11-5 Atmospheric inversion caused by thermal radiation.

In the first place, fog makes it possible to convert SO_3 to H_2SO_4. Second, fog sits in valleys and prevents the sun from warming the valley floor and breaking inversions, often prolonging air pollution episodes.

11.2 ◆ Major Air Pollutants

11.2.1 Particulates

Particulate pollutants can be classified by several means.

- *Dust* is defined as solid particles that are:
  a. entrained by process gases directly from the material being handled or processed, e.g., coal, ash, and cement;
  b. direct offspring of a parent material undergoing a mechanical operation, e.g., sawdust from woodworking;
  c. entrained materials used in a mechanical operation, e.g., sand from sandblasting.
- Dusts from grain elevators and coal-cleaning plants typify the last class of particulate. Dust consists of relatively large particles. Cement dust, for example, is about 100 μ in diameter. (In air-pollution control parlance, a micrometer (μm) is often referred to as a micron (μ). This usage is adopted here.)

- A *fume* is also a solid particle, frequently a metallic oxide, formed by the condensation of vapors by sublimation, distillation, calcination, or chemical reaction processes. Examples of fumes are zinc and lead oxide resulting from the condensation and oxidation of metal volatilized in a high-temperature process. The particles in fumes are quite small, with diameters from 0.03 to 0.3 μ.

- An entrained liquid particle formed by the condensation of a vapor and perhaps by chemical reaction is called a *mist*. Mists typically range from 0.5 to 3.0 μ in diameter.

- *Smoke* is made up of entrained solid particles formed as a result of incomplete combustion of carbonaceous materials. Although hydrocarbons, organic acids, sulfur oxides, and nitrogen oxides are also produced in combustion processes, only the solid particles resulting from the incomplete combustion of carbonaceous materials are called smoke. Smoke particles have diameters from 0.05 to approximately 1 μ.

- Finally, a *spray* is a liquid particle formed by the atomization of a parent liquid. Sprays settle out under gravity.

Approximate size ranges of the various types of air pollutants are shown in Figure 11–6.

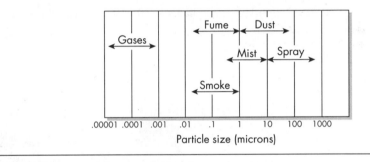

FIGURE 11–6 Definition of air pollutants by size.

11.2.2 Measurement of Particulates

The measurement of particulates is historically done using the *high-volume sampler* (or "hi-vol"). The high-volume sampler (Figure 11–7) operates much like a vacuum cleaner by simply forcing more than 2000 cubic meters (70,000 ft³) of air through a filter in 24 hours. The analysis is gravimetric; the filter is weighed before and after, and the difference is the particulates collected.

The air flow is measured by a small flow meter, usually calibrated in cubic feet of air per minute. Because the filter gets dirty during the 24 hours of operation, less air goes through the filter during the latter part of the test than in the beginning and the air flow must therefore be measured at both the start and end of the test period and the values averaged.

e • x • a • m • p • l • e **11.2**

Problem A clean filter is found to weigh 10.00 g. After 24 hours in a hi-vol, the filter plus dust weighs 10.10 g. The air flow at the start and end of the test is 60 and 40 ft³/minute, respectively. What is the particulate concentration?

Solution

$$\text{Weight of the particulates (dust)} = (10.10 - 10.00)g \times 10^6 \, \mu g/g$$

$$= 0.1 \times 10^6 \, \mu g$$

Average air flow $= (60 + 40)/2 = 50 \, ft^3/min$

Total air through the filter $= 50 \, ft^3/min \times 60 \, min/hr \times 24 \, hr/day \times 1 \, day$

$$= 72,000 \, ft^3$$

$$= 72,000 \, ft^3 \times 28.3 \times 10^{-3} \, m^3/ft^3$$

$$= 2038 \, m^3$$

Total suspended particulates $= (0.1 \times 10^6 \, \mu g)/2038 \, m^3 = 49 \, \mu g/m^3$

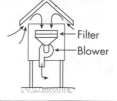

FIGURE 11-7 Hi-vol sampler.

The particulate concentration measured in this manner is often referred to as *total suspended particulates* (TSP) to differentiate it from other measurements of particulates.

Another widely used measure of particulates in the environmental health area is of *respirable particulates*, or those particulates that would be respired into lungs. These are generally defined as being less than 0.3 μ in size, and the measurements are with stacked filters. The first filter removes only particulates > 0.3 μ, and the second filter, having smaller spaces, removes the small respirable particulates.

The EPA has also recognized that the measurement of particulates can be badly skewed if a few really large particles happen to fall into the sampler. To get around this problem, they now measure particles smaller than 10 μ only. Designated symbolically as PM_{10}, for "particulate matter less than 10 microns," this measure is used in the ambient air quality standards.

Woodburning stoves are apparently particularly effective emitters of small particulates, of the PM_{10} variety. In one Oregon city, the concentration of PM_{10} concentrations reached above 700 $\mu g/m^3$, whereas the national ambient air quality standard (see below) is only 150 $\mu g/m^3$. Cities where the atmospheric conditions prevent the dispersal of wood smoke have been forced to pass woodburning bans during nights when the conditions could lead to high particulate levels in the atmosphere.

Interestingly, no continuous particulate measuring devices have yet been developed and accepted. The problem, of course, is that the measurement must be gravimetric, and it is difficult to construct a device that continuously weighs minute quantities of dust. Some indirect devices are used for estimating particulates, the most notable being the *nephelometer*, which actually measures light scatter, the assumption being that an atmosphere that contains particulates also scatters light. Unfortunately, different sizes of particulates scatter light differently and atmospheric moisture (fog) also interferes with the passage of light, but this should not be measured as particulates.

11.2.3   Gaseous Pollutants

In the context of air pollution control, gaseous pollutants include substances that are gases at normal temperature and pressure as well as vapors of substances that are liquid or solid at normal temperature and pressure. Among the gaseous pollutants of greatest importance in terms of present knowledge are carbon monoxide, hydrocarbons, hydrogen sulfide, nitrogen oxides, ozone and other oxidants, and sulfur oxides. Carbon dioxide should be added to this list because of its potential effect on climate. These and other gaseous air pollutants are listed in Table 11–1. Pollutant concentrations are commonly expressed as micrograms per cubic meter ($\mu g/m^3$).

11.2.4   Measurement of Gases

While the units of particulate measurement are consistently in terms of micrograms per cubic meter, the expression of the concentration of gases can be as

TABLE 11-1   Some Gaseous Air Pollutants

Name	Formula	Properties of Importance	Significance as Air Pollutant
Sulfur dioxide	SO_2	Colorless gas, intense choking odor, highly soluble in water to form sulfurous acid H_2SO_3	Damage to property, health, and vegetation
Sulfur trioxide	SO_3	Soluble in water to form sulfuric acid H_2SO_4	Highly corrosive
Hydrogen sulfide	H_2S	Rotten egg odor at low concentrations, odorless at high concentrations	Highly poisonous
Nitrous oxide	N_2O	Colorless gas, used as carrier gas in aerosol bottles	Relatively inert. Not produced in combustion
Nitric oxide	NO	Colorless gas	Produced during high-temperature, high-pressure, combustion. Oxidizes to NO_2
Nitrogen dioxide	NO_2	Brown to orange gas	Major component in the formation of photochemical smog
Carbon monoxide	CO	Colorless and odorless	Product of incomplete combustion. Poisonous
Carbon dioxide	CO_2	Colorless and odorless	Formed during complete combustion. Greenhouse gas

TABLE 11–1 *continued*			
Name	*Formula*	*Properties of Importance*	*Significance as Air Pollutant*
Ozone	O_3	Highly reactive	Damage to vegetation and property. Produced mainly during the formation of photochemical smog
Hydrocarbons	C_xH_y or HC	Many	Some hydrocarbons are emitted from automobiles and industries; others are formed in the atmosphere
Methane	CH_4	Combustible, odorless	Greenhouse gas
Chlorofluorocarbons	CFC	Nonreactive, excellent thermal properties	Depletes ozone in upper atmosphere

either parts per million (ppm) on volume to volume basis or as micrograms per cubic meter. The conversion from one to the other is

$$\mu g/m^3 = \frac{M \times 1000}{24.5} \times ppm$$

where M = molecular weight of the gas. This equation is applicable for conditions of 1 atmosphere and 25°C. For 1 atmosphere and 0°C (273°K), the constant becomes 22.4.

e • x • a • m • p • l • e **11.3**

Problem A stack gas contains carbon monoxide (CO) at a concentration of 10% by volume. What is the concentration of CO in $\mu g/m^3$? (Assume 25°C and 1 atmosphere pressure.)

Solution Since 1% by volume is 10,000 ppm, 10% by volume is 100,000 ppm. Since the molecular weight of CO is 28, the concentration in micrograms per cubic meter is

$$\frac{28 \times 1000}{24.5} \times 100,000 = 114 \times 10^6 \, \mu g/m^3$$

◆

The earliest gas-measurement techniques almost all involve the use of a *bubbler*, shown in Figure 11–8. The gas is literally bubbled through the liquid, which either reacts chemically with the gas of interest or into which the gas is dissolved. Wet chemical techniques are then used to measure the concentration of the gas.

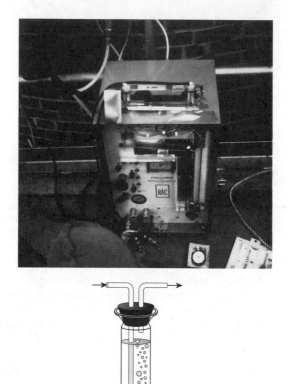

FIGURE 11–8   A typical bubbler for measuring gaseous air pollutants.

A simple (but now seldom used) bubbler technique for measuring SO_2 is to bubble air through hydrogen peroxide, so that the following reaction occurs:

$$SO_2 + H_2O_2 \rightarrow H_2SO_4$$

The amount of sulfuric acid formed can be determined by titrating the solution with a base of known strength.

One of the better third-generation methods of measuring SO_2 is the colorimetric pararosaniline method, in which SO_2 is bubbled into a liquid containing tetrachloromercurate (TCM). The SO_2 and TCM combine to form a stable complex. Pararosaniline is then added to this complex, with which it forms a colored solution. The amount of color is proportional to the SO_2 in the solution, and the color is measured with a spectrophotometer. (See Chapter 6 for ammonia measurement—another example of a colorimetric technique).

11.2.5 Measurement of Smoke

Air pollution has historically been associated with smoke—the darker the smoke, the more pollution. We now of course know that this isn't necessarily true, but many regulations (e.g., for municipal incinerators) are still written on the basis of smoke density.

The density of smoke has for many years been measured on the *Ringlemann scale*, devised in the late 1800s by Maxmilian Ringlemann, a French professor of engineering, running from 0 for white or transparent smoke to 5 for totally black opaque smoke. The test is conducted by holding a card such as shown in Figure 11-9 and comparing the blackness of the card to the smoke. A fairly dark smoke is thus said to be "Ringlemann 4," for example.

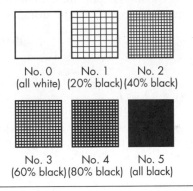

No. 0 (all white)   No. 1 (20% black)   No. 2 (40% black)

No. 3 (60% black)   No. 4 (80% black)   No. 5 (all black)

FIGURE 11-9 The Ringlemann scale.

11.2.6 Visibility

One of the obvious effects of air pollutants is the reduction in visibility. Loss of visibility is often defined as the condition when, in bright daylight, it is just possible to identify a large object such as a building, or at night, to be able to see a moderately bright light. This is of course a vague definition of visibility, but it is useful, especially for defining the limits of visibility.

Reductions in visibility can occur due to natural air "pollutants" such as terpenes from pine trees (that's why the Smokey Mountains are smokey) or from human-produced emissions. Many constituents can cause attenuated visibility, such as water droplets (fog) and gases (NO_2), but the most effective reduction in visibility is by small particulates. Particulates reduce visibility both by adsorbing the light and by scattering the light. In the first case, the light does not enter the eye of the viewer, and in the second case the scattering reduces the contrast between light and dark objects. Mathematically, the change in the intensity of a an object illuminated by a beam of intensity I, a distance x from the observer, is attenuated as

$$\frac{dI}{dx} = -\sigma I$$

where I = intensity of the light beam as viewed by the observer

   x = distance from the observer

   σ = constant that takes into account the atmospheric condition, often called the *extinction coefficient*

Integrated, this equation reads

$$\ln\left(\frac{I}{I_0}\right) = -\sigma x$$

where I_0 = is the intensity of the beam when the distance x approaches zero.

When the light intensity is reduced to about 2% of the unattenuated light, this is the lower limit of visibility for most people. If this value is substituted in the above equation, the distance x is then the limit of visibility, L_V, or

$$\ln(0.02) = -\sigma L_V$$

$$L_V = \frac{3.9}{\sigma}$$

As a rough approximation, the extinction coefficient is directly and linearly proportional to the concentration of particulates (for atmospheres containing less than 70% moisture). An approximate expression for visibility is then

$$L_V = \frac{1.2 \times 10^3}{C}$$

where L_V = the limit of visibility, km

   C = the concentration of particulates, $\mu g/m^3$

The constant 1.2×10^3 takes into account the conversion factor if L_V is in kilometers and C is micrograms per cubic meter. Remember that this is an approximate relationship and should be used only for atmospheres with less than 70% moisture.[1]

11.3 ◆ Sources and Effects of Air Pollution

Unwanted constituents in air, or air pollution, can have a detrimental effect on human health, on the health of other creatures, on the value of property, and on the quality of life. Some of the pollutants of human health concern are formed and emitted through natural processes. For example, naturally occurring particulates include pollen grains, fungus spores, salt spray, smoke particles from forest fires, and dust from volcanic eruptions. Gaseous pollutants from natural sources include carbon monoxide as a breakdown product in the degradation of hemoglobin, hydrocarbons in the form of terpenes from pine trees, hydrogen sulfide resulting from the breakdown of cysteine and other sulfur-containing amino acids by bacterial action, nitrogen oxides, and methane.

People-made sources of pollutants can be conveniently classified as stationary combustion, transportation, industrial process, and solid-waste disposal sources. The principal pollutant emissions from stationary combustion processes are particulate pollutants, as fly ash and smoke, and sulfur and nitrogen oxides. Sulfur oxide emissions are, of course, a function of the amount of sulfur present in the fuel. Thus, combustion of coal and oil, both of which contain appreciable amounts of sulfur, yields significant quantities of sulfur oxide.

Much of the knowledge of the effects of air pollution on people comes from the study of acute air pollution episodes. The two most famous episodes occurred in Donora, Pennsylvania, and in London, England. In both episodes, the pollutants affected a specific segment of the public—those individuals already suffering from diseases of the cardiorespiratory system. Another observation of great importance is that it was not possible to blame the adverse effects on any one pollutant. This observation puzzled the investigators (industrial hygiene experts), who were accustomed to studying industrial problems in which one could usually relate health effects to a specific pollutant. Today, after many years of study, we believe that the health problems during the episodes were attributable to the combined action of a particulate matter (solid or liquid particles) and sulfur dioxide, a gas. No one pollutant by itself, however, could have been responsible.

Except for these episodes, scientists have little information from which to evaluate the health effects of air pollution. Laboratory studies with animals are of some help, but the step from a rat to a person (anatomically speaking) is quite large.

Four of the most difficult problems in relating air pollution to health are unanswered questions concerning: (1) the existence of thresholds, (2) the total body burden of pollutants, (3) the time versus dosage problem, and (4) synergistic effects of various combinations of pollutants.

Threshold.   The existence of a threshold in health effects of pollutants has been debated for many years. As discussed in Chapter 1, there are several dose–response curves possible for a dose of a specific pollutant (e.g., carbon monoxide) and the response (e.g., reduction in the blood's oxygen-carrying capacity). One possibility is that there will be no effect on human metabolism until a critical concentration (the threshold) is reached. However, some pollutants can produce a detectable response for any finite concentration. Nor do these curves have to be linear. In air pollution, the most likely dose–response relationship for many pollutants is nonlinear, without an identifiable threshold but also a minimal response up to a higher concentration, at which point the response becomes severe. The problem is that for most pollutants the shapes of these curves are unknown.

Total Body Burden.   Not the entire dose of pollutants comes from air. For example, although a person breathes in about 50 μg/day of lead, the daily intake of lead from water and food is about 300 μg/day. In the setting of air quality standards for lead it must therefore be recognized that most of the lead intake is from food and water.

Time Versus Dosage.   Most pollutants require time to react, and the time of contact is as important as the level. The best example of this is the effect of carbon monoxide, as illustrated in Figure 11–10. CO reduces the oxygen-carrying capacity of the blood by combining with the hemoglobin and forming carboxyhemoglobin. At about 60% carboxyhemoglobin concentration, death results from lack of oxygen. The effects of CO at sublethal concentrations are usually reversible. Because of the time-response problem, ambient air quality standards are set at maximum allowable concentrations for a given time.

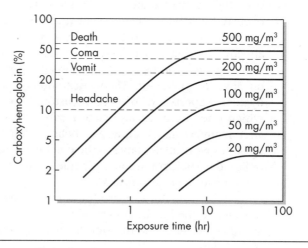

FIGURE 11–10 Effect of carbon monoxide on health. (Adapted from W. Agnew, *Proceedings of the Royal Society*, A307, 153, 1968.)

Synergism.   Synergism is defined as an effect that is greater than the sum of the parts. For example, black lung disease in coal miners occurs only when the miner is also a cigarette smoker. Coal mining by itself, or cigarette smoking by itself, will not cause black lung, but the synergistic action of the two puts miners who smoke at high risk.

The major target of air pollutants is the respiratory system, pictured in Figure 11–11. Air (and entrained pollutants) enter the body through the throat and nasal cavities and pass to the lungs through the trachea. In the lungs, the air moves through bronchial tubes to the alveoli, small air sacks in which the gas transfer takes place. Pollutants are either absorbed into the bloodstream or moved out of the lungs by tiny hair cells called cilia, which are continually sweeping mucus up into the throat. The respiratory system can be damaged by both particulate and gaseous pollutants.

Particles greater than 0.1 μ will usually be caught in the upper respiratory system and swept out by the cilia. However, particles less than 0.1 μ in diameter can move into the alveoli, where there are no cilia, and they can stay there for a long time and cause damage to the lung. Droplets of H_2SO_4 and lead particles represent the most serious forms of particulate matter in air.

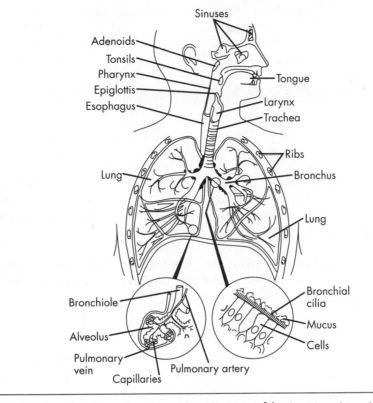

FIGURE 11–11   The respiratory system. (Courtesy of the American Lung Association.)

The effects of carbon monoxide are potentially deadly. Human hemoglobin (Hb) has a carbon monoxide affinity 210 times greater than its affinity for oxygen, and thus CO combines readily with hemoglobin to form carboxyhemoglobin (COHb). The formation of COHb reduces the hemoglobin available to carry oxygen, and this can cause death by asphyxiation.

Ozone (O_3) in the atmosphere is an eye irritant and can cause damage to materials such as rubber in automobile tires. Ozone is usually a part of urban air and photochemical smog.

A number of diseases are thought to be causally related to air pollution, including lung cancer, emphysema, and asthma. There is overwhelming evidence that air pollution can increase the risk of these diseases, especially when tied synergistically to cigarette smoking, yet the actual cause-and-effect relationship is not medically proven. It is incorrect to assert that air pollution, no matter how bad (or cigarettes, for that matter), *cause* lung cancer or other respiratory diseases.

Sulfur dioxide is perhaps the most insidious air pollutant. It acts as an irritant and restricts air flow and slows the action of the cilia. Since it is highly soluble, it should be readily removed by the mucus membrane, but studies have shown that it can invade the deep reaches of the lung by first adsorbing onto tiny particles, then using this transport to reach the deep lung—a classic case of synergistic action.

The amount of sulfur oxide produced can be calculated if the sulfur content of the fuel is known, as the following example illustrates.

e • x • a • m • p • l • e   11.4

Problem   During a two-week-long major air pollution episode in London in 1952, an estimated 25,000 metric tons of coal was burned that had an average sulfur content of about 4%. The mixing depth (height of the inversion layer or cap over the city, which prevented the pollutants from escaping) was about 150 m, over an area of 1200 km². If initially there was no SO_2 in the atmosphere (a conservative assumption), what is the expected SO_2 concentration at the end of two weeks?

Solution   Use the "box model" for calculating the concentration. That is, consider the volume over London a black box, and use the well-worn materials balance equation:

$$\begin{bmatrix} \text{rate of } SO_2 \\ \text{ACCUMULATED} \end{bmatrix} = \begin{bmatrix} \text{rate of } SO_2 \\ \text{IN} \end{bmatrix} - \begin{bmatrix} \text{rate of } SO_2 \\ \text{OUT} \end{bmatrix}$$
$$+ \begin{bmatrix} \text{rate of } SO_2 \\ \text{PRODUCED} \end{bmatrix} - \begin{bmatrix} \text{rate of } SO_2 \\ \text{CONSUMED} \end{bmatrix}$$

The rate of SO_2 out was zero, since nothing escaped from under this *haze hood*. The rate produced and consumed is also assumed to be zero if it is assumed that sulfur oxide was not created or destroyed in the atmosphere. The rate of SO_2 concentration increasing in the box is constant, or

$$\begin{bmatrix} \text{rate of } SO_2 \\ \text{ACCUMULATED} \end{bmatrix} = \begin{bmatrix} \text{rate of } SO_2 \\ \text{IN} \end{bmatrix} - 0 + 0 - 0$$

$$\frac{dA}{dt} = k$$

or

$$A = A_0 + kt$$

where A = the mass of SO_2.

The sulfur emitted is calculated as = 25,000 metric tons of coal burned per week × 0.04 = 1000 metric tons of sulfur emitted per week. Sulfur has an atomic weight of 32 and sulfur dioxide has a molecular weight of 64. The sulfur dioxide emitted per week is (64/32) × 1000 = 2000 metric tons per week.

If the initial concentration, A_0, is assumed to be zero,

$$A = 0 + 2000 \text{ (tons/week)} \times 2 \text{ weeks} = 4000 \text{ metric tons of } SO_2$$

accumulated in the atmosphere over London.

The volume into which this is mixed is

$$150 \text{ m} \times 1200 \text{ km}^2 \times 10^6 \text{ m}^2/\text{km}^2 = 180,000 \times 10^6 \text{ m}^3$$

The concentration of SO_2 at the end of two weeks should have been

$$\frac{4000 \text{ metric tons} \times 10^6 \text{ g/metric ton} \times 10^6 \text{ }\mu\text{g/g}}{180,000 \times 10^6 \text{ m}^3} = 22,000 \text{ }\mu\text{g/m}^3$$

◆

The actual peak concentration of SO_2 during the 1952 London episode was less than 2000 $\mu\text{g/m}^3$. Yet the calculation above shows more than 10 times that value. Where did all the SO_2 go? The answer lies in the assumption used in the calculation that no SO_2 was consumed. In fact, there is a continuous scavenging of SO_2 from the atmosphere by contact with buildings, vegetation, wildlife, and human beings. Most important, sulfur oxides seem to be a major constituent of poorly buffered precipitation, more commonly known as acid rain.

11.3.1 Sulfur and Nitrogen Oxides and Acid Rain

One way in which SO_2 is removed from the atmosphere is the formation of acid rain. Normal, uncontaminated rain has a pH of about 5.6, but acid rain can be as low as pH 2 or even below. Acid rain formation is a complex process, and the dynamics are not fully understood. In its simplest terms, SO_2 is emitted from the combustion of fuels containing sulfur, the reaction being

$$S + O_2 \xrightarrow{\text{heat}} SO_2$$

$$SO_2 + O \xrightarrow{\text{sunlight}} SO_3$$

$$SO_3 + H_2O \longrightarrow H_2SO_4$$

H_2SO_4 is of course sulfuric acid. Sulfur oxides do not literally produce sulfuric acid in the clouds, but the idea is the same. The precipitation from air containing high concentrations of sulfur oxides is poorly buffered and readily drops its pH.

Nitrogen oxides, emitted mostly from automobile exhaust, but also from any other high-temperature combustion, contribute to the acid mix in the atmosphere. The chemical reactions that apparently occur with nitrogen are

$$N_2 + O_2 \longrightarrow 2NO$$

$$NO + O_3 \longrightarrow NO_2 + O_2$$

$$NO_2 + O_3 + H_2O \longrightarrow 2HNO_3 + O_2$$

where HNO_3 is of course nitric acid.

The effect of acid rain has been devastating. Hundreds of lakes in North America and Scandinavia have become so acidic that they no longer can support fish life. In a recent study of Norwegian lakes, more than 70% of the lakes having a pH of less than 4.5 contained no fish, while nearly all lakes with a pH of 5.5 and above contained fish. The low pH not only affects fish directly, but contributes to the release of potentially toxic metals such as aluminum, thus magnifying the problem.

In North America, acid rain has already wiped out all fish and many plants in 50% of the high mountain lakes in the Adirondacks. The pH in many of these lakes has reached such levels of acidity as to replace the native plants with acid-tolerant mats of algae.

The deposition of atmospheric acid on freshwater aquatic systems prompted EPA to suggest a limit of from 10 to 20 kg SO_4^{-2} per hectare per year. If "Newton's law of air pollution" is used (what goes up must come down), it is easy to see that the amount of sulfuric and nitric oxides emitted is vastly greater than this limit. For example, just for the state of Ohio alone, the total annual emissions are 2.4×10^6 metric tons of SO_2 per year. If all this is converted to SO_4^{-2} and is deposited on the state of Ohio, the total would be 360 kg per hectare per year.[2]

But not all of this sulfur falls on the folks in Ohio, and much of it is exported by the atmosphere to places far away. Similar calculations for the sulfur emissions for the northeastern United States indicates that the rate of sulfur emission is 4 to 5 times greater than the rate of deposition. Where does it all go?

The Canadians have a ready and compelling answer. They have for many years blamed the United States for the formation of most of the acid rain that invades across the border. Similarly, much of the problem in Scandinavia can be traced to the use of tall stacks in Great Britain and the lowland countries of continental Europe. For years British industry simply built taller and taller stacks as a method of air pollution control, reducing the immediate ground level concentration, but emitting the same pollutants into the higher atmosphere. The air quality in the United Kingdom improved, but at the expense of acid rain in other parts of Europe.

Pollution across political boundaries is a particularly difficult regulatory problem. The big stick of police power is no longer available. Why *should* the U.K. worry about acid rain in Scandinavia? Why *should* the Germans clean up the Rhine before it flows through The Netherlands? Why *should* Israel stop taking more than its share of water out of the Dead Sea, which it supposedly shares with Jordan? Laws are no longer useful, and threatened retaliation is unlikely. What forces are there to encourage these countries to do the right thing? Is there such a thing as "international ethics"?

11.3.2   Photochemical Smog

An important approach to classification of air pollutants is that of *primary* and *secondary pollutants*. A primary pollutant is defined as one that is emitted as such to the atmosphere, whereas secondary pollutants are actually produced in the atmosphere by chemical reactions. The components of automobile exhaust are particularly important in the formation of secondary pollutants. The well-known and much discussed Los Angeles smog is a case of secondary pollutant formation. Table 11–2 lists in simplified form some of the key reactions in the formation of photochemical smog.

TABLE 11–2	Simplified Reaction Scheme for Photochemical Smog			

NO_2
Nitrogen
    dioxide
	+ Light	→	NO
			Nitric
			    oxide
				+ O
				Atomic
				    oxygen

O	+ O_2 Molecular oxygen	→	O_3 Ozone	
O_3	+ NO	→	NO_2	+ O_2
O	+ HC Hydro- carbon	→	$HCO°$ Radical	
$HCO°$	+ O_2	→	$HCO_3°$ Radical	
$HCO_3°$	+ HC	→	Aldehydes, ketones, etc.	
$HCO_3°$	+ NO	→	$HCO_2°$ Radical	+ NO_2
$HCO_3°$	+ O_2	→	O_3	+ $HCO_2°$
$HCO_x°$ radical	+ NO_2	→	Peroxyacetyl nitrates	

The reaction sequence illustrates how nitrogen oxides formed in the combustion of gasoline and other fuels and emitted to the atmosphere are acted upon by sunlight to yield ozone (O_3), a compound not emitted as such from sources and hence considered a secondary pollutant. Ozone in turn reacts with hydrocarbons to form a series of compounds that includes aldehydes, organic acids, and epoxy compounds. The atmosphere can be viewed as a huge reaction vessel wherein new compounds are being formed while others are being destroyed.

The formation of photochemical smog is a dynamic process. Figure 11–12 is an illustration of how the concentrations of some of the components vary during the day. Note that as the morning rush hour begins the NO levels increase, followed quickly by NO_2. As the latter reacts with sunlight, O_3 and other oxidants are produced. The hydrocarbon level similarly increases at the beginning of the day and then drops off in the evening.

The reactions involved in photochemical smog remained a mystery for many years. Particularly baffling was the formation of high ozone levels. As seen from the first three reactions in Table 11–2, for every mole of NO_2 reacting to make atomic oxygen and hence ozone, one mole of NO_2 is created from reaction with the ozone. All of these reactions are fast so that if these are the only reactions occurring in photochemical smog, it should be impossible to build up high ozone levels. How, then, could the ozone concentrations build to such high levels?

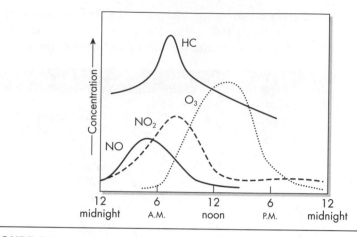

FIGURE 11–12 Formation of photochemical smog during a sunny 24-hour period.

One answer is that NO enters into other reactions, especially with various hydrocarbon radicals, and thus allows excess ozone to accumulate in the atmosphere (the 7th reaction in Table 11–2). In addition, some hydrocarbon radicals react with molecular oxygen and also produce ozone.

The chemistry of photochemical smog is still not clearly understood. This was witnessed by the attempt to reduce hydrocarbon emissions to control ozone levels. The thinking was that if HC is not available, the O_3 will be used to oxidize NO to NO_2, thus using the available ozone. Unfortunately, this control strategy was a failure, and the answer seems to be that all primary pollutants involved in photochemical smog formation must be controlled.

11.3.3   Ozone Depletion

Ozone (O_3) is an eye irritant at usual urban levels, but urban ozone should not be confused with stratospheric ozone which occurs 7 to 10 miles above the earth's surface. Stratospheric ozone acts as an ultraviolet radiation shield. Chlorofluorocarbons (CFCs), emitted mostly from refrigeration units and spray cans, can apparently alter this shield and increase the risk of skin cancer as well as change global ecology in unpredictable ways, including contributing to global warming.

Two of the most important CFCs are trichlorofluoromethane, $CFCl_3$, and dichlorodifluoromethane, CF_2Cl_2, both of which are inert and nonwater soluble, and therefore do not wash out of the atmosphere. They drift into the upper atmosphere and are eventually destroyed by shortwave solar radiation, releasing chlorine, which can react with ozone. The depletion of ozone can allow the ultraviolet radiation to pass through unimpeded, and this can have a serious potential effect on the incidence of skin cancer. While basking in the sun is not a good idea in the first place, doing so when the ozone layer is not effective in screening out much of the ultraviolet rays is even more risky.

11.3.4   Global Warming

The earth acts as a reflector to the sun's rays, receiving the radiation from the sun, reflecting some of it into space (called *albedo*), and adsorbing the rest, only to reradiate this into space as heat. In effect the earth acts as a wave converter, receiving the high-energy, high-frequency radiation from the sun and converting most of it into low-energy, low-frequency heat to be radiated back into space. In this manner, the earth maintains a balance of temperature, so that

$$\begin{bmatrix} \text{energy from the sun} \\ \text{IN} \end{bmatrix} = \begin{bmatrix} \text{energy radiated back to space} \\ \text{OUT} \end{bmatrix}$$

Unfortunately, some gases such as methane (CH_4) and carbon dioxide (CO_2) adsorb radiation at wavelengths approximately the same as the heat radiation trying to find its way back to space. Because the radiation is adsorbed in the atmosphere by these gases, the temperature of the atmosphere increases, heating the earth. The system works exactly like a greenhouse in that light energy (shortwave, high-frequency radiation) passes through the greenhouse glass, but the long-wavelength, low-frequency heat radiation is prevented from escaping. The gases that adsorb the heat energy radiation are properly referred to as *greenhouse gases* since they in effect cause the earth to heat up just like a greenhouse.

Although it is fairly easy to measure the concentrations of greenhouse gases such as CO_2, CH_4, and N_2O, and to show that the levels of these gases have been increasing during the past 50 years, it is another matter to argue from empirical evidence that the earth's temperature is actually increasing. The temperature of the earth undergoes continual change, with fluctuation frequencies ranging from a few years to thousands of years. Even if it were possible to measure accurately the earth's temperature, there would be no proof that the change is being caused by the higher concentration of the greenhouse gases.

There seems to be growing consensus, however, that even though it is not possible to prove without doubt that global warming is occurring, the net effect can be so devastating to the earth that it would be imprudent to simply sit back and wait for the irrefutable proof. By then the change could be so immense that the effect might be irreversible.

The most remarkable thing about global temperature might be that it has been so constant over millions of years. Somehow the earth has been able to develop the thermal conditions suitable for the creation and support of life. Speculating on this idea, James Lovelock suggested the Gaia hypothesis, that the earth is actually a single entity (mother nature), which lives much like any other creature and has to adapt to changing conditions as well as to fight off diseases.[3] The name Gaia comes from the Greek name for the nurturing earth goddess. Some Gaians have interpreted this notion in its broadest and most spiritual sense, taking the view that the earth really is one organism, albeit an unusual one; and it has many of the characteristics of other organisms. Others (including James Lovelock, who has been a little embarrassed about this Gaia business) see the hypothesis as nothing but a feedback–response model and do not ascribe mysticism to it.

But suppose humans are simply one part of a whole living organism, the earth, much like brain cells are a part of the animal. If this is true, then it makes no sense to destroy one's own body, and therefore it makes no sense for humans to destroy the rest of the creatures that co-inhabit the earth. To think of the earth as a living organism suggests a spiritual rather than a scientific approach, especially if it is taken literally.

If we accept this notion, some interesting extensions can result. One could speculate (and this is pure whimsy!) that the earth (Gaia) is still developing and is going through adolescent stages. Most notably, it has not settled on the carbon balance. Millions of years ago, most of the carbon on earth used to be in the atmosphere as carbon dioxide, and the preponderance of CO_2 promoted the development of plants. But as the plants grew, they soon started to rob the atmosphere of carbon dioxide, replacing it with oxygen. This change in atmospheric gases promoted the growth of animals, which converted the oxygen back to carbon dioxide, once again seeking a balance. Unfortunately, Gaia made a mistake and did not count on the animals dying in such huge numbers, trapping the carbon in deep geologic deposits, which eventually became coal, oil, and natural gas.

How to get this carbon out? What is needed is some semi-intelligent being with opposable digits. So Gaia invented humans! It can therefore be argued that our sole purpose on earth is to dig up the carbon deposits as rapidly as possible and to liberate the carbon dioxide. This is the long-sought-for "meaning of life."

Unfortunately, Gaia again messed up and did not count on the humans becoming so prolific and destructive, and especially did not count on humans inventing nuclear power, which eliminates their sole purpose for existing. The sheer numbers of humans (especially if they are not going to dig up the carbon) become a problem, much as a pathogenic bacterium causes an infection in an organism. It is not a coincidence that the growth of the human population on earth is not at all unlike the growth of cancer cells in a malignancy.

So next Gaia has to limit the number of humans, and she will do so by developing "antihumatics" (as in antibiotics), which will kill off a sufficient number of humans and once again bring the earth to a proper balance. The increase in highly resistant strains of bacteria and viruses may be the first indications of a general culling of the human population.

The application of the Gaia hypothesis to environmental ethics leads to existentialism—the notion that life is meaningless (except as a burner of fossil fuel) and that one should not concern oneself with the environment. Grim stuff.

11.3.5   Other Sources of Air Pollutants

Nitrogen oxides are formed by the thermal fixation of atmospheric nitrogen in high-temperature processes. Accordingly, almost any combustion operation will produce nitric oxide (NO). Other pollutants of interest from combustion processes are sulfur oxides, organic acids, aldehydes, ammonia, and carbon monoxide. The amount of carbon monoxide emitted relates to the efficiency of the combustion operation, that is, a more efficient combustion operation will oxidize more of the carbon present to carbon dioxide, reducing the amount of carbon monoxide emitted. The quantity of sulfur oxides emitted depends entirely on the amount of sulfur in the fuel. Of the fuels used in stationary combustion, coal contains the largest fraction of sulfur while natural gas contains practically no sulfur. As an added advantage, the particulate emissions from the combustion of natural gas are almost nil.

Transportation sources, particularly automobiles using the internal combustion engine, constitute a major source of air pollution. Particulate emissions from the automobile include smoke and lead particles, the latter usually as halogenated compounds. Smoke emissions, as in any other combustion operation, are due to the incomplete combustion of carbonaceous material. However, lead emissions relate directly to the addition of tetraethyl lead to the fuel as an antiknock compound. Gaseous pollutants from transportation sources include carbon monoxide, nitrogen oxides, and hydrocarbons. Hydrocarbon emissions result from incomplete combustion and evaporation from the crankcase, carburetor, and gasoline tank.

Pollutant emissions from industrial processes reflect the ingenuity of modern industrial technology. Thus, nearly every imaginable form of pollutant is emitted in some quantity by some industrial operation.

Although solid waste disposal operations need not be a major source of air pollutants, many communities still permit backyard burning of solid waste. Other communities use incinerators for solid-waste management. Burning, whether in the backyard or in incinerators, is an attempt to reduce the volume of waste, but may produce instead a variety of difficult-to-control pollutants. Inefficient systems contribute to many undesirable odors, as well as carbon monoxide, small amounts of nitrogen oxides, organic acids, hydrocarbons, aldehydes, and smoke.

11.3.6 Indoor Air

A sticky problem for the Environmental Protection Agency (EPA) has been control of indoor air quality. Does EPA's mandate include the indoor environment, and should it in effect become a health agency, or should it limit its concern to outdoor air? Pragmatically, if EPA does not address the problems of indoor air, no other federal agency seems ready to do so.

Indoor air quality is of importance to health simply because we spend so much time indoors and the quality of the air we breathe is seldom monitored. Contaminated indoor air can cause any number of health problems, including eye irritation, headache, nausea, sneezing, dermatitis, heartburn, drowsiness, and many other symptoms. These problems can arise as a result of breathing harmful pollutants such as

- Asbestos—from fireproofing and vinyl floors
- Carbon monoxide—from smoking, space heaters, stoves
- Formaldehyde—from carpets, ceiling tile, paneling

- Particulates—from smoking, fireplaces, dusting
- Nitrogen oxides—from kerosene stoves, gas stoves
- Ozone—from photocopying machines
- Radon—diffused from the soil
- Sulfur dioxide—from kerosene heaters
- Volatile organics—from smoking, paints, solvents, cooking

Although the air exchange and cleaning in office buildings and many apartments is controlled, private residences usually depend on natural ventilation to provide air exchange.

Most houses are in fact poorly sealed, and air leakage takes place at many locations including through doors and windows, exhaust vents, and chimneys. In warmer weather some homes of course have forced ventilation, with whole house fans or individual window fans.

In calculating the ventilation of any enclosure such as a home or office, engineers use the concept of *air change*, defined as

$$A = \frac{q}{V}$$

where A = number of air changes per hour, hr^{-1}
$\quad q$ = flow rate, m^3/hr
$\quad V$ = volume of enclosure, m^3

e · x · a · m · p · l · e 11.5

Problem   A small room is to be used for a copy machine, and there is concern that the ozone level may be too high unless the room is ventilated. The volume of the room is 700 ft^3, and it is recommended that the number of air changes per hour be 30. What flow of air must the fan deliver?

Solution   From the above equation,

fan air handling capacity = 30/hr × 700 ft^3 = 21,000 ft^3/hr

The emission of indoor air pollutants can be handily analyzed using the "black box" technique, mainly because an enclosure literally is a box. Pollutants are emitted within the box and are completely and ideally mixed, while clean air is brought in and contaminated air is flushed out. The materials balance in terms of the pollutants is

$$
\begin{bmatrix} \text{rate of} \\ \text{pollutants} \\ \text{ACCUMULATED} \end{bmatrix} = \begin{bmatrix} \text{rate of} \\ \text{pollutants} \\ \text{IN} \end{bmatrix} - \begin{bmatrix} \text{rate of} \\ \text{pollutants} \\ \text{OUT} \end{bmatrix} \\ + \begin{bmatrix} \text{rate of} \\ \text{pollutants} \\ \text{PRODUCED} \end{bmatrix} - \begin{bmatrix} \text{rate of} \\ \text{pollutants} \\ \text{CONSUMED} \end{bmatrix}
$$

If steady state is assumed, then the first term is zero. The second term can be assumed zero if the incoming air is clean, and since the pollutants are not consumed, the last term drops out. Hence,

$$0 = 0 - [\text{rate OUT}] + [\text{rate PRODUCED}] - 0$$

The rate of pollutants leaving is equal to

$$Q = CAV$$

where Q = rate of pollutant leaving the room, mg/hr
C = concentration of pollutants in the room, mg/m³
A = air change rate, number per/hour
V = volume of the enclosure, m³

The rate of pollutants being produced is called the *source strength*, S, expressed as milligrams of pollutants emitted per hour. The concentration of the pollutant in the room and in the exiting air is

$$C = S/(AV)$$

---

e • x • a • m • p • l • e   11.6

Problem   Smokers in a room (10 ft × 10 ft × 8 ft) are smoking so that there is on average one cigarette burning at all times. If the emission rate of a cigarette is 86 mg/hr of CO, and if the ventilation is 0.2 change per hour, what is the level of CO in the room?

Solution

$$C = (86 \text{ mg/hr})/(800 \text{ ft}^3 \times 0.2/\text{hr}) = 0.537 \text{ mg/ft}^3$$

This converts to 19,000 μg/m³. Is this a concentration that would be of concern? (Refer to Figure 11–10, on page 338.)

◆

If the pollutant is decaying or being removed by adhering to surfaces, then the "CONSUMED" term cannot be assumed to be zero. The decay or removal of the pollutant can be assumed to be first order, or proportional to the concentration in the enclosure. Similarly, if the ventilation air is not perfectly clean and contains some of the pollutant, the "IN" term cannot be zero. The steady-state equation then must read

$$0 = [\text{rate IN}] - [\text{rate OUT}] + [\text{rate PRODUCED}] - [\text{rate CONSUMED}]$$

$$0 = C_0 AV - CAV + S - KCV$$

or

$$C = \frac{S/V + C_0 A}{A + K}$$

where C_0 = ambient air quality, air used for ventilation, mg/m³

A = air change rate, per hour

V = volume of enclosure

C = concentration in the enclosure, or exhaust concentration, mg/m³

S = source strength, mg/hr

K = rate of decay constant, per hour

One of the most insidious indoor air pollutants is secondary smoke from smoking cigarettes and pipes. Most of the particulates in a cigarette are emitted into the room without being inhaled by the smoker, and these are then inhaled by everyone else in the room. Cigarette smoke also contains high amounts of CO, and when several persons are smoking in a room, the CO level can become high enough to affect performance. One of the smallest rooms we commonly live in is our car, and cigarette smokers can significantly affect the CO level in a car. Smokers exhale high levels of CO even when they are not smoking, and this has been blamed for the "sleepy driver syndrome" in commuter cars.

Another important and troublesome indoor air pollutant is *radon*, a naturally emitted gas entering homes through the basement, well water, and even building materials. Radon and its radioactive daughters are part of the natural decay chain beginning with uranium and ending with lead. The decay products of radon—polonium, lead, and bismuth—are easily inhaled and can readily find their way into lungs. The most important health effect of radon therefore is lung cancer. Radon control is discussed further in Chapter 14.

There are no effective means of correlating the inhalation of radon gas and incidence of lung cancer, and all studies depend on very high exposure rates such as to uranium miners. Relatively speaking, however, the risk of radon in the home is considerable when compared to other potential sources of cancer. Because radon is a carcinogen, there is no threshold, and therefore there is no "safe level" of radon in the home. Nevertheless, the EPA suggests that residents living in homes that contain 4 pCi/L (pico Curies per liter) due to radon should consider taking corrective action, and at 8 pCi/L, such action is recommended. Mitigating action usually involves the sealing of the basement and other soil contact areas, and the ventilation of basements to prevent the radon gas from seeping upstairs into the living quarters.

It is estimated that the risk of contracting lung cancer due to living for 70 years in a house containing 1.5 pCi/L is about 0.3 percent.[4] This is a high risk when compared to other regulated carcinogens, which are in the range of 1×10^{-4} percent. Radon exposure in fact ranks as the first potential problem in EPA's list of problem areas, with an expectation of between 5000 and 20,000 lung cancers annually from exposure to radon and its progeny in homes.

Although this may seem to represent a major problem, the risk of harm by radon still pales when compared to the voluntary risks we routinely impose on ourselves, such as driving cars, drinking alcohol, and smoking. In fact, living all of one's life in a home with between 10 and 20 pCi/L represents an equivalent risk of smoking one pack of cigarettes per day.

11.4 ◆ Air Quality Standards

Historically, local communities have been afraid to enact strict air pollution ordinances for fear of driving away industry. While there are cheaper and less restrictive locations, industrial plants could threaten to leave town, taking much-needed jobs with them. In the United States, only federal legislation has been able to prevent this type of blackmail.

But even with the present federal legislation, industry has the option to move "offshore," to the Caribbean or to Latin America, where pollution-control regulations are not as stringent. This seems to be a clearly unethical trend, transporting pollution to other countries less able to resist such contamination.

However, the United States grew rich in part because we were able to produce better products at cheaper prices, and one reason the cost was low was because we did not bother with pollution control. The water and air were free, and there was "so much of it" that it made little sense to clean up. Nature seemed to do a pretty good job by itself.

Now countries considerably poorer than the United States are saying that they would like to become just as wealthy and would like to use their water and air the same way—that is, pollute it. They welcome American industry that promises to build plants in their countries and decline to implement strict pollution regulations.

What should be the ethical stance of American industry in this case? If industry declines to invest in the poorer countries, these countries are deprived of income and benefits, and the gap between poor and rich nations will increase. If, however, industry invests in these countries and builds manufacturing facilities, industry must have some advantage such as minimal pollution-control regulations to make it worth their while, and hence the air and water will be polluted to what we would consider unacceptable levels. What ethical responsibility, if any, does industry have to implement American pollution-control standards for facilities located in countries that do not require such strict standards?

11.4.1 Air Quality Legislation in the United States

The history of air quality standards has been long and often tumultuous. At first, all air pollution was thought of as smoke pollution, and the first legislation governing smoke was passed in Los Angeles in 1905. Los Angeles sits in a basin that is frequently covered by a thermal inversion, and even from the days of the Conquistadors, it was evident that smoke from campfires rose only to a set level and remained there.

By 1943, the number of automobiles in Los Angeles had increased to where they were experiencing the world's first photochemical smog. This phenomenon was thought at first to be due to industries, and unique to Los Angeles. It took many years for both of these myths to be exposed. Angelenos simply did not want to believe that their much-beloved cars were the cause of such putrid air. It became obvious that something had to be done about automobile emissions, but this was a classic uphill battle for scientists and regulators. The automobile lobby, which includes the automobile manufacturers, the oil industry, and the people who wanted cheap personal transportation, engaged in a now-notorious campaign of foot-dragging, denial, and even obfuscation. They launched public relations and lobbying campaigns to prevent or slow down any tampering with cars or fuels. When it finally became evident to all that smog in Los Angeles was indeed due mostly to internal-combustion engines, the industry insisted that this was a special case for Los Angeles, and that national legislation was not necessary. Photochemical smog could not happen anywhere else, they claimed.

Finally, as it became evident that both California and the United States were about to pass "technology-driving" legislation restricting auto emissions, the American auto industry bemoaned their inability to meet these goals and warned that cars would become undrivable. Much of their bombast was blunted when Honda came out with a vehicle that not only met all of the stringent exhaust requirements, but also got 40 miles to a gallon of gas!

The 1963 Clean Air Act, a major piece of legislation that for the first time anywhere set both emission and ambient air quality limits, was passed in great part due to the efforts of Senator Edmund Muskie of Maine. The act required the federal government to set ambient air quality standards for seven major air pollutants. These were not implemented until 1970, and the states were given until 1975 to meet them. So as not to drag out this gruesome tale, suffice it to say that many states have yet to meet these standards, and the city of Los Angeles may have to go to extreme measures to reduce its photochemical smog.

The 1990 amendments to the Clean Air Act add more than 180 hazardous materials (now known as *air toxics*) to the ambient air quality standards and requires significant cuts in the emission of sulfur and nitrogen oxides, precursors of acid rain. The act also extends the deadline for meeting the ambient ozone standard until the year 2007—which makes it 44 years since the original 1963 Clean Air Act!

11.4.2   Emission and Ambient Air Quality Standards

The emission standards are regulated (and for the most part set) by the individual state air quality offices. As an example of an emission standard, an incinerator might not be allowed to exceed emissions of X $\mu g/m^3$ of particulates at a specified CO_2 minimum level. The latter is necessary since particulate concentrations can be reduced by simply diluting with excess air. But since air contains negligible CO_2, the minimum CO_2 level prevents emission standard attainment by simple dilution.

The EPA also has the authority to set national emission performance standards for hazardous air pollutants. The list of pollutants is growing daily, and once a pollutant is *listed*, the EPA can enforce the emission standards for larger facilities. At present, listed are chemicals such as asbestos, benzene, ethylene glycol, methanol, phenol, styrene, vinyl chloride, and hundreds more.

The second type of air quality standards parallel the stream standards in water quality. These standards specify the minimum quality of ambient air. Included in these standards are particulates, sulfur oxide, carbon monoxide, photochemical oxidants, hydrocarbons, nitrogen oxides, and lead. As data become available, other standards will be developed.

The ambient air quality standards are of two kinds: primary and secondary. Primary standards are intended to relate to human health, whereas secondary standards address such problems as corrosion, animal health, visibility, and so on. Table 11–3 is a listing of the present national ambient air quality standards (NAAQS).

TABLE 11-3 National Ambient Air Quality Standards		
Pollutant	*Primary Standard* $(\mu g/m^3)$	*Secondary Standard* $(\mu g/m^3)$
Particulate matter less than 10 μ in diameter		
24-hour average	150	150
annual geometric mean	50	50
Sulfur dioxide		
3-hour maximum		1300
24-hour average	365	60
annual geometric mean	80	260
Nitrogen oxides		
annual geometric mean	100	100
Carbon monoxide		
1-hour maximum	10,000	10,000
8-hour maximum	40,000	40,000
Ozone		
1-hour maximum	210	210
Hydrocarbons		
3-hour maximum	160	160
Lead		
quarterly arithmetic mean	1.5	1.5

Areas in the United States where the national ambient air quality standards are exceeded more than once or twice a year for any of the pollutants are known as *non-attainment areas* for those pollutants. In such areas, air pollution control programs must be initiated to bring the area back into compliance, and industries contemplating expansion must show how they can improve the air quality by reducing emissions. In some areas, either automobiles will be required to change fuels to reduce emissions or travel restrictions on the use of private automobiles will be initiated. Since the internal-combustion engine is the single largest contributor of air pollution, a replacement that uses hydrogen or electricity would be a significant contribution to cleaner air.

◆ Abbreviations

A = air change per hour
C = concentration, $\mu g/m^3$
K = decay constant, per hour
q = air flow rate, m^3/hr

Q = pollutant emission from an enclosure, mg/hr
S = source strength, $\mu g/hr$
V = volume, m^3

◆ Problems

11-1   A 1974 car is driven an average of 1000 miles/month. The EPA 1974 emission standards are 3.4 g/mi for HC and 30 g/mi of CO.

   a.   How much CO and HC would be emitted during the year?

   b.   How long would it take to exceed a lethal concentration of CO in a common double-car garage, 20 × 25 × 7 ft?

11-2   A 2.5% level of CO in hemoglobin (COHb) has been shown to cause impairment in time-interval discrimination. The level of CO on crowded city streets sometimes hits 10 $\mu g/m^3$. An approximate relationship between CO and COHb (after prolonged exposure) is

$$COHb(\%) = 0.5 + [0.16 \times 10(CO \text{ concentration in } \mu g/m^3)]$$

   a.   What level of COHb would a traffic cop be subjected to during a working day directing traffic on a city street?

   b.   Might this affect the cop's attitude? (See Figure 11-10.)

11-3   Photochemical smog is a serious problem in many large cities.

   a.   Draw a graph showing the concentration of NO, NO_2, HC, and O_3, in the Los Angeles area during a sunny, smoggy day.

   b.   Draw another graph showing how the same curves appear on a cloudy day. Explain the difference.

   c.   The only feasible way of reducing the formation of photochemical smog in Los Angeles seems to be to prevent automobiles from entering the city. Draw the same curves as they might appear if *all* cars were banned from LA streets.

   d.   Would this ever happen? Why or why not?

11-4   Give three examples of synergism in air pollution.

11-5   If SO_2 is so soluble in water, how can it get to the deeper reaches of the lung without first dissolving in the mucus?

11-6   A hi-vol clean filter weighs 18.0 g and the dirty filter weighs 18.6 g. The initial and final air flows are 70 and 40 ft^3/min.

a. What volume of air went through the filter in 24 hours?

b. What was the concentration of particulates in the air?

11-7 A high-volume sampler draws air in at an average rate of 70 ft^3/min. If the particulate reading is 200 μg/m^3, what was the weight of the dust on the filter?

11-8 Research and report on one of the classical air pollution episodes such as Meuse Valley, London, or Donora.

11-9 What does Ringlemann 5 tell you about a smoke being emitted from a chimney?

11-10 The data for a hi-vol are as follows:

> Clean filter: 20.0 g
>
> Dirty filter: 20.5 g
>
> Initial air flow: 70 ft^3/min
>
> Final air flow: 50 ft^3/min
>
> Time: 24 hours

a. What volume of air was put through the filter? (Answer in cubic feet.)

b. How is the air flow in a hi-vol measured?

c. What is the weight of the particulates collected?

d. If half of the collected particulates are <10 μ, does this air meet national ambient air quality standards?

11-11 If the primary ambient air quality standard for nitrogen oxides (as NO$_2$) is 100 μg/m^3, what is this in ppm? (Assume 25°C and 1 atmosphere pressure.)

11-12 The concentration of carbon monoxide in a smoke-filled room can reach as high as 500 ppm.

a. What is this in μg/m^3? (Assume 1 atmosphere and 25°C.)

b. What effect would this have on people who are sitting around having a political discussion for 4 hours?

11-13 Figure 11-13 shows schematically the global average energy flow between space, the atmosphere, and the earth's surface. The units in this figure are in watts per square meter of surface area.

a. Using space, the atmosphere, and the earth as black boxes, check if the numbers represent valid balances.

b. Suppose the earth experienced the eruption of several large volcanos, spewing dust into the atmosphere. How might these numbers change, and what effect would such an event have on the earth's temperature?

c. Since most of the greenhouse gases are emitted from industrialized nations in the northern hemisphere, what reasoning might the director of an environmental protection agency in an equatorial country present to argue for the curtailment of greenhouse gas emissions? Write a short radio commentary for the environmental protection agency director, one that he could read for the National Public Radio in the United States. Remember that

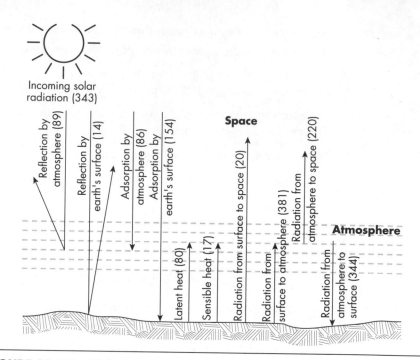

FIGURE 11-13   Global average energy flows. Units are watts per square meter of surface area. (Adapted from J. Harte, *Consider a Spherical Cow*, Wm. Kaufmann, Menlo Park, CA 1985.)

emotionalism will not win nearly the public support that a solid ethical argument will.

11-14   A wood-burning stove operates for 4 hours in a room measuring 5 × 5 × 2.5 meters, with a ventilation of 0.5 air changes per hour. The CO level reaches 5 mg/m³ and remains there. Assume that the ventilation air has negligible CO and that CO does not decay.

   a.   At what rate does the stove emit CO?

   b.   If this stove is used in a small hut measuring 2 × 3 × 3 meters, what would be the CO concentration in the hut?

   c.   Is burning wood any better environmentally than using natural gas or oil? Analyze the use of wood as alternative energy for heating based on presumed environmental impact and cost.

11-15   The CEO of a prominent environmental consulting firm, Steven Fisher of Brown & Caldwell, is quoted as saying, "We have to realize that all of the work in the environmental area is really being done because of fear. That in turn leads to public pressure, legislation, and then enforcement of that legislation."[5]

     Fear of what? Do you think he is right? If so, is this fear well founded? If not, what is driving environmental legislation and enforcement? Consider these questions in light of the history of air pollution control legislation in the United

States, and write a one-page discussion of why you think this legislation was passed and why it is taking EPA so long to enforce the regulations.

11-16   The rate of coal excavation is decreasing by about 1.2% per year, while the use of oil and gas is increasing by about 3% per year. Estimate how long it will take to double the rate of carbon emissions.

11-17   The particulate concentration in an urban atmosphere is 160 $\mu g/m^3$.

   a.   What would be the visibility at the airport?

   b.   Suppose the measurement was made early in the morning when the atmosphere was 90% saturated with moisture. Would the visibility have been lower or higher then?

11-18   A pilot reports that the visibility at the airport is 3 miles.

   a.   If it is a dry day, what might be the concentration of the particulates in this city?

   b.   Does this concentration exceed the National Ambient Air Quality Standards for particulates?

11-19[6]   Metals are frequently coated with thin films of light hydrocarbon oil to prevent oxidation of the raw metal feedstock during shipping and storage. The operations is accomplished by immersing the metal pieces in a vapor and having the solvent condense on the surface. Unfortunately, about 90% of the solvent eventually escapes into the atmosphere.

At one time, trichlorethylene (TCE) was the most widely used solvent for degreasing and metal cleaning. TCE, however, is a major contributor to photochemical smog, acting as one of the hydrocarbons that reacts with nitric oxides, thereby allowing high concentrations of ozone to build up. In the past few years 1,1,1-trichloroethane (TCA) has been substituted for TCE. The substitution is not without problems, however, since TCA has potential for both stratospheric ozone depletion and global warming. The following table illustrates a comparison of TCE and TCA relative to their environmental effects.

	Smog Formation Potential per Ton	Global Warming Potential per Ton	Ozone Depletion Potential per Ton
TCE	350	6.9	negligible
TCA	3	390	1000

The increasingly stringent regulation of TCE, a photochemically reactive chlorinated solvent, encouraged industry to convert to TCA. The results indicate that while TCE plays a relatively minor role in photochemical smog formation, its contribution to global warming and to the depletion of stratospheric ozone is relatively significant. A dilemma arises from the conflicting criteria used in choosing a solvent. For instance, should we replace a photochemically reactive solvent that breaks down easily in the troposphere with a solvent that is persistent and makes its way up into the stratosphere? Once there, the solvent will absorb infrared radiation and contribute to global warming or release a chlorine atom, which when participates in the catalytic destruction of the ozone layer.

Write a one-page paper outlining your criteria for making the decision between TCE and TCA. You may wish to look back in Chapter 1 to remind yourself of the various ways engineers make decisions.

◆ Endnotes

1. This discussion is based in part on R. D. Ross, ed., *Air Pollution and Industry,* Van Nostrand Reinhold, New York (1972), and K. Wark, and C. F. Warner, *Air Pollution,* Harper & Row, New York (1981).

2. Stephen Schwartz, "Acid Deposition: Unraveling a Regional Phenomenon," *Science* 243 (February 1989).

3. James Lovelock, Gaia: A New Look at Life on Earth, Oxford Univ. Press, New York, (1979).

4. A. Nero, "The Indoor Radon Story," Technology Review (January 1986).

5. Quoted in "The Environmental Age," *Engineering News Record,* 228:25:22 (June 1992).

6. This problem is based on a similar problem in David Allen, N. Bakshani, and Kirsten Sinclair Rosselot, *Pollution Prevention: Homework and Design Problems for Engineering Curricula,* American Institute of Chemical Engineers and other societies (1991). Used with permission.

chapter
12

AIR QUALITY
CONTROL

The easiest way to control air pollution is to eliminate the source of the pollution. Surprisingly, this is often also the most economical solution to an air pollution problem. Decommissioning a sludge incinerator and placing the sludge on dedicated land, for example, may be a great deal more economical than installing air-cleaning equipment for the incinerator. In other cases, a modification of the process, such as switching to natural gas instead of coal in an electrical power plant, will eliminate the immediate air pollution problem. Most often, however, control is achieved by some form of air treatment similar in concept to water treatment. In this chapter, some alternatives available for treating emissions are discussed, followed by a review of the last control strategy—dispersion.

12.1 ◆ Treatment of Emissions

Selection of the correct treatment device requires matching characteristics of the pollutant with features of the control device. It is important to remember that the sizes of air pollutants range many orders of magnitude, and it is therefore not reasonable to expect one device to be effective and efficient for all pollutants. In addition, the types of chemicals in emissions often will dictate the use of some devices. For example, a gas containing a high concentration of SO_3 could be cleaned by water sprays, but the resulting sulfuric acid might present serious corrosion problems.

The various air pollution control devices are conveniently divided into those applicable for controlling particulates and those used for controlling gaseous pollutants. The reason, of course, is the difference in the size. Gas molecules have diameters of about 0.0001 μ; particulates range from 0.1 μ and up.

Any air pollution control device can be treated as a black box separation device. The removal of the pollutant is calculated using the principles of separation covered in Chapter 3. If a gas (such as air) contains an unwanted constituent (such as dust), the pollution control device is designed to remove the dust. The recovery efficiency of any separating device is expressed by equation 3.4, reprinted here.

$$R_1 = \frac{x_1}{x_0} \times 100$$

where R_1 = recovery, %

$\quad x_1$ = amount of pollutant collected by the treatment device per unit time, kg/sec

$\quad x_0$ = amount of pollutant entering the device per unit time, kg/sec

Difficulties may arise in defining the x_0 term. Some dust, for example, may be composed of particles so small that the treatment device cannot be expected to remove this fraction. These very small particles should not in all fairness be included in x_0. Yet it is common practice to simply let x_0 equal the contaminants to be removed, such as total particulates.

e • x • a • m • p • l • e   **12.1**

Problem   An air pollution control device is to remove a particulate, which is being emitted at a concentration of 125,000 $\mu g/m^3$ at an air flow rate of 180 m^3/sec. The device removes 0.48 metric tons per day. What are the concentration of the emission and the recovery of collection?

Solution   A black box and a materials balance with regard to particulates is first set up and the old materials balance equation used once again:

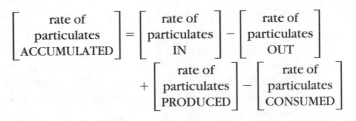

$$\begin{bmatrix} \text{rate of} \\ \text{particulates} \\ \text{ACCUMULATED} \end{bmatrix} = \begin{bmatrix} \text{rate of} \\ \text{particulates} \\ \text{IN} \end{bmatrix} - \begin{bmatrix} \text{rate of} \\ \text{particulates} \\ \text{OUT} \end{bmatrix}$$

$$+ \begin{bmatrix} \text{rate of} \\ \text{particulates} \\ \text{PRODUCED} \end{bmatrix} - \begin{bmatrix} \text{rate of} \\ \text{particulates} \\ \text{CONSUMED} \end{bmatrix}$$

Since this is in steady state, and there are no particulates produced or consumed, the equation reduces to

[rate of particulates IN] = [rate of particulates OUT]

Figure 12–1 shows the black box. The *particulates in* is the feed and is calculated as the flow rate times the concentration, which, as shown in Chapter 3, yields the mass flow rate:

$$(180 \ \text{m}^3/\text{sec}) \times (125{,}000 \ \mu\text{g/m}^3) \times (10^{-6} \ \mu\text{g/g}) = 22.5 \ \text{g/sec}$$

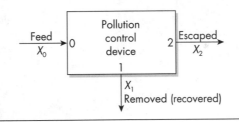

FIGURE 12–1   An air pollution control device as a black box.

The *particulates out* consists of the particles that escape and those that are collected. The latter is calculated as

$$\frac{0.48 \ \text{tons}}{\text{day}} \times 10^6 \times \frac{1 \ \text{hr}}{3600 \ \text{sec}} \times \frac{1 \ \text{day}}{24 \ \text{hr}} = 5.5 \ \text{g/sec}$$

The balance is

$$22.5 = 5.5 + [\text{particulates that escape}]$$

escaped particulates = 17 g/sec

$$\text{emission concentration} = \frac{(17 \ \text{g/sec} \times 10^6 \ \mu\text{g/g})}{180 \ \text{m}^3/\text{sec}}$$

$$= 94{,}000 \ \mu\text{g/m}^3$$

The recovery is

$$R = (x_1 \times 100)/x_0 = (5.5 \times 100)/22.5 = 24\%$$

12.1.1   Control of Particulates

The simplest devices for controlling particulates are *settling chambers* consisting of nothing more than wide places in the exhaust flue where larger particles can settle out, usually with a baffle to slow the emission stream. Obviously, only very large particulates (>100 μ) can be efficiently removed in settling chambers.

Possibly the most popular, economical, and effective means of controlling particulates is the *cyclone*. Figure 12-2 shows a simple schematic. The dirty air is blasted into a conical cylinder, but off centerline. This creates a violent

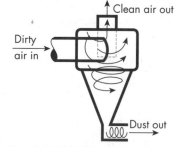

FIGURE 12-2   Cyclone used for dust collection.

swirl within the cone, and the heavy solids migrate to the wall of the cylinder, where they slow down due to friction, slide down the cone, and finally exit at the bottom. The clean air is in the middle of the cylinder and exits out the top.

Bag (or *fabric*) *filters* used for controlling particulates (Figure 12–3) operate like the common vacuum cleaner. Fabric bags are used to collect the dust, which must be periodically shaken out of the bags. The fabric will remove nearly all particulates, including submicron sizes. Bag filters are widely used in many industrial applications, but are sensitive to high temperatures and humidity.

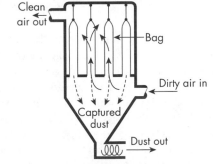

FIGURE 12–3   Bag filter used for control of particulate air pollutants.

The basic mechanism of dust removal in fabric filters is similar to the action of sand filters in water quality management. The dust particles adhere to the fabric due to entrapment and surface forces. They are brought into contact by impingement and/or Brownian diffusion. Since fabric filters commonly have an air space-to-fiber ratio of 1:1, the removal mechanism cannot be simple sieving.

The *spray tower* or *scrubber*, pictured in Figure 12-4, is an effective method for removing large particulates. However, scrubbers have two major drawbacks:

- They produce a visible plume, albeit only water vapor. The lay public seldom differentiates between a water vapor plume and any other visible plume, and hence public relations often dictate no visible plume.

- The waste is now in liquid form and some manner of water treatment is necessary.

High-efficiency scrubbers promote the contact between air and water by violent action in a narrow throat section into which the water is introduced. Generally, the more violent the encounter, and hence the smaller the gas bubbles or water droplets, the more effective the scrubbing.

Electrostatic precipitators are widely used in power plants, mainly because power is readily available. The particulate matter is removed by first being charged by electrons jumping from one high-voltage electrode to the other, and then migrating to the positively charged collecting electrode. The type of electrostatic precipitator shown in Figure 12-5 consists of a pipe with a wire hanging down the middle. The particulates collect on the pipe and must be removed by banging the pipe with a hammer. Electrostatic precipitators have no moving parts, require only electricity to operate, and are extremely effective in removing submicron particulates.

The efficiencies of the various control devices obviously vary widely with the particle size of the pollutants. Figure 12-6 shows approximate collection-efficiency curves, as a function particle size, for the various devices discussed.

12.1.2   Control of Gaseous Pollutants

The control of gases involves the removal of the pollutant from the gaseous emissions, a chemical change in the pollutant, or a change in the process producing the pollutant.

Wet scrubbers, as already discussed, can remove gaseous pollutants by simply dissolving them in the water. Alternatively, a chemical may be injected into the scrubber water, which then reacts with the pollutants. This is the basis for most SO_2-removal techniques, as discussed below.

Adsorption is a useful method when it is possible to bring the pollutant into contact with an efficient adsorber like activated carbon, as shown in Figure 12-7.

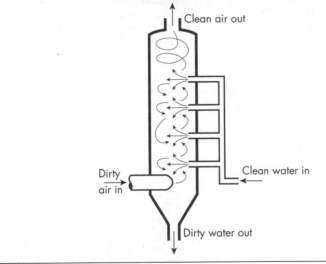

FIGURE 12-4   Spray tower or scrubber. In the photograph (courtesy of UnitedMcGill), the scrubber is the high round tower.

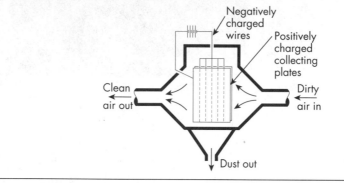

FIGURE 12–5   Electrostatic precipitator used for control of particulate air pollutants.

Incineration or *flaring* is used when an organic pollutant can be oxidized to CO_2 and water. A variation of incineration is *catalytic combustion*, whereby the temperature of the reaction is lowered by the use of a catalyst that mediates the reaction (Figure 12–8).

12.1.3   Control of Sulfur Oxides

As noted earlier, sulfur oxides (SO_2 and SO_3) are serious and yet ubiquitous air pollutants. The major source of sulfur oxides (or SO_x as they are often referred to in shorthand) is coal-fired power plants. The increasingly strict standards for SO_x control have prompted the development of a number of options and techniques for reducing emissions of sulfur oxides. Among these options are the following.

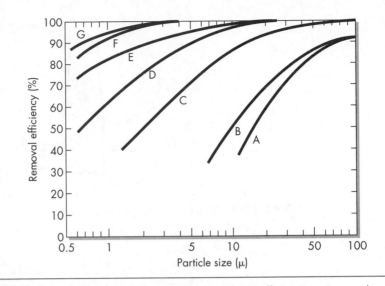

FIGURE 12-6   Comparison of approximate removal efficiencies. A = settling chamber, B = simple cyclone, C = high-efficiency cyclone, D = electrostatic precipitator, E = spray tower wet scrubber, F = venturi scrubber, G = bag filter. (Adapted from C. E. Lappe, "Processes Use Many Collection Types," *Chemical Engineering* 58:145, May 1951.)

Change to Low-Sulfur Fuel.   Natural gas and oil are considerably lower in sulfur than coal. However, uncertain and expensive supplies make this option risky.

Desulfurize the Coal.   Sulfur in coal is either organic or inorganic. The inorganic form is iron pyrite (FeS_2), which, since it occurs in discrete particles, can be removed by washing. The removal of the organic sulfur (generally about 60% of the total) requires chemical reactions and is most economically accomplished if the coal is gasified (changed into a gas resembling natural gas).

Tall Stacks.   A shortsighted method, albeit locally economical, of SO_2 control is to build incredibly tall smokestacks and disperse the SO_2. This option has

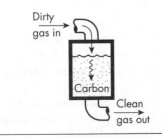

FIGURE 12-7   Adsorber for removing air pollutants.

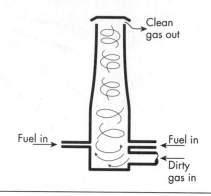

Clean
gas out

Fuel in

Fuel in

Dirty
gas in

FIGURE 12-8 Incinerator used for burning gaseous pollutants.

been employed in Great Britain and is in part responsible for the acid rain problem plaguing Scandinavia.

Flue-Gas Desulfurization.   The last option is to reduce the SO_2 emitted by cleaning the gases coming from the combustion, the so-called flue gases. Many systems have been employed, and a great deal of research is underway at present to make these processes more efficient.

The most widely used method of SO_2 removal is to contact the sulfur with lime. The reaction is

$$SO_2 + CaO \rightarrow CaSO_3$$

or if limestone is used,

$$SO_2 + CaCO_3 \rightarrow CaSO_4 + CO_2$$

Both the calcium sulfite and sulfate (gypsum) are solids that have low solubilities and can be separated in gravity settling tanks. The calcium salts thus formed represent a staggering disposal problem. It is possible to convert the sulfur to H_2S, H_2SO_4, or elemental sulfur and market these raw materials. Unfortunately, the total possible markets for these chemicals is far less than the anticipated production from desulfurization.

Before moving on to the next topic, it might again be useful to reiterate the importance of matching the type of pollutant to be removed with the proper process. Figure 12-9 reemphasizes the importance of particle size in the application control equipment.

Other properties, however, may be equally important. A scrubber, for example, removes not only particulates but gases that can be dissolved in the water. Thus SO_2, which is readily soluble, but not NO, which is poorly soluble, would be removed in a scrubber. The selection of the proper control technology is an important component of the environmental engineering profession.

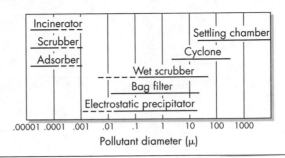

FIGURE 12-9 The effectiveness of various air pollution control devices depends on particle size.

12.2 ◆ Dispersion of Air Pollutants

Recall from the previous discussion of meteorology that atmospheric conditions primarily determine the dispersion of air pollutants. If the conditions are super-adiabatic, a great deal of vertical air movement and turbulence are produced, and dispersion is enhanced. The subadiabatic prevailing lapse rate is by contrast a very stable system. An inversion is an extreme subadiabatic condition, and the vertical air movement within an inversion is almost nil.

The effect of atmospheric stability of a plume can be illustrated as in Figure 12-10. A superadiabatic lapse rate produces atmospheric instability and a *looping* plume, whereas a neutral lapse rate evens out the plume, producing a *coning* plume. If the plume is emitted into an inversion layer, a *fanning* plume will result, a highly descriptive name since from above it can be seen that the plume fans out horizontally without any vertical dispersion. A particu-larly nasty situation is the *fumigation* condition, when an inversion cap is placed on the plume, but a superadiabatic lapse rate under the inversion causes mixing and high ground-level concentrations.

The distance a plume rises is also of importance if dispersion is to be attained. Although there have been no theoretical models that consistently predict plume rise, a number of empirical models have been suggested. Briggs developed a model that seems to effectively predict plume rise from power sta-tions.[1]

$$\Delta h = 2.6 \left(\frac{F}{\bar{u}S} \right)^{1/3}$$

$$F = \frac{gV_s d^2 (T_s - T_a)}{4(T_a + 273)}$$

$$S = \frac{g}{(T_a + 273)} \left(\frac{\Delta T_a}{\Delta z} + 0.01 \right)$$

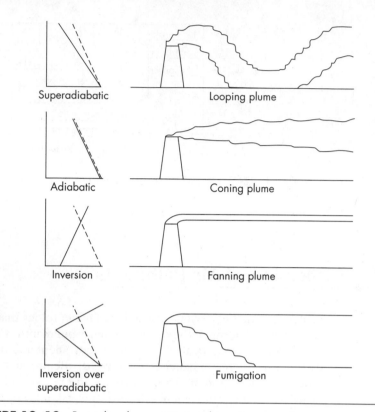

Superadiabatic   Looping plume

Adiabatic   Coning plume

Inversion   Fanning plume

Inversion over   Fumigation
superadiabatic

FIGURE 12-10 Prevailing lapse rates produce signature plumes.

where Δh = plume rise above top of stack, m
 $\bar{u}$ = average wind speed, m/sec
 $\Delta T/\Delta z$ = prevailing lapse rate, the change in temperature with eleva-
                tion, °C/m
 V_s = stack gas exit velocity, m/sec
 d = stack exit diameter, m
 g = gravitational acceleration, 9.8 m/sec^2
 T_a = temperature of the atmosphere, °C
 T_s = temperature of the stack gas, °C
 F = buoyancy flux, m^{4}sec^{-3}

e • x • a • m • p • l • e   12.2

Problem   A stack has an emission exiting at 3 m/sec through a stack diameter
of 2 m. The average wind speed is 6 m/sec. The air temperature at stack
elevation is 28°C and the temperature of the emission is 167°C. The atmosphere
is at neutral stability. What is the expected rise of the plume?

Solution   For neutral stability, $\Delta T/\Delta z = 0.01°C/m$.

$$\Delta h = 2.6\left(\frac{F}{6S}\right)^{1/3}$$

$$F = \frac{9.8(3)(2)^2(167 - 28)}{4(28 + 273)} = 13.6$$

$$S = \frac{9.8}{(28 + 273)}(0.01 + 0.01) = 6.51 \times 10^{-4}$$

$$\Delta h = 2.6\left(\frac{13.6}{6(6.51 \times 10^{-4})}\right)^{1/3} \cong 40 \text{ m}$$

◆

Dispersion is the process of spreading out the emission over a large area and thereby reducing the concentration of the specific pollutants. The plume spread or dispersion is in two dimensions: horizontal and vertical. It is assumed that the greatest concentration of the pollutants is in the plume centerline, that is, in the direction of the prevailing wind. The farther away from the centerline, the lower the concentration. If the spread of a plume in both directions is approximated by a Gaussian probability curve, as introduced in Chapter 2, the concentration of a pollutant at any distance x downwind from the source can be calculated as

$$C_{(x,y,z)} = \frac{Q}{2\pi\bar{u}\sigma_y\sigma_z} \exp\left(-\frac{1}{2}[(y/\sigma_y)^2 + (z/\sigma_z)^2]\right) \tag{12.1}$$

where $C_{(x,y,z)}$ = concentration at some point in the coordinate space, kg/m³
Q = source strength, or the emissions, kg/sec
$\bar{u}$ = average wind speed, m/sec
σ_y and σ_z = standard deviation of the dispersion in the y and z directions
y = distance crosswind horizontally, m
z = distance crosswind vertically, m

The coordinates are shown in Figure 12-11. Note that z is in the vertical direction, y is horizontal crosswind, and x is downwind.

The standard deviations are measures of how much the plume spreads. If σ_y and σ_z are large, the spread is great, and the concentration is of course low. The opposite is true if the spread is small. The dispersion is dependent on both atmospheric stability (as noted earlier) and the distance from the source. Figure 12-12 is one approximation for the dispersion coefficients. Atmospheric stability is denoted in Figure 12-12 by letters ranging from A to F. Table 12-1 is a key for selecting the proper stability condition. (These curves and the table are based on some data, but a lot of it is extrapolation. This is especially evident when one notes that under A stability category, it is possible to have a vertical standard deviation of 10,000 km. It gets pretty silly.)

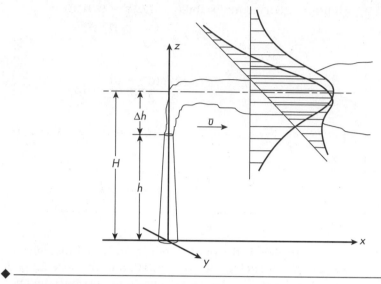

FIGURE 12-11   The Gaussian dispersion model.

A plume emitted from a stack has an effective height H, which is calculated as the stack height plus the plume rise Δh as calculated previously and as shown in Figure 12-11. The elevation of the plume centerline is then $z = H$, and the diffusion equation is

$$C_{(x,y,z)} = \frac{Q}{2\pi\bar{u}\sigma_y\sigma_z} \exp\left(-\frac{1}{2}[(y/\sigma_y)^2 + ((z - H)/\sigma_z)^2]\right) \qquad (12.2)$$

Equations 12.1 and 12.2 hold as long as the ground does not influence the diffusion. This is usually not a good assumption because the ground is not an efficient sink for the pollutants, and the levels must be higher at ground level due to the inability of the plume to disperse into the ground. The ground effect can be taken into account by assuming an imaginary mirror image source at elevation $z - H$, as shown in Figure 12-13. Taking the reflection of the ground into account, the dispersion equation reads as follows.

$$C_{(x,y,z)} = \frac{Q}{2\pi\bar{u}\sigma_z\sigma_y} \times \left[\exp\left(-\frac{1}{2}(y/\sigma_y)^2\right) \times \left(\exp\left(-\frac{1}{2}[(z + H)/\sigma_z]^2\right)\right.\right.$$
$$\left.\left. + \exp\left(-\frac{1}{2}[(z - H)/\sigma_z]^2\right)\right)\right] \qquad (12.3)$$

Equation 12.3 is the most general dispersion equation, taking into account the reflection of the ground and the emission of the pollutant at some effective stack height H. We can simplify this equation by making various assumptions. For example, if the measurement is taken at ground level and the plume is also emitted at ground level, both the z and H terms are zero, yielding the equation

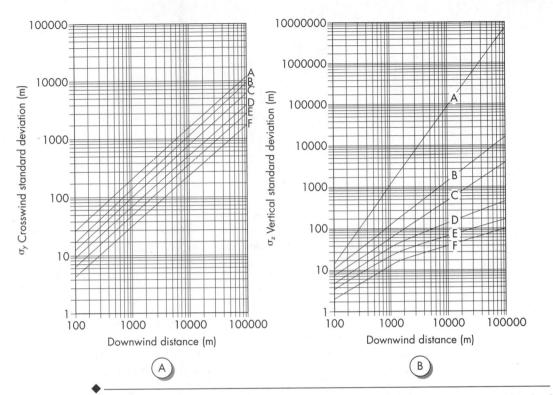

FIGURE 12-12 Dispersion coefficients. (Adapted from D. B. Turner, *Workbook of Atmospheric Dispersion Estimates*, U. S. Department of Health, Education and Welfare, Public Health Service, National Center for Air Pollution Control, Publication No. 999-AP-28.)

TABLE 12-1 Atmospheric Stability Key for Figure 12-12[3]

| Surface Wind Speed (at 10 m) (m/sec) | DAY | | | NIGHT | |
| | Incoming Solar Radiation (Sunshine) | | | Thinly Overcast or 4/8 Low Cloud | 3/8 Cloud |
	Strong	Moderate	Slight		
<2	A	A–B	B	—	—
2–3	A–B	B	C	E	F
3–5	B	B–C	C	D[a]	E
5–6	C	C–D	D	D	D
>6	C	D	D	D	D

[a] The neutral category, D, should be assumed for overcast conditions day or night.

Source: F. Pasquil, "The Estimation of the Dispersion of Windborne Material," *Meteorology Magazine* 90:1063 (1961).

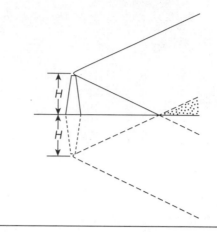

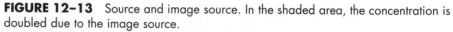

FIGURE 12-13 Source and image source. In the shaded area, the concentration is doubled due to the image source.

$$C_{(x,y,0)} = \frac{Q}{2\pi\bar{u}\sigma_z\sigma_y} \exp\left(-\frac{1}{2}(y/\sigma_y)^2\right) \qquad \text{at } z = 0$$

If the emission is at ground level ($H = 0$), the pollutant is measured on its centerline ($y = 0$), and the measurement is also at ground level ($z = 0$), then

$$C_{(x,0,0)} = \frac{Q}{\pi\bar{u}\sigma_z\sigma_y} \qquad (y = 0, H = 0, \text{ and } z = 0)$$

e • x • a • m • p • l • e 12.3

Problem   Given a sunny summer afternoon with average wind, $\bar{u} = 4$ m/sec, emission $Q = 0.01$ kg/sec, and the effective stack height $H = 20$ m, find the ground level concentration at 200 m from the stack.

Solution   From Figure 12-12, at 200 m, $\sigma_y = 36$ and $\sigma_z = 20$ for unstable superadiabatic strong solar radiations (Table 12-1), the atmospheric conditions are type B, and noting that maximum concentrations occur on the plume centerline, at $y = 0$. Using equation 12.3

$$C_{(200,0,0)} = \frac{0.01}{2(3.14)(4)(36)(20)} \times \left[\exp\left(-\frac{1}{2}(0/36)^2\right)\right.$$

$$\left. \times \left(\exp\left(-\frac{1}{2}[(0 - 20)/20]^2\right) + \exp\left(-\frac{1}{2}[(0 + 20)/20]^2\right)\right)\right]$$

$$C_{(200,0,0)} = 6.7 \times 10^{-7} \text{ kg/m}^3$$

or

$$C_{(200,0,0)} = 670 \ \mu\text{g/m}^3$$

Finally, it should be pointed out that the accuracy of this plume dispersion analysis is very poor. Air pollution modelers are usually pleased to find their models predicting concentrations to within an order of magnitude!

12.3 ◆ Control of Moving Sources

Although many of the previously mentioned control techniques can apply to moving sources as well as to stationary ones, one very special moving source— the automobile—deserves special attention. Although the automobile has many potential sources of pollution, there are only a few important points requiring control (Figure 12-14).

- Evaporation of hydrocarbons (HCs) from the fuel tank
- Evaporation of HCs from the carburetor
- Emissions of unburned gasoline and partially oxidized HCs from the crankcase
- The NO_x, HCs, and CO from the exhaust

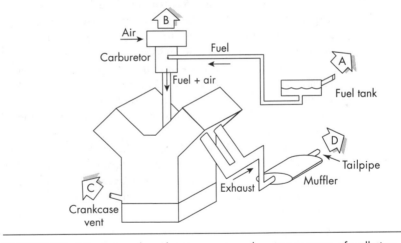

FIGURE 12-14   Internal-combustion engine, showing sources of pollution.

The evaporative losses from the gas tanks have been reduced by the use of gas tank caps that prevent the vapor from escaping. Losses from carburetors have been reduced by use of activated-carbon canisters that store the vapors emitted when the engine is turned off and the hot gasoline in the carburetor vaporizes. When the car is restarted, the vapors can be purged by air and burned in the engine (Figure 12-15).

The third source of pollution, the crankcase vent, has been eliminated by closing off the vent to the atmosphere and recycling the blow-by gases into the intake manifold. The positive crankcase ventilation (PCV) valve is a small check valve that prevents the buildup of pressure in the crankcase.

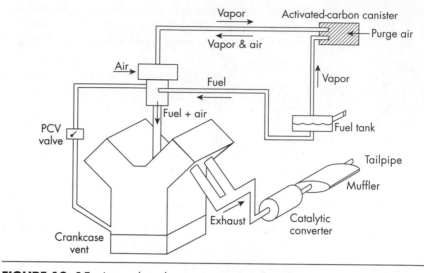

FIGURE 12–15   Internal-combustion engine, with pollution control devices.

The most difficult control problem is the exhaust, which accounts for approximately 60% of the HCs and almost all the NO_x, CO, and lead. One immediate problem is how to measure these emissions. It is not as simple as sticking a sampler up the tailpipe, since the quantity of pollutants emitted changes with the mode of operation. The effect of operation on emissions is illustrated in Table 12–2. Note that when the car is accelerating the combustion is efficient (low CO and HCs) and the high compression produces a lot of NO_x. However, decelerating results in low NO_x and very high HCs due to partially burned fuel.

Because of these difficulties the EPA has instituted a standard test for measuring emissions. This test procedure includes a cold start, acceleration and cruising on a dynamometer to simulate a load on the wheels, and a hot start.

Emission control techniques for the internal-combustion automobile engine include the following:

TABLE 12–2   Effect of Engine Operation on Automotive Exhaust Characteristics, Shown as Fraction of Idling Emissions

	Component		
	CO	HCs	NO_x
Idling	1.0	1.0	1.0
Accelerating	0.6	0.4	100
Cruising	0.6	0.3	66
Decelerating	0.6	11.4	1.0

- Tuning the engine to burn fuel efficiently
- Installation of catalytic reactors
- Engine modifications

A tune-up can have a significant effect on emission components. For example, a high air : fuel ratio (a lean mixture) will reduce both CO and HCs, but will increase NO_x. Typical emissions resulting from changing the air : fuel ratio are shown in Figure 12–16. A well-tuned car is the first line of defense for controlling automobile emissions, regardless of what other devices and/or processes are used.

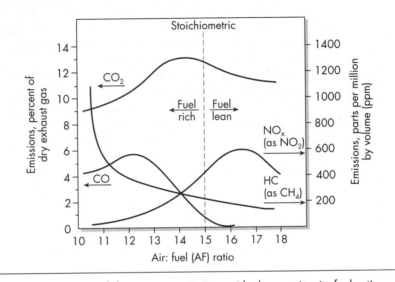

FIGURE 12–16 Variability in auto emissions with changes in air : fuel ratio.

The second control strategy, now used in all gasoline-powered cars sold in the United States, is the catalytic converter, which oxidizes the CO and HCs to CO_2 and H_2O. The most popular catalyst is platinum, which can be fouled by some gasoline additives such as lead, thus prompting the elimination of leaded gasoline. The second problem with catalytic converters is that the sulfur compounds in gasoline are oxidized to particulate SO_3, thus increasing the sulfur levels in urban environments and the global acid rain problem.

The greatest advance in engine development, however, has been the complete redesign of engines so as to produce less emission. For example, the geometric configuration of the cylinder in which the combustion occurs is important since complete combustion requires that all the gasoline ignite together and burn as a steady flame for the very short time required. Cylinders that have nooks and crannies in which the air–gasoline mixture can hide will produce partially combusted emissions such as hydrocarbons and carbon monoxide. A second advance has been fuel injectors that accurately measure the exact amount of gasoline needed by the engine and pump this in, avoiding the problem of the engine sucking in too much fuel from the carburetor and then emitting it in the exhaust.

Even with extensive investment in engineering, however, it will be difficult to manufacture a totally clean internal-combustion engine. Electric cars are clean but can store only limited power; thus their range is limited. In addition, the electricity used to power such vehicles must be generated in power plants, thereby creating more pollution.

The diesel engines used in trucks and buses also are important sources of pollution. There are, however, fewer of these vehicles in operation than gasoline-powered cars, and the emissions may not be as harmful owing to the nature of diesel engines. The main problem associated with diesel engines is the visible smoke plume and odors, two characteristics that have led to considerable public irritation with diesel-powered vehicles. But from a public health viewpoint, diesel exhaust does not constitute the problem that the exhaust from gasoline-powered cars does. Diesel engines in passenger cars can, in fact, readily meet strict emission control standards.

One of the most destructive effects of automobile emission is the deterioration of buildings, statuary, and other materials. Athens, Greece, for example, has one of the highest levels of photochemical smog, given their plethora of sunlight and incredible number of nonregulated automobiles. As a result, the buildings of the Acropolis are rapidly deteriorating, as are other remnants of the ancient Greek civilization. Ironically, the theft of many of the most valuable pieces by the British around the turn of the century has resulted in the saving of these treasures. The clean, controlled air in the British Museum is far better than the putrid atmosphere in Athens. But does this make the theft any less reprehensible?

Do we, in fact, have a duty to preserve things? What is worth preserving? And who decides? The things that may or may not be worth preserving are either constructed by humans such as buildings, or natural things and places such as the Grand Canyon or the Gettysburg Battlefield. We may decide to preserve such landmarks for one of two reasons: to keep the structure or thing in existence purely for its own sake, or to not destroy it because we enjoy looking at it.[2] There is a distinction of course between just enjoying the landmark and using it, as one would a bridge or building. If there is a use for the landmark, based on economics or other human desires, then there is no question that the landmark has some value. But some things do not have utility, such as some old buildings and art. Is it possible to develop an argument for the preservation of mere things, irrespective of their utility to humans?

One such argument was advanced by Christopher Stone in his provocative book *Should Trees Have Standing?*[3] Stone argues that if corporations, municipalities, and ships are considered legal entities, with rights and responsibilities, is it not reasonable to grant similar rights to trees, forests, and mountains? He does not suggest that a tree should have the same rights as humans, but that a tree should be able to be represented in court (to have standing) just as corporations are. If this argument is valid, then buildings and other inanimate objects can similarly have legal rights, and their "interests" can be represented in court. It should then be possible to sue the City of Athens on behalf of the Acropolis.

But the argument by Stone for granting legal rights to trees and other objects is not very strong. Perhaps the only reason we would prefer not to destroy landmarks such as historic buildings and natural wonders is that these things are components of our physical environment and are necessary to a historical grounding of our civilization. The Acropolis in Athens is a bunch of rocks. Not a single roof remains. The land would be worth millions if developed for condominiums. Yet we strive to protect it, and an international project has been hard at work treating the stone to prevent further deterioration. We do this because it is important to our heritage, and we want to preserve it for future generations. While it is often uncertain how we should act toward these future generations, it is quite obvious that if we allowed landmarks such as the Acropolis to be destroyed by air pollution from automobiles, future generations would not benefit from our inaction.

◆ Abbreviations

C	= concentration of pollutants, $\mu g/m^3$	$\bar{u}$	= average wind speed, m/sec
d	= stack exit diameter, m	V	= stack gas velocity, m/sec
F	= buoyancy factor (see equation)	x	= distance downwind, m
		y	= distance crosswind, m
H	= effective stack height, m	z	= distance vertically, m
HC	= hydrocarbon	Δh	= plume rise above top of stack, m
NO_x	= nitrogen oxide	$\Delta T/\Delta z$	= prevailing lapse rate, °C/m
Q	= source strength, or the rate of emissions, kg/sec	σ_y	= standard deviation crosswind
R	= recovery, %	σ_z	= standard deviation vertically
T_a	= temperature of the atmosphere at stack height, °C		
T_s	= temperature of the stack gas, °C		

◆ Problems

12-1 Taking into account cost, ease of operation, and ultimate disposal of residuals, what type of control device would you suggest for the following emissions?

a. A dust with particle range of 5–10 μ?

b. A gas containing 20% SO_2 and 80% N_2?

c. A gas containing 90% HC and 10% O_2?

12-2 A stack emission has the following characteristics: 90% SO_2, 10% N_2, no particulates. What treatment device would you suggest and why?

12-3 An industrial emission has the following characteristics: 80% N_2, 15% O_2, and 5% CO_2 and no particulates. You are called in as a consultant to advise on the type of air pollution control equipment required. What would be your recommendation? How much would you charge for your time?

12-4 A whiskey distillery has hired you as a consultant to design the air pollution equipment for their new plant, to be built in a residential area. A major environmental effect of a whiskey distillery is the odor produced by the process, but many people think that this odor is pleasant (it's kind of a sweet musty aroma). How would you handle this assignment?

12-5 Given the following temperature soundings:

Elevation (m)	Temperature (°C)
0	20
50	15
100	10
150	15
200	20
250	15
300	20

What type of plume would you expect if the exit temperature of the plume is 10°C and the smoke stack is

a. 50 m tall?

b. 150 m tall?

c. 250 m tall?

12-6 Consider a prevailing lapse rate that has these temperatures: ground = 21°C, 500 m = 20°C, 600 m = 19°C, 1000 m = 20°C. If a parcel of air is released at 500 m and at 20°C, would it tend to sink, rise, or remain where it is? If a stack is 500 m tall, what type of plume would you expect to see?

12-7 A power plant burns 1000 tons of coal per day, 2% of which is sulfur, and all this is emitted from the 100-m stack. For a wind speed of 10 m/sec, calculate

a. The maximum ground-level concentration of SO_2, 10 km downwind from the plant

b. The maximum ground-level concentration and the point at which this occurs for stability categories A, C, and F. (Part b should be done with a spreadsheet. Insert, values of σ_y and σ_z as required.)

12-8 The odor threshold of H_2S is about 0.7 $\mu g/m^3$. If an industry emits 0.08 g/sec of H_2S out of a 40-m stack during an overcast night with a wind speed of 3 m/sec, estimate the area (in terms of x and y coordinates) where H_2S would be detected. (This is much too tedious to do by hand. Write a computer program and either use curvefitting to read values of σ_y and σ_z or construct a table and use this in your program.)

12-9 A power plant emits 300 kg/hr of SO_2 into a neutral atmosphere with a wind speed of 2 m/sec. The σ_y and σ_z values are assumed to be 100 m and 30 m, respectively.

a. What will be the ground-level concentration of SO_2, expressed in micrograms per cubic meter if the town is 2 km off the plume centerline and the plant has a stack that produces an effective stack height of 20 m?

b. What will be the ground-level concentration if the power plant installs a 150-m stack?

12–10 A temperature sounding balloon feeds back the following data:

Elevation (m above ground level)	Temperature (°C)
0	20
20	20
40	20
60	21
80	21
100	20
120	17
140	16
160	14
200	12

What type of plume would you expect to see out of a stack 70-m high? Why?

12–11 A coal-fired power plant burns 1000 metric tons (1 metric ton = 1000 kg) of coal per day, at 2% sulfur.

a. What is the SO_2 emission rate?

b. What is the SO_2 concentration 1.0 km downwind ($x = 1000$ m), 0.1 km off centerline in y direction ($y = 100$ m), and at ground level ($z = 0$) if the wind speed is 4 m/sec and its a cloudy day? The effective stack height is 100 m. Ignore reflection.

c. What is the ground-level concentration if the reflection of the ground is included?

d. If the source is at ground level ($H = 0$), and it is necessary to measure the concentration at ground level ($z = 0$), what is the concentration (include reflection)?

12–12 A stack is 1000 ft tall and emits smoke at 90°F. The ground-level temperature is 80°F and the temperature at 2000 ft is 100°F (assume a straight-line lapse rate between these two temperatures).

a. Draw a picture of the expected plume and name the type of plume.

b. If in the above situation the plume temperature is 92°F, how high would the plume rise (assume zero stack velocity and perfect adiabatic conditions)?

12–13 Suppose you were asked to design equipment for controlling emissions from an industry. For the three emissions below, draw block diagrams showing which equipment you might choose to obtain about 90% efficiency.

Emission 1:   Particle size range—70 to 200 μ
No gaseous emissions

Temperature—200°F
No space limitations

Emission 2:   Particle size range 0.1 to 200 μ
No gaseous emissions
Temperature—200°F
No space limitations
Visible plume not acceptable

Emission 3:   Particle size—5 to 40 μ
Gaseous emission—SO_2
Temperature—1200°F
Contaminated water not acceptable (no treatment facilities available)
Severe space limitations

12–14   A furniture manufacturer emits air pollution that consists of mostly particulate material of diameter 10 μ and bigger.

a.   As the consulting engineer, what type of treatment system would you suggest for this factory?

b.   The collected pollutant is to be disposed of in the local landfill. After the facility is constructed and the system is placed into operation, you notice that the material headed for the landfill has a strange organic odor to it. You ask the plant manager and she tells you it's probably xylene, a highly toxic chemical. You tell the plant manager that this material is potentially dangerous to the workers and it's probably a hazardous waste and should not be placed in the landfill. She tells you to mind your own business. You have been fully paid for the job and are no longer officially retained as their engineer. What do you do? Be realistic with your answer, not idealistic.

12–15   Assume that a power plant is emitting 400 tons per day of particulates and that a town is 10 km north of the plant. The wind is 25% of the time from the south, 25% of the time from the north, and 50% of the time from the west. There is no east wind. The wind speed is always 5 km/hr. (How's that for simplification?) Assume neutral stability (C) condition for all west winds, unstable conditions for north winds (A), and inversion conditions for south winds (F). Calculate the average ambient particulate air quality in the town. Does the air quality in the town meet the EPA National Ambient Air Quality Standards?

12–16[4]   A sewage sludge incinerator has a scrubber for removing the particulate material from its emissions, but the scrubber has been acting up, and a small pilot bag filter is being tested for applicability for treating the particulate emissions. The experimental setup is to split the emissions from the incinerator so that the 200 m^3/sec total flow is split with 97% going to the scrubber and only 3% going to the pilot bag filter. The particulate concentration of the untreated emission is 125 mg/m^3. The solids collected at the baghouse are collected hourly and average out to 2.6 kg/hr. The water does not have any particulates in it when it enters the scrubber at a flow rate of 2000 L/min. There is negligible evaporation in the scrubber, and the dirty scrubber water is found to carry solids at a rate of 52 kg/hr.

    a.   What is the efficiency of each method of air pollution control?

    b.   Regardless of the outcome in the efficiency calculations, is there any reason why the scrubber would still be superior as a pollution control device?

12-17   What is the particulate removal efficiency of a cyclone if it has to remove particles of diameter

    a.   100 μ

    b.   10 μ

    c.   1 μ

    d.   0.1 μ

12-18   A new coal-fired power plant is to burn a coal of 3% sulfur and a heating value of 11,000 Btu/lb. What is the minimum efficiency that a SO_2 scrubbing device will need to have to meet the new source emission standard of 1.2 lb $SO_2/10^6$ Btu?

12-19   Why can't the gasoline engine be tuned so that it results in the minimum production of all three pollutants, CO, HCs, and NO_x simultaneously?

12-20   A paper mill has a plant in a valley, and it wants to build a stack to push the plume centerline over the mountain so it can reduce the sulfur dioxide concentration in its own valley. The mountain is 4 km away, and its elevation is 3400 feet. The valley floor, where the plant is, is at an elevation of 1400 feet. If the emission temperature is 200°C and the prevailing temperature in the valley is 20°C, the wind velocity is 2 m/sec and a prevailing lapse rate of 0.006°C/m can be assumed, how high must the stack be to achieve their objective? (Note: You will have to assume some numbers, and each solution will be unique.)

12-21   If a car is designed to burn ethanol (C_2H_5OH), what would the stoichiometric air : fuel ratio be?

12-22   A fertilizer manufacturer emits hydrogen fluoride (HF) at a rate of 0.9 kg/sec from a stack with an effective height of 200 m. The average wind speed is 4.4 m/sec and the stability is category B.

    a.   What is the concentration of HF, in micrograms per cubic meter at ground level; on plume centerline; at 0.5 km from the stack?

    b.   Calculate also the ground-level concentration at 1.0, 1.5, 2.0, 2.5, 3.0, 3.5, and 4.0 km from the stack. Plot the results as concentration vs. distance.

12-23   For the conditions in Problem 12-24, draw a three-dimensional picture of the ground-level HF concentration.

12-24   The concentration of SO_2 at ground level is not allowed to exceed 80 $\mu g/m^3$. If the wind speed is 4 m/sec on a clear day and if the source emits 0.05 kg SO_2/sec, what must the effective stack height be to meet this requirement?

12-25   A chemical plant expects to emit 200 kg/day of a highly odoriferous but nondangerous gas. Downwind from the plant at a distance of 2000 m is a small town. The plant management decides to build a stack that will be tall enough so that the plant will not attain concentration levels of more than 10 $\mu g/m^3$ in the town. How tall will the stack have to be? Assume the gas exit velocity of 5 m/sec, the gas exit temperature of 150°C, the stack diameter of 1.5 m, the ambient temperature of 20°C, a wind velocity of 4 m/sec, and a prevailing lapse rate of 0.004°C/m.

12–26   A major catalog auto parts company advertises a book with the title of *How to Bypass Emission Controls.* For a measly $7.95 (plus shipping and handling) you can find how to get from 14% to 140% better gas mileage, increase acceleration and performance, run cooler and smoother, and have longer engine life. The book includes easy-to-follow instructions for both amateurs and professionals.

It is not illegal to tamper with your own car. You may not pass the emission inspections in most states, but modifying your car is not against the law.

When permission was asked from the mail order company to reproduce the ad in this book, they refused permission. If it is perfectly legal, and they are providing a public service, why would they not allow their ad to be reproduced?

Reconstruct the management board meeting where this request was discussed. Make up characters such as the company CEO, the VP for marketing, the legal counsel, and so on. Create a dialogue that may have occurred, ending with the decision not to allow the ad to be reproduced in this book. How did ethics come into the discussion (if at all)?

◆ Endnotes

1. G. A. Briggs, *Plume Rise*, AEC Critical Review Series, TID-25075 (1969); as modified and discussed in K. Wark and C. F. Warner, *Air Pollution*, Harper & Row, New York (1981).
2. M. P. Golding and M. H. Golding, "Why Preserve Landmarks? A Preliminary Inquiry," in Goodpaster and Jayne (eds.), *Ethics and the Problems of the 21st Century*, University of Notre Dame Press, South Bend, IN (1979).
3. Christopher D. Stone, *Should Trees Have Standing?* Avon Books, New York (1975).
4. This problem is credited to William Ball, Johns Hopkins University.

chapter
13

SOLID WASTE

Four pounds per day doesn't sound like much, until it is multiplied by the total number of people in the United States. Suddenly 880,000,000 pounds of garbage *per day* sounds like what it is, a humongous lot of trash.

What is to be done with these solid residues from our "effluent" society? The search for an answer represents a monstrous challenge to the engineering profession.

Common ordinary household and commercial waste, called *refuse* or sometimes *municipal solid waste (MSW)*, is the subject of this chapter. Technically, refuse is made up of *garbage*, which is food waste, and *rubbish*, almost everything else in your "garbage" can. *Trash* is larger items such as old refrigerators, tree limbs, mattresses, and other bulky items that are not commonly collected with the household refuse. A very important subcategory of solid waste, called *hazardous waste*, is covered in the next chapter.

The municipal solid waste problem can be separated into three steps:

- Collection and transportation of household, commercial and industrial solid waste
- Recovery of useful fractions from this material
- Disposal of the residues into the environment

13.1 ◆ Collection of Refuse

In the United States and in most other countries, solid waste from households and commercial establishments is collected by trucks. Sometimes these are open-bed trucks that carry trash or bagged refuse, but more often these vehicles are *packers*, trucks that use hydraulic rams to compact the refuse to reduce its volume and make it possible for the truck to carry larger loads (Figure 13–1). Commercial and industrial collections are facilitated by the use of containers that are either emptied into the truck using a hydraulic mechanism or carried by the truck to the disposal site (Figure 13–2). Recently, specialized vehicles for collecting separated materials such as newspaper, aluminum cans, and glass bottles have become commonplace (Figure 13–3).

Household collection of mixed refuse is usually by a packer truck with three workers, one driver and two loaders. These workers bring the refuse from the backyard garbage cans to the truck and then drive the full truck to the disposal area. The entire operation is a study in inefficiency and hazardous work conditions. The safety record of solid waste collection personnel is by far the worst of any group of workers (three times as bad as coal miners, for example).

Various modifications to this collection method have been implemented to cut collection cost and reduce accidents, including the use of compactors and garbage grinders in the kitchen and rolling-can systems for homes. The rolling-green-can system has revolutionized garbage collection. In most operations, the householder is given a large plastic can, about the volume of two normal garbage cans, and is asked to roll the can to the curb every week for collection. The truck is equipped with a hydraulic lift that empties the green can into the truck (Figure 13–4). Invariably, the green-can system has saved communities money, has significantly reduced injuries, and has been widely accepted by the citizens.

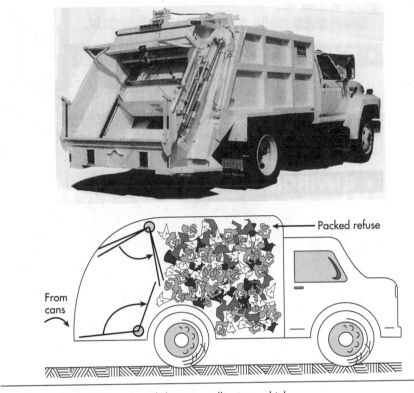

Packed refuse

From cans

FIGURE 13-1 Domestic solid waste collection vehicle.

Other alternate systems have been developed for collecting refuse, one especially interesting one being a system of underground pneumatic pipes. The pneumatic collection system at Disney World in Florida has collection stations scattered throughout the park that receive the refuse, and the pneumatic pipes deliver the waste to a central processing plant.

The selection of a proper route for collection vehicles, known as *route optimization*, can result in significant savings to a city. The problem of route optimization was first addressed in 1736 by the famous Swiss mathematician, Leonard Euler (1707–1783). He was asked to design a parade route for Königsberg such that the parade would not cross any bridge over the River Pregel more than once and would return to its starting place. Euler's problem is pictured in Figure 13–5A.

Not only did Euler show that such a route was impossible for the king's parade, but he generalized the problem by specifying what conditions are necessary to establish such a route, now known as a *Euler's tour*. The objective of truck routing is to create a Euler's tour, where a street is traversed only once and *deadheading*, traveling twice down the same street, is eliminated. A Euler's tour is also known as a *unicoursal route*, since the traveler courses each street only once.

Travel takes place along specific *links* (streets and bridges in Euler's problem) that connect *nodes* (intersections). A Euler's tour is possible only

FIGURE 13–2 Commercial solid waste collection vehicle.

if an even number of links enters all the nodes. The nodes in Figure 13–5B are A through D, and *all* of them have an odd number of links, so the parade the king wanted was not possible. Euler's principle is illustrated by Example 13.1.

e • x • a • m • p • l • e **13.1**

Problem   A street network is shown in Figure 13–6A on page 391. If refuse from homes along these streets is to be collected by a vehicle that intends to travel down each street only once (solid waste is collected along both sides of the street, a situation that would typically occur in a residential neighborhood), is a Euler's tour possible? If refuse is to be collected only on the block face (refuse is collected on only one side of a street, a situation that would typically occur in heavy traffic with large streets), is a Euler's tour possible in this second case?

Solution The street network is reduced to a series of links and nodes in Figure 13-6B for the case where the truck travels down a street only once. Note that eight of the nodes have an odd number of links, and thus a Euler's tour is impossible.

In the second case, the links are shown in Figure 13-6C, and all the nodes have an even number of links, indicating that a unicoursal route is possible. But what is it?

◆

FIGURE 13-3 Collection vehicles for separated materials.

FIGURE 13-4 Hydraulic lifts are used to empty the green cans into the collection vehicle.

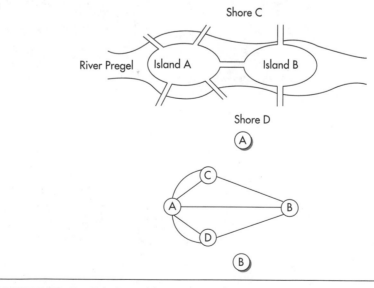

FIGURE 13-5 Euler's problem in Königsberg.

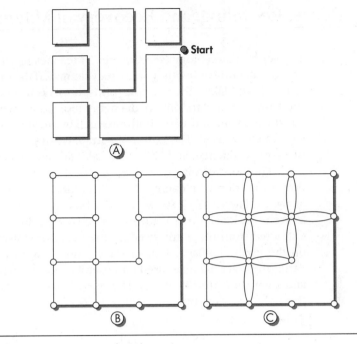

FIGURE 13–6 Routing of trucks.

Computer programs are available for developing the most efficient route possible, but often these are not used because

- They take too much time to write and debug the computer program, or the available software is not appropriate for the specific situation;
- It's possible to develop a very good solution (maybe not the absolute optimal solution) by commonsense means;
- The collection crews will change the routes around to suit themselves anyway!

Commonsense routing is sometimes called *heuristic routing*, which means the same thing. Some sensible rules-of-thumb, when followed, will go a long way toward producing the best collection solution. For example:

- Try to always make right-hand turns.
- Try to travel in long straight lines.
- Try not to leave a one-way street as an exit from a node.

Most of these rules are, as the name suggests, common sense. Tremendous savings can often be realized by seemingly minor modifications to the collection system since this is usually the greatest cost center for refuse management.

13.2 ◆ Reuse, Recycling, and Recovery of Materials from Refuse

The science of ecology (see Chapter 7) teaches us that if dynamic ecosystems are to remain healthy, they must recycle materials. In simple ecosystems such as ponds and lakes, for example, phosphorus is used during photosynthesis by the aquatic plants to build high-energy molecules, which are then used by the aquatic animals, and when both produce waste and die, phosphorus is released so it can be reused. The health of natural ecosystems can be measured by their diversity, resilience, and ability to maintain homeostasis (steady state).

The flow of materials through the human ecosystem is not unlike the flow of nutrients or energy through natural ecosystems and can be similarly analyzed. Figure 13–7 shows a black box that represents human society, just as we use a black box to express an ecosystem. In an ecosystem, nutrients are extracted from the earth, used by the living organisms, and then redeposited on the earth. Similarly, human society uses raw materials extracted from the earth, these are manufactured into useful products to be used by human beings, and then they are discarded. The mass balance shows the flow of materials through this black box.

$$\begin{bmatrix} \text{rate of} \\ \text{materials} \\ \text{ACCUMULATED} \end{bmatrix} = \begin{bmatrix} \text{rate of} \\ \text{materials} \\ \text{IN} \end{bmatrix} - \begin{bmatrix} \text{rate of} \\ \text{materials} \\ \text{OUT} \end{bmatrix} \\ + \begin{bmatrix} \text{rate of} \\ \text{materials} \\ \text{PRODUCED} \end{bmatrix} - \begin{bmatrix} \text{rate of} \\ \text{materials} \\ \text{CONSUMED} \end{bmatrix}$$

If a raw material such as iron is considered, at steady state the amount of iron ore extracted from the earth must equal the amount of iron discarded as ferrous materials.

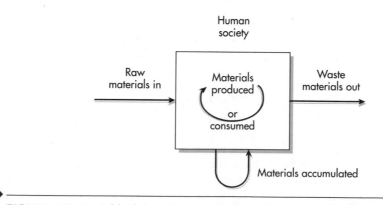

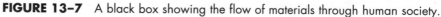

FIGURE 13–7   A black box showing the flow of materials through human society.

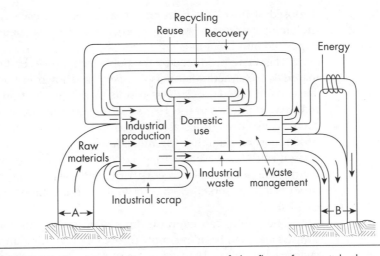

FIGURE 13–8 A graphic representation of the flow of materials through human society.

Figure 13-8 is a more detailed representation of materials flow through human society. The width of the bands is intended to indicate the mass rate of flow (e.g., millions of tons per year). The wider the band, the larger the flow. All the materials originate from the earth, and the amount of material extracted is represented in the figure by the letter A. These are the *raw materials* such as iron ore, oil, and other primary materials that feed industry. These materials are extracted and fed into the manufacturing sector for the production of useful goods.

Not all the extracted materials can be used, however, and some become *industrial waste* that must be disposed of into the environment. Some more material can become *industrial scrap*, and can be used by the same industry or shipped to some other industry through a waste exchange. The primary distinction between scrap and other types of secondary materials is that scrap never enters the public sector—it is never used by the public and thus the public does not have a decision to make about its eventual destination.

The materials in the form of manufactured goods flow from industry to the public. The public uses these materials, and then discards some of them after their usefulness is finished (e.g., batteries) or when the public is sufficiently wealthy to choose to use other products (e.g., CDs replacing LP records). The material the public chooses to classify as waste becomes *domestic waste*, as shown in the figure, and this material is given to the waste management sector.

Some organic waste can be first used for energy conversion before its disposal. Combustion of this waste produces electric power or steam that would have otherwise been produced by the combustion of nonreplenishable fossil fuels or the use of nuclear energy. The materials returned to the environment, expressed as tons per year, are the same, however. An atom of carbon is still disposed of into the environment whether it is carbon as carbon dioxide in the emissions from a waste-to-energy facility or as cellulose in the newspaper deposited into a landfill.

The waste produced by the public can, however, be processed and useful materials reclaimed. Materials *recovery*, as shown in the figure, uses domestic waste as the feed stock. This results in the production of materials useful to industry. Promoted in the 1970s as "resource recovery," this process includes such diverse operations as the recovery of steel from old automobiles to the production of compost for nurseries. Today a central processing facility is known as a *Materials Recovery Facility*, or MRF (pronounced "Murph"). The main feature of materials recovery in a MRF is that the recovered materials are produced from mixed domestic waste and are then reintroduced into industrial use.

The public can exercise three alternate means for getting rid of its unwanted material. Besides simply throwing everything into a mixed-waste receptacle, each individual can *reuse* the products and put the products to secondary, often imaginative use. For example, the use of paper shopping bags for the disposal of trash is a secondary use for the paper bags. Reuse can also include extending the life of products such as buying automobile tires that last for 60,000 miles instead of 30,000 miles. The tires are in effect "reused" for the extra 30,000 miles.

A third alternative is for the individual to separate the materials before they are discarded. The materials separation is at the source of the waste such as the kitchen, and this operation is often termed *source separation*. The separation into different types of materials by the person who decides to produce the waste, and the collection and subsequent reintroduction of this material into the public sector, is known as *recycling*. Note that recycling is not source separation. Source separation is only the first step in the recycling process that must include collection, processing, transport, and eventual sale to an industry that then uses the material. The recycling process depends on the willingness of industry to purchase the recycled material, and no amount of separation at the household level will make this operation viable without the development of dependable markets.

In summary, raw materials are extracted by industry for the production of goods for public use. Industry produces waste that becomes either industrial waste for disposal or industrial scrap. The public has a choice of simply giving the unwanted materials to the waste managers, reusing the products, or recycling. Recovery occurs generally out of the public's control and involves the processing of mixed waste for the production of useful materials.

Figure 13-8 is useful in showing the effect of reuse, recycling, and recovery. In all three cases, an increase in materials flows through these loops will reduce waste. For example, the use of paper shopping bags for trash disposal presupposes that the public chooses not to purchase plastic bags for trash disposal. If this occurs, fewer plastic bags will be produced and sold, and the total amount of waste destined for disposal will likewise be reduced. Glass bottles can be separated from refuse, collected, and remanufactured into new glass bottles, thereby increasing the recycling loop. Finally, mixed refuse can be hand sorted and corrugated cardboard removed and recovered, increasing materials flow in the recovery loop.

Manufacturers can enhance the feasibility of recycling and recovery of material by consciously producing products that are simple and inexpensive to recover or recycle or that can be reused. At some point, society may choose to mandate such rules, but voluntary efforts might be far more effective than governmental dictates.

In all three methods, reuse, recycling, and recovery, the primary goal is purity. For example, the daily refuse from a city of 100,000 would contain perhaps 200 tons per day of paper. Secondary paper sells for about $20/ton (it fluctuates greatly), so that the income to the community would be about $4000 per day, or about $1.5 million per year! So why isn't every community recovering the paper from its refuse and selling it? The answer is elementary; because processing the refuse to recover the used paper costs more than producing paper from trees.

The obvious solution is to never dirty up the paper (and other materials that might have market value) in the first place. Recycled paper is clean and has a market value. But recycling programs often rely on voluntary cooperation, and such cooperation can be fickle. Try as we might, it has been almost impossible to get the public to separate more than 25% of the material before collection.

One way to increase this percentage might be to adopt laws that *require* people to separate their refuse. But this is not a successful approach in a democracy, since public officials advocating unpopular regulation can be removed from office. Source separation in totalitarian regimes, however, is easy to implement, but this is hardly a compelling argument to abolish democracy. A further complication is that in some inner cities it is difficult to convince people to put refuse in trash cans, much less convince them to separate the refuse into the various components.

Theoretically, vast amounts of materials can be reclaimed from refuse, but this is not an easy task regardless of how it is approached. In recycling, a person about to discard an item must first identify it by some characteristics and then manually separate the item into a separate bin. In the recovery operation, the search for purity requires a mechanical identification and separation of materials. The separation in both cases relies on some readily identifiable characteristic or properties of the specific material that distinguish it from all others. This characteristic is known as a *code*, and this code is used to separate the material from the rest of the mixed refuse using a *switch*, as originally introduced in Chapter 3.

In recycling, the code is usually simple and visual. Anyone can identify newspapers from aluminum cans. But sometimes confusion can occur, such as identifying aluminum cans from steel cans, or newsprint from glossy magazines, especially if the glossy magazines are wrapped up in the Sunday paper.

The most difficult operation in recycling is the identification and separation of plastics. Because mixed plastic has few economical uses, plastic recycling is economical only if the different types of plastic are separated from each other. However, most people cannot distinguish one type of plastic from another. The

FIGURE 13-9   Identification markings on plastics.

plastics industry has responded by marking most consumer products with a code that identifies the type of plastic, as shown in Figure 13–9. Plastics that can be recycled are all common products used in everyday life, some of which are listed in Table 13–1.

Theoretically, a person about to discard an unwanted plastic item only has to look at the code and separate the various types of plastics accordingly. In fact, there is almost no chance that a domestic household will have seven different waste receptacles for plastics, nor are there enough of these to be economically collected and used. Typically only the most common types of plastic are recycled, including PETE (polyethylene terephthalate), the material out of which the two-liter soft drink bottles are made, and HDPE (high-density polyethylene), the white plastic used for milk bottles.

Mechanical recovery operations have a chance of succeeding if the material presented for separation is clearly identified by a code and if the switch is then sensitive to that code. Currently, no such technology exists. It is impossible, for example, to mechanically identify and separate all the PETE soft drink bottles from refuse. In fact, most recovery operations employ *pickers*, persons who identify the most readily separable materials such as corrugated cardboard and HPDE milk cartons before the refuse is mechanically processed.

TABLE 13-1 Common Types of Plastics that May Be Recycled			
Code Number	Chemical Name	Nickname	Typical Uses
1	Polyethylene terephthalate	PETE	Soft drink bottles
2	High-density polyethylene	HDPE	Milk cartons
3	Polyvinyl chloride	PVC	Food packaging, wire insulation, and pipe
4	Low-density polyethylene	LDPE	Plastic film used for food wrapping, trash bags, grocery bags, and baby diapers
5	Polypropylene	PP	Automobile battery casings and bottle caps
6	Polystyrene	PS	Food packaging, foam cups and plates, and eating utensils
7	Mixed plastic		Fence posts, benches, and pallets

Most items in refuse are not made of a single material, and to be able to use mechanical separation, these items must be separated into discrete pieces consisting of a single material. A common ''tin can,'' for example, contains steel in its body, zinc on the seam, a paper wrapper on the outside, and perhaps an aluminum top. Other common items in refuse provide equally challenging problems in separation.

One means of producing single-material pieces and assisting in the separation process is to decrease the particle size of refuse by grinding up the larger pieces. This will increase the number of particles and achieve many ''clean'' or single-material particles. The size-reduction step, although not strictly materials separation, is commonly a second step in a MRF, after the picking process.

Size reduction is followed by various other processes such as air classification (which separates the light paper and plastics) and magnetic separation (which separates iron and steel). Various unit operations used in a typical refuse processing plant are shown in Figure 13-10.

The recovery of materials, although it sounds terribly attractive, is still a marginal option. The most difficult problem faced by engineers designing such facilities is the availability of firm markets for the recovered product. The markets can be quite volatile, and secondary-material prices can fluctuate wildly. One example is the secondary-paper market.

Paper industry companies are what is known as *vertically integrated*, meaning that the company owns and operates all the steps in the papermaking process. They own the lands on which the forests are grown, they do their own logging, and they take the logs to their own paper mill. Finally, the company markets the finished paper product to the public. This is schematically shown in Figure 13-11.

Suppose a paper company finds that it has a base demand of 100 million tons of paper. It then adjusts the logging and pulp and paper operations to meet this demand. Now suppose there is a short-term fluctuation of 5 million tons that has to be met. There is no way the paper company can plant the trees necessary to meet this immediate demand, nor are they able to increase the capacity of the pulp and paper mills on such short notice. What they do then is to go to the secondary-paper market and purchase the secondary fiber to meet the incremental demand. If several large paper companies find that they have an increased demand, they will all try to purchase the secondary paper, and suddenly the demand will increase the price of waste paper greatly. When either the demand decreases or the paper company has been able to expand its capacity, it no longer needs the secondary paper, and the price of waste paper plummets. Because paper companies purchase waste paper *on the margin*, secondary-paper dealers are always in either a boom or bust situation, and the price is highly variable. When paper companies that use only secondary paper to produce consumer products increase their production (due to the demand for recycled or recovered paper), these extreme fluctuations are dampened out.

One product that *always* has a market is energy. Since refuse is about 80% combustible material, it can be burned as is or processed to produce a *refuse-derived fuel*. A cross section of a typical waste-to-energy facility is shown

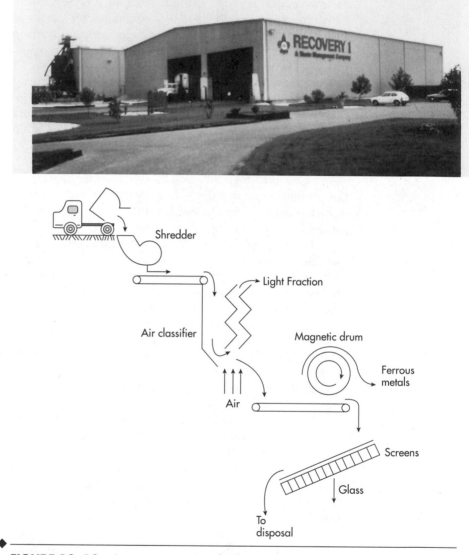

FIGURE 13-10   A resource recovery facility producing various marketable products from municipal solid waste.

in Figure 13-12. The refuse is dumped from the collection trucks into a pit that mixes and equalizes the flow over the 24-hour period since such facilities must operate around the clock. A crane lifts the refuse from the pit and places it in a chute that feeds the furnace. The grate mechanism moves the refuse, tumbling it and forcing in air from the bottom as well as the top as the combustion takes place. The hot gases produced from the burning refuse is cooled with a bank of tubes filled with water. As the gases are cooled, the water is

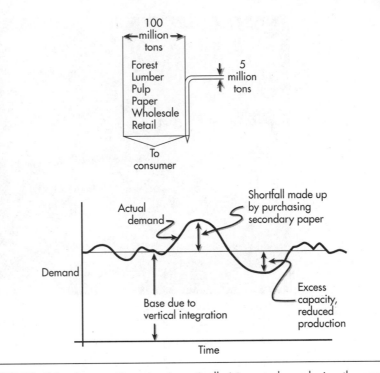

FIGURE 13–11   A paper company is vertically integrated, producing the paper sold to consumers from trees. Short-term fluctuations in demand are met by the purchase of secondary paper.

heated, producing low-pressure steam. The steam can be used for heating and cooling, or for producing electricity in a turbine. The cooled gases are then cleaned by electrostatic precipitators or other efficient particulate control devices and discharged through a stack.

Because solid waste can be combusted as is, and because it also can be processed in many ways before combustion, there might be confusion as to what exactly is being burned. The American Society for Testing and Materials has developed a scheme for classifying solid waste destined for combustion, or refuse-derived fuel (RDF). These designations are as follows:

RDF-1   Unprocessed MSW.

RDF-2   MSW shredded but no separation of materials.

RDF-3   Organic fraction of shredded MSW. This is usually produced in a MRF or from source-separated organics such as newsprint.

RDF-4   Organic waste produced by a MRF that has been further shredded into a fine almost-powder form, sometimes called "fluff."

RDF-5   Organic waste produced by a MRF that has been densified by a pelletizer or a similar device. These pellets often can be fired with coal in existing furnaces.

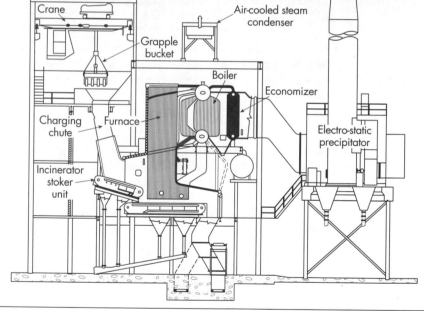

FIGURE 13–12   A waste-to-energy facility.

RDF-6   Organic fraction of the waste that has been further processed into a liquid fuel, such as oil.

RDF-7   Organic waste processed into a gaseous fuel.

At the present time, RDF-4 is seldom used because of the extra cost of processing. RDF-6 was attempted some years ago, but the full-scale facility failed to operate properly. Organic MSW can be processed into a gas by anaerobic digestion. Though the process is deceptively (and seductively) simple and appears to use common components, all experiences to this date have been failures. The most difficult problems are the preprocessing of the waste to achieve a high-quality organic fraction, the mixing of the high-solids digesters, and the presence of toxic material that can severely hinder the operation of the anaerobic digester.

One reason such waste-to-energy facilities have not found greater favor is the concern with the emissions, but this seems to be a misplaced concern. All studies have shown that the risk of municipal solid waste combustion facilities on human health is negligible, far below the risks resulting from the combustion of gasoline in automobiles, for example.

Of particular concern to many people is the production of "dioxin" in waste combustion. "Dioxin" is actually a combination of many members of a family of organic compounds called polychlorinated dibenzodioxins (PCDD). Members of this family are characterized by a triple-ring structure of two benzene rings connected by a pair of oxygen atoms (Figure 13–13). A related family of organic chemicals are the polychlorinated dibenzofurans (PCDF), which have a similar structure except that the two benzene rings are connected by only one oxygen. Since any of the carbon sites are able to attach to either a hydrogen or a chlorine atom, the number of possibilities is great. The sites that are used for the attachment of chlorine atoms are identified by number, and this signature identifies the specific form of PCDD or PCDF. For example, 2,3,7,8-tetrachloro-dibenzo-*p*-dioxin (or 2,3,7,8-TCDD in shorthand) has four chlorine atoms at the four outside corners, as shown in Figure 13–13. This form of dioxin, especially toxic to laboratory animals, is often identified as a primary constituent of contaminated pesticides and emissions from waste-to-energy plants.

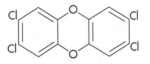

FIGURE 13–13   2,3,7,8-tetrachloro-dibenzo-*p*-dioxin.

All the PCDD and PCDF compounds (referred to here as "dioxins") have been found to be extremely toxic to animals, with the LD_{50} for guinea pigs being about 1 μg/kg body weight. Neither PCDD or PCDF compounds have found any commercial use and are not manufactured. They do occur, however, as contaminants in other organic chemicals. Various forms of "dioxins" have been found in pesticides (such as Agent Orange, widely used during the Vietnam War) and in various chlorinated organic chemicals such as chlorophenols.

Curiously, recent evidence has not borne out the same level of toxicity to humans, and it seems less likely that "dioxins" are actually as harmful as they might seem from laboratory studies. A large chemical spill in Italy was expected to result in public health disaster, based on the extrapolations from animal experiments, but thus far this does not seem to have materialized. Nevertheless, "dioxins" are able, at very low concentrations, to disrupt normal metabolic processes, and this has caused the EPA to continue to place severe limitations on the emission of "dioxins" from incinerators.

There is little doubt that waste-to-energy plants emit trace amounts of "dioxins," but nobody knows for sure how the "dioxins" originate. It is possible that some of them are in the waste, and simply not combusted, being emitted with the off-gases. However, it also is likely that *any* combustion process that has even trace amounts of chlorine produces "dioxins" and that these are simply an end product of the combustion process. The presence of trace quantities of "dioxins" in emissions from wood stoves and fireplaces seems to confirm this view.

It might be well to remember here that the two sources of risk, incinerators vs. fireplaces, are clearly unequal. The effect of the latter on human health is greater than the effect of incinerator emissions. But the fireplace is a *voluntary* risk, whereas the incinerator is an *involuntary* risk. As noted in Chapter 1, people are willing to accept voluntary risks 1000 times higher than involuntary risks, and they are therefore able to vehemently oppose incinerators while enjoying a romantic fire in the fireplace.

Although the volume of the refuse is reduced by more than 90% in waste-to-energy facilities, the remaining 10% still has to be disposed of somehow, as well as the materials such as old refrigerators that cannot be incinerated. A landfill is therefore necessary even if the refuse is combusted, and a waste-to-energy plant is therefore not an ultimate disposal facility. A landfill for ash is a great deal simpler and smaller than a landfill for refuse, and the problem with siting landfills has resulted in many more waste-to-energy facilities because the volume reduction significantly extends the lives of existing landfills.

13.3 ◆ Ultimate Disposal of Refuse

The disposal of solid wastes is a misnomer. Our present practices amount to nothing more than hiding the wastes well enough so they cannot be readily found.

The only two realistic options for disposal are in the oceans (or other large bodies of water) and on land. The former is at present forbidden by federal law and is becoming similarly illegal in most other developed nations. Little else needs to be said of ocean disposal, except perhaps that its use was a less than glorious chapter in the annals of public health and environmental engineering.

13.3.1   Sanitary Landfills

The placement of solid waste on land is called a *dump* in the United States and a *tip* in Great Britain (as in "tipping"). The dump is by far the least expensive means of solid waste disposal, and thus was the original method of choice for almost all inland communities. The operation of a dump is simple and involves nothing more than making sure that the trucks empty at the proper spot. Volume is often reduced by setting the dumps on fire, thus prolonging dump life.

Rodents, odor, air pollution, and insects at the dump, however, can result in serious public health and aesthetic problems, and alternate methods for disposal are necessary. Larger communities can afford to use an incinerator for volume reduction, but smaller towns cannot afford such capital investment, and this has led to the development of the *sanitary landfill*.

The sanitary landfill differs markedly from open dumps in that the latter are simply places to dump wastes, whereas sanitary landfills are engineered, designed, and operated according to acceptable standards. The basic principle of a landfill operation is to prepare a site with liners to deter pollution of groundwater, deposit the refuse in the pit, compact it with specially built heavy machinery with huge steel wheels, and cover the material with earth at the conclusion of each day's operation (Figure 13–14). Siting and developing a proper landfill requires planning and engineering design skills.

Even though the tipping fees paid for the use of landfills is charged on the basis of weight of refuse accepted, landfill capacity is measured in terms of volume, not weight. Engineers designing the landfills first estimate the total volume available to them and then estimate the density of the refuse as it is deposited and compacted in the landfill. The density of refuse increases markedly as it is first generated in the kitchen and then finally placed into the landfill. Table 13–2 is a crude estimate of the density of household refuse.

An added complication in the calculation of landfill volume is the need for the daily dirt cover. The more dirt is placed on the refuse, the less volume is available for the refuse itself. Commonly, engineers estimate that the volume occupied by the cover dirt is one-fourth of the total landfill volume.

e • x • a • m • p • l • e **13.2**

Problem   Imagine a town where 10,000 households each fill one 80-gallon container of refuse per week. To what density would a 20-cubic yard packer truck have to compact the refuse to be able to collect all the households during one trip?

Solution   There is of course (what else) a mass balance involved here. Imagine the packer truck as a black box, and the refuse goes in at the households and out at the landfill.

$$[\text{mass IN}] = [\text{mass OUT}]$$

$$V_L C_L = V_P C_P$$

where V and C are the volume and density of the refuse, and subscripts L and P denote loose and packed refuse. Assume that the density in the cans is 200 lb/yd^3.

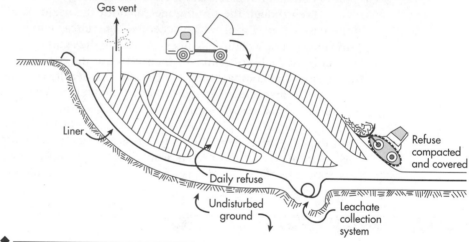

FIGURE 13–14   A sanitary landfill.

TABLE 13–2 Approximate Density of Municipal Solid Waste	
Location/Condition	*Density (lb/yd³)*
As generated, in the kitchen trash can	100
In the "garbage" can	200
In the packer truck	800
In the landfill, as placed	1000
In the landfill, after compaction by overlaying refuse	1200

$$[(10{,}000 \text{ households}) \ (80 \text{ gal/household})$$

$$\times \ (0.00495)] \ [200 \text{ lb/yd}^3] = (20 \text{ yd}^3)C_P$$

$$C_P = 39{,}600 \text{ lb/yd}^3(!)$$

Clearly impossible. Note that the 0.00495 is the conversion factor from gallons to cubic yards.

◆

Sanitary landfills are not inert. The buried organic material decomposes anaerobically, producing various gases, such as methane and carbon dioxide, and liquids that have extremely high pollutional capacity when they enter the groundwater. Liners made of either impervious clay or synthetic materials such as plastic are used to try to prevent the movement of leachate into the groundwater. Figure 13–15 shows how a synthetic landfill liner is installed in a prepared pit. The seams have to be carefully sealed, and a layer of soil placed on the liner to prevent landfill vehicles from puncturing it.

Synthetic landfill liners are useful in capturing most of the leachate, but they cannot be perfect. No landfill is sufficiently tight that groundwater contamination by leachate is totally avoided. Wells have to be drilled around the landfill to check for groundwater contamination from leaking liners, and if such contamination is found, remedial action is necessary.

The use of plastic liners has substantially increased the cost of landfills to the point where a modern landfill costs nearly as much per ton of refuse as a waste-to-energy plant. And of course the landfill never disappears—it will be there for many years to come, limiting the use of the land for other purposes.

Modern landfills also require the gases to be collected and either burned or vented to the atmosphere. The gases are about 50% carbon dioxide and 50% methane, both of which are greenhouse gases. In the past, when gas control in landfills was not practiced, the gases were known to cause problems with odor, soil productivity, and even explosions. Now the larger landfills use the gases for running turbines for the production of electricity for sale to the power company. Smaller landfills simply vent the gases to the atmosphere.

No matter what the propaganda, sanitary landfills are at best an interim solution to our solid waste problem. Environmental concerns will soon dictate materials and/or energy recovery as the disposal method of choice for solid waste management.

Obviously, we must attack the solid waste problem from both ends—reducing the total quantity of wastes by making materials more recyclable, and developing more environmentally acceptable disposal methods. We are still many years away from the development and use of fully recyclable or biodegradable materials. The only truly disposable package available today is the ice cream cone.

There is a third way we can affect the quantity and content of the solid waste stream: carefully select the materials and products we use and hence have to throw away. There is a lot to be said for being selective in the type of packaging accepted for various products. For example, foam plastic wrappers

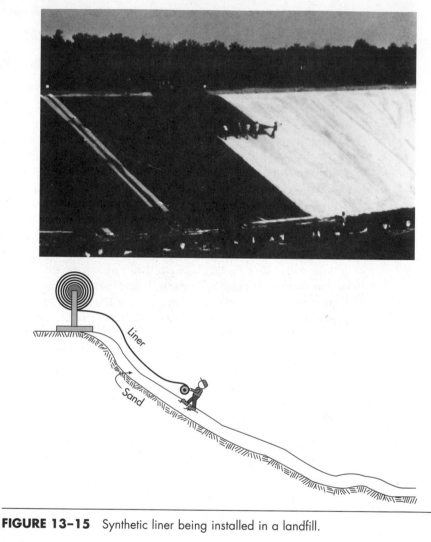

FIGURE 13-15   Synthetic liner being installed in a landfill.

for fast foods have little utility. A piece of paper works just as well, as many fast food chains discovered once their customers started complaining (and in some cases passing local ordinances prohibiting such packaging). Rejecting useless bags at stores when a bag is not needed is not bad manners. We can all do a lot of little things to make a big impact on the quantity and composition of the solid waste stream.

The question of course is, why should we? Consider the utility of a person rejecting an unnecessary bag at a store. The effort is significant (you sometimes have to argue) and the convenience is less since now the purchase is not carried as conveniently. The bag would not have cost anything personally to dispose of if the collection is a municipal service. What then is the benefit to you to reject the bag? Essentially none at all. The quantity of landfill space you personally save by your actions is negligible, and your choice of purchases that reduce the quantity of waste will not benefit you at all.

So the question is, why ought you to do something that is not to your measurable benefit?

The answer might be that "it is the right thing to do," and this may be perfectly satisfying to you. It makes you feel a part of the community, all pulling together to achieve some good end. You could argue that if you did not do this, you would have no basis for expecting others to do the same, and then there would be no way to achieve anything by community action. But if someone says, "Prove to me that I ought to participate in such altruistic activity," you have to admit that there is no such proof. There are arguments, but all of them depend on the basic goodness of humans. When that is no longer the norm, then there is no hope for human civilization.

13.3.2 Life Cycle Analysis

One means of getting a handle on questions of material and product use is to conduct what has become known as a *life cycle analysis*. Such an analysis is a holistic approach to pollution prevention by analyzing the entire life cycle of a product, process, or activity, encompassing raw materials, manufacturing, transportation, distribution, use, maintenance, recycling, and final disposal. In other words, life cycle analysis should yield a complete picture of the environmental impact of a product.

Life cycle analyses are performed for several reasons, including the comparison of products for purchasing and a comparison of products by industry. In the former case, the total environmental effect of say glass returnable bottles could be compared to the environmental effect of nonrecyclable plastic bottles. If all the factors going into the manufacture, distribution, and disposal of both types of bottles are considered, one container might be shown to be clearly superior. In the case of comparing the products of an industry, we might determine if the use of phosphate builders in detergents is more detrimental than the use of substitutes, which have their own problems in treatment and disposal.

One problem with such studies is that they are often conducted by industry groups or individual corporations, and (surprise!) the results often promote their own product. For example, Proctor & Gamble, the manufacturer of a popular brand of disposable baby diapers, found in a study done for them that the cloth diapers consume three times more energy than the disposable kind. But a study by the National Association of Diaper Services found that disposable diapers consume 70% more energy than cloth diapers. The difference was in the accounting procedure. If one uses the energy contained in the disposable diaper as recoverable in a waste-to-energy facility, then the disposable diaper is more energy efficient.[1]

Life cycle analyses also suffer from a dearth of data. Some of the information critical to the calculations is virtually impossible to obtain. For example, something as simple as the tonnage of solid waste collected in the United States is not readily calculable or measurable. Even if the data *were* there, the procedure suffers from the unavailability of a single accounting system. Is there an optimal level of pollution or must all pollutants be removed 100% (a virtual impossibility)? If there is air pollution and water pollution, how must these be compared?

A simple example of the difficulties in life cycle analysis would be in finding the solution to the great coffee cup debate—whether to use paper coffee cups or polystyrene coffee cups. The answer most people would give is not to use either, but instead to rely on the permanent mug. But there nevertheless are times when disposable cups are necessary (e.g., in hospitals), and a decision must be made as to which type to choose.[2] So let's use life cycle analysis to make a decision.

The paper cup comes from trees, but the act of cutting trees results in environmental degradation. The foam cup comes from hydrocarbons such as oil and gas, and this also results in adverse environmental impact, including the use of nonrenewable resources. The production of the paper cup results in significant water pollution, with 30 to 50 kg of BOD per cup produced while the production of the foam cup contributes essentially no BOD. The production of the paper cup results in the emission of chlorine, carbon dioxide, reduced sulfides, and particulates, while the production of the foam cup results in none of these. The paper cup does not require chlorofluorocarbons, but neither do the newer foam cups ever since the CFCs in polystyrene were phased out. The foam cups, however, contribute from 35 to 50 kg per cup of pentane emissions, while the paper cup contributes none. The recyclability of the foam cup is much higher than the paper cup since the latter is made from several materials, including the plastic coating on the paper. They both burn well, although the foam cup produces 40,000 kJ/kg while the paper cup produces only 20,000 kJ/kg. In the landfill, the paper cup degrades into CO_2 and CH_4, both greenhouse gases, while the foam cup is inert. Since it is inert, it will remain in the landfill for a very long time, while the paper cup will eventually (but very slowly!) decompose. If the landfill is considered a waste storage receptacle, then the foam cup is superior, since it does not participate in the reaction, while the paper cup produces gases and probably leachate. If, however, the landfill is thought of as a treatment facility, then the foam cup is highly detrimental.

So which cup is better for the environment? If you wanted to do the right thing, which cup should you use? This question, like so many others in this book, is not an easy one to answer.

13.4 ◆ Integrated Solid Waste Management

The EPA has developed a national strategy for the management of solid waste, called the Integrated Solid Waste Management (ISWM). The intent of this plan is to assist local communities in their decision making by encouraging those strategies that are the most environmentally acceptable. The EPA ISWM strategy suggests that the list of the most to least desirable solid waste management strategies should be as follows:

- Source reduction
- Recycling
- Combustion
- Landfilling

That is, when an integrated solid waste management plan is implemented for a community, the first means of attacking the problem should be by *source reduction*. This is an unfortunate term because it is both incorrect and misleading. One is not reducing sources but rather reducing the amount of waste coming from a source and thus the term really should be "waste reduction."

On a household level it is possible through judicious buying to significantly reduce the amount of waste generated in each household. For example, buying produce unwrapped instead of on plastic trays with shrink-wrap is one simple way of reducing waste. Purchasing products that have a longer life is a second way of achieving waste reduction. Buying automobile tires that last 80,000 miles instead of those that wear out at 40,000 reduces the production of waste tires by 50%. In addition, getting your name off mailing lists is a big help because it reduces junk mail. This option of solid waste management is even more effective on the industrial and commercial level.

Recycling is the second option and should be undertaken when most of the waste-reduction options have been implemented. Unfortunately, the EPA confuses recycling with recovery and groups them together as meaning any technique that results in the diversion of waste. Recycling is of course the collection and processing of the separated waste, the result being new consumer products. Recovery is the separation of mixed waste, also with the end result of producing new raw materials for industry.

The third level of the ISWM plan is solid waste combustion, which really should mean all methods of treatment. The idea is to take the solid waste stream and to transform it into a nonpolluting product. This may be by combustion, but other thermal and chemical treatment methods may eventually prove just as effective.

Finally, if all of the above techniques have been implemented and/or considered, and there is still waste left over (which there will be), the final solution is landfilling. There really is no alternative to landfilling (except disposal in deep water—which is now illegal), and therefore every community must develop some landfilling alternative.

Although this ISWM strategy is useful, it can lead to problems if taken literally. Some communities cannot, as hard as they might try, implement recycling. In other communities, the only option right from the start is landfilling. In others, the intelligent thing to do would be to provide one type of treatment for one kind of waste (e.g., combustion for yard waste) and one for another (e.g., landfilling for refuse). This is where engineering judgment comes into play and where the solid waste engineer really earns his or her salary. We have to juggle all the options and integrate these with the special features (economics, history, politics) of the community. The engineer is, after all, there to serve the needs of the people.

◆ Abbreviations

C	= density of refuse, lb/yd³	PETE	= polyethylene terephthalate
HDPE	= high-density polyethylene	PP	= polypropylene
LDPE	= low-density polyethylene	PS	= polystyrene
PCDD	= polychlorinated dibenzodi-	PVC	= polyvinyl chloride
	oxins (commonly called	RDF	= refuse-derived fuel
	"dioxins")	V	= volume of refuse, yd³
PCDF	= polychlorinated dibenzo-		
	furans		

◆ Problems

13–1   The street plan in Figure 13–6A is analyzed as Figure 13–6C if the collection has to be by *blockfaces*, that is, the truck must travel along each block to collect and the crew cannot cross the street. Since it is possible to construct a Euler's tour, find one solution.

13–2   Using the street networks in Figure 13–16, design the most efficient route possible if the truck is

  a.   collecting only one side of a street at a time

  b.   collecting streets both sides at the same time

13–3   The objective of this assignment is to evaluate and report on the solid waste you personally generate. For one day, selected at random, collect *all* the solid waste you would have normally discarded. This includes food, newspapers,

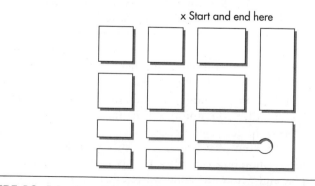

x Start and end here

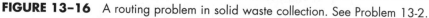

FIGURE 13–16   A routing problem in solid waste collection. See Problem 13-2.

beverage containers, and so on. Using a scale, weigh your solid waste and report it as follows:

Component	Weight	Percent of Total Weight
Paper		
Plastics		
Aluminum		
Steel		
Glass		
Garbage		
Other	_____	_____
		100

Your report should include a data sheet and a discussion. Answer the following questions.

a. How do your percentage and total generation compare to national averages?

b. What in your refuse might have been reusable (as distinct from recoverable), and if you had reused it, how much would this have reduced the refuse?

c. What in your refuse is *recoverable*? How might this be done?

13-4 One of the costliest aspects of municipal refuse collection is the movement of the refuse from the household to the truck. Suggest a method by which this can be improved. Originality counts heavily, practicality far less.

13-5 One of the unexpected results of the reunification of Germany is the "Trabant Problem." The Trabant, or "Trabi" as it was popularly known, was the communists' answer to the Volkswagen, a cheap vehicle powered by a notoriously inefficient and dirty two-stroke engine. The body was made out of a cellulose filler covered by a phenol formaldehyde resin that produces a plastic that cannot be melted down. Along with freedom for the East Germans comes the problem of what to do with the millions of Trabi hulks that litter the countryside and clog up the solid waste disposal system (Figure 13-17). The only solution seems to be to burn them, but this produces air pollution that Eastern Europe can ill afford.

Suggest how you would approach the "Trabi Problem" if suddenly you were placed in charge of this project. List all the feasible alternatives, and discuss each, finally choosing the one you favor.

13-6 Plastic bags at food stores have become ubiquitous. Often recycling advocates point to the plastic bags as the prototype of waste and pollution, as stuff that clogs up our landfills. In retaliation, plastic bag manufacturers have begun a public relations campaign to promote their product. On one of the flyers (printed on paper) they say:

> The [plastic] bag does not emit toxic fumes when properly incinerated. When burned in waste-to-energy plants, the resulting by-products from combustion are carbon dioxide and water vapor, the very same by-products that you and I produce when we breathe. The bag is inert in landfills where it does not contribute to leaching bacterial or explosive gas prob-

FIGURE 13–17   One method of disposing of a Trabi hulk. See Problem 13-5. (Photo courtesy of *Chemical Engineering*. Used with permission.)

lems. The bag photodegrades in sunlight to the point that normal environmental factors of wind and rain will cause it to break into very small pieces, thereby addressing the unsightly litter problem.

Critique this statement. Is all of it true? What part is not? Is anything misleading? Do you agree with their evaluation? Write a one-page response.

13-7   The siting of landfills is a major problem for many communities. This is often an exasperating job for engineers because the public is so intimately involved. A prominent environmental engineer, Glenn Daigger of CH2M Hill, is quoted as follows:

Environmental matters are on the front page today because we, the environmental industry, are not meeting people's expectations. They're telling us that accountability and quality are not open questions that ought to be considered. It's sometimes difficult to grasp in the face of all the misinformation out there: Ultimately, we're responsible for accommodating the public's point of view, not the other way around.[3]

Do you agree with him? Should the engineer "accommodate the public's point of view" or should the engineer impose his or her own point of view on the public, since the engineer has a much better understanding of the problem? Write a one-page summary of what you believe should be the engineer's role in the siting of a landfill for a community.

13-8   The recycling symbols shown in Figure 13-18 are taken from various forms of packaging.

a.   What purpose do you suppose the container manufacturers had in mind when placing the symbols on their packages?

FIGURE 13-18 Recycling symbols off consumer packaging. See Problem 13-8.

   b. Was this use ethical in all cases? Argue your conclusion.

   c. Cut out another recycling symbol from a package and discuss its merits.

13-9 A high school has a student body of 660 students. Studies have shown that, on average, each student is expected to contribute 0.2 lb of solid waste from all sources except the cafeteria, which contributes 320 lb daily. If the density of refuse in a dumpster is 200 lb/yd^3, and if trash pickup occurs only once a week, how many dumpsters does the school need if each dumpster can hold 20 yd^3?

13-10 Figure 13-19 shows an isolated development that is to be serviced by refuse pickup. Numbers on the blocks indicate the number of households that need to be picked up on that block. Assume each household is 4 persons, and that the waste generation is 3 lb/capita/day. The density of refuse in the old-style packer truck is only about 400 lb/yd^3.

   a. If the pickup is to occur once a week, how many trucks would be needed?

   b. If the trucks entered and exited the development at the point shown by *E*, what would be one efficient routing for these trucks? (Count the number of deadheads as an estimate of efficiency.)

   c. If the old trucks were abandoned and new ones achieving compaction to 1000 lb/yd^3 were purchased, how would the collection system be changed?

13-11 How do you think the composition of municipal solid waste will change in 20 years? Why?

13-12 A commercial establishment has a private hauler who weighs the solid waste transported to the landfill. The data for 10 consecutive weeks are shown below.

Week No.	Refuse (lb)
1	540
2	620
3	920
4	410
5	312
6	820
7	540
8	420
9	280
10	780

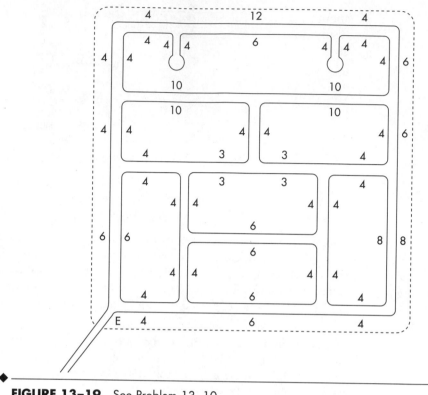

FIGURE 13-19   See Problem 13-10.

a.   What is the average waste generation for this establishment?

b.   What might be the *lowest* week in the entire year, based on these data?

c.   If the dumpster is able to handle 2000 lb, will it ever be full? (This is a curve ball. Careful with your answer.)

13-13   A solid waste incinerator has a maximum operating capacity of 100 tons/day. The facility receives MSW during the work week, but not during the weekends or on holidays. Typically, it receives the following daily loads:

Day	Received (tons/day)
Mondays	180
Tuesdays	160
Wednesdays	150
Thursdays	120
Fridays	80
Saturdays	0
Sundays	0

How large should the receiving pit be to hold enough waste to operate through a three-day weekend? (There is no one right answer to this problem. Think about what you would do if you were the engineer having to make this decision.) Justify your answer.

13-13 Suppose you are the city engineer of a small community, and the town council tells you to design a "recycling" program that will achieve at least a 50% diversion from the landfill. What would be your response to the town council? If you agree to try to achieve such a diversion, how would you do it? For this problem, you are to prepare a formal response to the town council, including a plan of action and what would be required for its success. This response would include a cover letter addressed to the council and a report of several pages, all bound in a report format with title page and cover.

13-14 Design, using flow diagrams, how you would produce RDF 1, RDF 2, RDF 3, and RDF 4 as defined by the ASTM standards for refuse-derived fuel.

◆ Endnotes

1. "Life Cycle Analysis Measures Greeness, But Results May Not Be Black and White," *Wall Street Journal* (28 February 1991).
2. Martin B. Hocking, "Paper versus Polystyrene: A Complex Choice," *Science* 251 (1 February 1991).
3. Quoted in the "The New Environment Age," *Engineering News Record*, 228:25 (22 June 1992).

HAZARDOUS WASTE

W hat is so amazing, in retrospect, is the total nonchalance with which the American people tolerated the disposal of highly toxic wastes ever since industry first started producing such materials. Open dumps, huge waste lagoons, blatant dumping into waterways . . . all this existed for decades. Since the early 1980s, however, the public has become conscious of the consequences of such indiscriminate dumping. Since then we have made heroic efforts to clean up the most odious examples of environmental insults and health dangers and to regulate industry so as to prevent future problems.

14.1 ◆ Defining Hazardous Waste

A *hazardous substance* is defined by the EPA as any substance that

> because of its quantity, concentration, or physical, chemical, or infectious characteristics may cause, or significantly contribute to, an increase in mortality; or cause an increase in serious irreversible, or incapacitating reversible illness; or pose a substantial present or potential hazard to human health and the environment when improperly treated, stored, transported, or disposed of, or otherwise managed.

Hazardous waste is a name given to material that, when intended for disposal, meets one of two criteria:

1. It contains one or more of the *criteria pollutants*, or those chemicals that have been explicitly identified as hazardous. There are at present more than 50,000 chemicals thus identified.

2. The waste can be defined (by laboratory tests) to be at least one of the following:

   - Flammable
   - Corrosive
   - Reactive
   - Toxic

Flammable materials are defined as those liquids with flash points below 60°C, or those materials that are "easily ignited and burn vigorously and persistently." Corrosive materials are those that, in an aqueous solution, have pH values outside the range of 2.0 to 12.5, or any liquid that exhibits corrosivity to steel at a rate greater than 6.35 mm per year. Reactive wastes are classified as unstable and able either to form toxic fumes or to explode. The greatest difficulty in defining hazardous waste comes from establishing what is and what is not toxic.

Toxicity is almost impossible to measure. Toxic to what animals (or plants), at what concentrations, or over what time periods?

The EPA defines toxicity in terms of four criteria:

   - Bioconcentration
   - LD_{50}
   - LDC_{50}
   - Phytotoxicity

Bioconcentration is the ability of a material to be retained in animal tissue to the extent that organisms higher on the trophic level will have increasingly higher concentrations of this chemical. Many pesticides, for example, will reside in the fatty tissues of animals and will not break down very fast. As the smaller creatures are eaten by the larger ones, the concentration in the fatty tissues of the larger organisms can reach toxic levels for them. Of most concern are aquatic animals and birds such as seals and pelicans, which feed on fish, or eagles, falcons, condors, and other carnivorous birds.

LD_{50} is a measure of how much of a certain chemical is needed to kill half of a group of test specimens such as mice. Mice in a toxicity study are fed progressively higher doses of the poison until half of them die, and this dose is then known as the lethal dose (50%), or LD_{50}. The lower the amount of the toxin used to kill 50% of the specimens, the higher the toxic value of the chemical. Some chemicals, such as dioxin, for example, or PCBs, show incredibly low levels of LD_{50}s, suggesting that these chemicals are extremely dangerous to small animals and other test species.

There are, however, some serious problems with such a test. The tests are conducted on a laboratory specimen such as a mouse, and the amount of that toxin that would produce death in humans is extrapolated based on body weight, as illustrated in Example 14.1.

e • x • a • m • p • l • e **14.1**

Problem   A toxicity study on the resistance of mice to a new pesticide has been conducted, with the following results:

Amount Ingested (mg)	Fraction of Mice That Died After 4 Hours
0 (control)	0
0.01	0
0.02	0.1
0.03	0.1
0.04	0.3
0.05	0.7
0.06	1.0
0.07	1.0

What is the LD_{50} of this pesticide for a mouse? What is the LD_{50} for a typical human being weighing 70 kg? Assume a mouse weighs 20 g.

Solution   The data are plotted as in Figure 14–1, and the point at which 50% of the mice die is identified. The mouse LD_{50} is therefore about 0.042 mg. The human lethal dose is estimated as

$$[70,000 \text{ g}/20 \text{ g}] \times 0.042 \text{ mg} = 147 \text{ mg}$$

◆

This is a shaky conclusion, however. First, the physiology of a person is quite different from that of a mouse, and a person may be able to ingest either more or less of a toxin before showing an adverse effect. Second, the effect

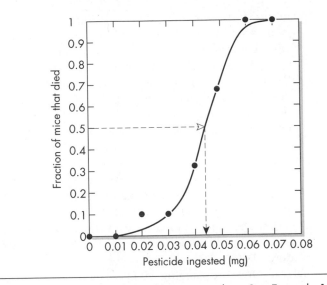

y-axis: Fraction of mice that died

x-axis: Pesticide ingested (mg)

FIGURE 14–1   Calculation of LD_{50} from mouse data. See Example 14.1.

measured is an acute effect, not a long-term (chronic) effect. Thus toxicity of chemicals that affect the body slowly, over years, is not measured, since the mouse experiments are done in hours. Finally, the chemical being investigated may act synergistically with other toxins, and this technique assumes that there is only one adverse effect at a time. As an example of such problems, consider the case of dioxin (see Chapter 13), which has been shown to be extremely toxic to small laboratory animals. All available epidemiological data, however, show that humans appear to be considerably more resilient to dioxins than the data would suggest.

The LDC_{50} is the *concentration* at which some chemical is toxic, and this is used where the amount ingested cannot be measured, such as in the aquatic environment or in evaluating the quality or air. Specimens such as goldfish are placed in a series of aquariums and increasingly higher concentrations of a toxin are administered. The fraction of fish dying within a given time is then recorded, and the LDC_{50} calculated. As a rough guideline, a waste is considered toxic if it is found to have a LD_{50} of less than 50 mg/kg body weight, or if the LDC_{50} is less than 2 mg/kg.

Finally, a chemical is considered toxic if it exhibits phytotoxicity, or toxicity to plants. Thus all herbicides are, by this definition, toxic materials, and they must be disposed of and treated as hazardous wastes.

A final criterion for being hazardous is if the material is radioactive. Radioactive wastes are, however, handled separately and are governed by separate rules and regulations. Radioactive waste disposal is covered in Section 14.3 of this chapter.

One concern in hazardous waste disposal is the speed with which the chemical can be set free to produce toxic effects in plants or animals. For example, one oft-used method of hazardous waste disposal is to mix the waste with a slurry consisting of cement, lime, and other materials. When the mixture is allowed to harden, the toxic material is safely buried inside the block of concrete from which it cannot escape and cause trouble.

Or can it? This question is also relevant in cases such as the disposal of incinerator ash. Many of the toxic materials such as heavy metals are not destroyed during incineration and escape with the ash. If these metals are safely tied to the ash and cannot leach into the groundwater, then there seems to be no problem. If, however, they leach into the water when the ash is placed in a landfill, then the ash would have to be treated as a hazardous waste and disposed of accordingly.

How then is it possible to measure the rate at which such potential toxins can escape the material in which it is at present embedded? As a crude approximation of such potential leaching, the EPA uses an extraction procedure, where the waste is mixed with water and shaken for a number of hours. This process has become known as the Toxicity Characteristics Leaching Procedure (TCLP).

Once the leaching has taken place, the leachate is analyzed for the possible hazardous materials. The EPA has determined that there is a list of contaminants that constitute possible acute harm, and that the level of these contaminants must not be exceeded in the leachate. Table 14-1 includes a few such chemicals. Many of the numbers on this list are EPA Drinking Water Standards multiplied by 100 to obtain the leaching standard. Critics of the test point out that calculating toxicity on the basis of drinking water standards is compounding a series of potential errors. The drinking water standards are, after all, based on scarce data, and often set on the basis of expediency (see Chapter 8). How is it possible then to multiply these spurious numbers by 100 and with a straight face say that something is or is not toxic?

The TCLP also is a test with a lot of loose ends. The conditions such as the water pH and temperature are of course important, as is the turbulence of the mixing and the condition of the solid placed into the mixer. The financial

TABLE 14-1 A Few Examples of EPA's Maximum Allowable Concentrations in Leachates From the Toxicity Characteristics Leaching Procedure (TCLP) Test

Contaminant	Allowable Level (mg/l)
Arsenic	5.0
Cadmium	1.0
Carbon tetrachloride	0.5
Chromium	0.5
Lead	5.0
Lindane	0.4
Pentachlorophenol	100
Trichloroethylene	0.5
Vinyl chloride	0.2

implications to an industry can be staggering if one of their leaching tests shows contaminant levels exceeding the allowable concentrations. The test result that mean so much to an industrial firm often has a feeble epidemiological base. Unfortunately, until we can come up with something better, this is all we've got. Yes it is conservative, but this technique gives us a crude handle on the possible detrimental effects of hazardous wastes.

In summary, a waste can be listed as a hazardous waste if it flunks any of the tests that would keep it from being so classified. Using this criterion, the EPA has developed a list of chemicals it considers toxic. The list is long and growing as new chemicals and waste streams are identified. Once a chemical or process stream is "listed," it needs to be treated as a hazardous waste and is subject to all the applicable regulations. Getting "delisted" is a difficult and expensive process, and the burden of proof is on the petitioner. This is, in a way, a situation where the chemical is considered guilty until proven innocent.

If a waste is "listed," its disposal becomes a true headache since only very few hazardous waste disposal facilities are available. Usually long and costly transportation is necessary, forcing the industry either to dispose of the waste surreptitiously (such as the 200 miles of roadway in North Carolina contaminated by PCBs when the drivers just opened the drain valves and drained the trucks instead of making the long trips to the disposal facilities) or to redesign their plant so as not to *create* waste. The latter requires skill and capital, and more than one industry has had to close when one or both were unavailable.

14.2 ◆ Hazardous Waste Management

Hazardous wastes are controlled in the United States by the EPA. This mandate is under one of several pieces of legislation, most notably the Toxic Substances Control Act (TOSCA), the Comprehensive Environmental Response, Compensation, and Liability Act (CERCLA), and the Resource Conservation and Recovery Act (RCRA). TOSCA is aimed at preventing the creation of wastes that may eventually prove damaging or difficult to dispose of safely, whereas RCRA addresses the disposal of hazardous materials by establishing standards for secure landfills and treatment processes. CERCLA is directed at correcting the mistakes of the past by cleaning up old hazardous waste sites and is usually referred to as the Superfund Act since it set up a large trust fund paid for by chemical and extractive industries for the purpose of providing resources to clean up abandoned sites. The EPA uses these funds to address acute problems caused by improper hazardous waste disposal or accidental discharge and the cleanup of old sites.

14.2.1   Cleanup of Old Sites

As part of the Superfund program, the EPA has established a National Priorities List of hazardous waste sites that are in need of immediate cleanup and are eligible for funding under CERCLA. Estimates of the number of these sites vary, but the latest is that there are about 425,000 sites in need of some degree of cleanup, at a cost of about $700 billion (!). The National Priority List now includes 2200 sites that are in immediate need of attention. As of this writing, only about 40 sites have been cleaned up, so there is a long way to go.

Because of the large number of potential Superfund sites, the EPA has developed a ranking system so that the worst situations can be cleaned up first, and the less critical can wait until funds, time, and personnel become available. This Hazard Ranking System (HRS) tries to incorporate the more sensitive effects of a hazardous waste area, including potential adverse health effects, the potential for flammability or explosions, and the potential for direct contact. These three modes are calculated separately and then used to obtain the final HRS score. Being conscious of public opinion, the scores tend to be overwhelmed in situations where human health is in danger, especially for explosive materials stored in densely populated areas. The potential for groundwater contamination in areas where water supplies are from groundwater sources also receives priority status. This technique minimizes the criticism often faced by EPA that they have not done enough to clean up the Superfund sites. They have, in fact, tackled the most difficult and most sensitive sites first, and should be commended for the rapid response to a problem that has been festering for generations.

In addition to the EPA list, the Department of Energy has 110 sites that require cleanup, and the Department of Defense has a staggering 17,000 sites in need of attention, of which perhaps 500 need extensive work. Much of this is the legacy of nuclear weapon testing and manufacturing, since many of these facilities were strictly secret and the contracting operators could do almost anything they wished with the waste.

The type of work conducted at these sites depends on the severity and extent of the problem. In some cases where there is an imminent threat to human health, the EPA can authorize a *removal action*, which results in the hazardous material being removed to safe disposal or treatment. In less acute instances, the EPA authorizes *remedial action* for hazardous waste sites, which may consist of the removal of the material or, more often, the stabilization of the site so that it is less likely to cause health problems. As with the notorious Love Canal in Niagara Falls, the remedial action of choice was to essentially encapsulate the site and monitor all groundwater and air emissions. The cost of excavating the canal would have been outrageous and there was no guarantee that the eventual disposition of the hazardous materials would have been any safer than leaving it there. In addition, the actual act of removal and transport might have resulted in significant human health problems. Remedial action therefore simply implies that the site has been identified and action taken to minimize the risk of the hazardous material causing human health problems.

The most common situation is where some chemical has already contaminated the groundwater and remediation is necessary. For example, a dry cleaner may accidentally (or purposefully) discharge its waste cleaning fluid and this may find its way into a drinking water well for a nearby residence. Once the problem is detected, the first question is whether or not the contamination is life-threatening or if it poses a significant threat to the environment. Then a series of test are run, using monitoring wells or soil samples, to determine the geology of the area and the size and shape of the plume or range of the contaminated area.

Depending on the seriousness of the situation, several remedial action options are available. If there is no threat to life, and if it can be expected that the chemical will eventually metabolize into harmless end products, then one solution is to do nothing. In most cases, this is not so, and direct intervention is necessary.

Containment is used where there is no need to remove the offending material, and/or if the cost of removal is prohibitive, as was the case at Love Canal. Containment is usually the installation of slurry walls, deep trenches filled with bentonite clay or some other highly nonpermeable material, and continuous monitoring for leakage out of the containment. With time, the offending material might slowly biodegrade or chemically change to nontoxic form, or new treatment methods may become available for detoxifying this waste.

Extraction and treatment is the pumping of contaminated groundwater to the surface for either disposal or treatment, or the excavation of contaminated soil for treatment. Sometimes air is blasted into the ground and the contaminated air collected.

The characteristics of a hazardous chemical often determine its location underground and therefore dictate the remediation process. If the chemical is immiscible with water and is lighter than water, it most likely will float on top of the aquifer, as shown in Figure 14-2A. Pumping this chemical out of the ground is relatively simple. If the chemical readily dissolves in water, however, then it can be expected to be mixed in a contaminated plume, as shown in Figure 14-2B. In such situations it will be necessary to contain the plume either by barrier walls and pumping low concentrations of the waste to the surface for treatment, or by sinking discharge wells to reverse the flow of water. A third possibility is that the chemical is denser than water and is immiscible, in which case it would be expected to lie on an impervious layer somewhere under the aquifer, as shown in Figure 14-2C. If the plume is to be pumped to the surface for treatment, wells are drilled around the contaminated site so that the groundwater flow is reversed by injection and pumping, preventing groundwater from leaving the area, as discussed in Chapter 9. Once the contaminated water is extracted, it must be treated, and the choice of treatment obviously depends on the problem. If the contamination is from hydrocarbons such as trichloroethylene, it is possible to remove this with activated carbon.

Some soils may be so badly contaminated that the only option is to excavate the site and treat the soil. This is usually the case with PCB contamination, because no other method seems to work well. The soil is dug out and usually incinerated to remove the PCBs, and then returned to the site.

In-situ treatment of the contaminated soil involves the injection of either bacteria or chemicals that will destroy the offending material. If heavy metals are of concern, these can be tied up chemically, or "fixed" so that they will not leach into the groundwater. Organic solvents and other chemicals can be degraded by injecting freeze-dried bacteria. The past few years have seen an

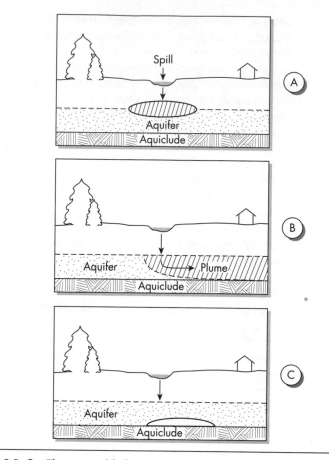

FIGURE 14–2 Three possible locations for underground hazardous waste.

amazing discovery of microorganisms able to decompose materials that were previously thought to be refractive or even toxic.

14.2.2   Treatment of Hazardous Materials

The treatment of materials deemed hazardous is obviously specific to the material and the specifics of the situation. There are a number of alternatives that engineers may consider in such treatment operations.

  If the hazardous material is organic and is readily biodegradable, most often the least expensive and most dependable treatment is biological, using techniques described in Chapter 10. The situation becomes interesting, however, when the hazardous material is an anthropogenic compound (created by people). Since these combinations of carbon, hydrogen, and oxygen are new to nature, there may not be any microorganisms that can use them as an energy source. In some cases, it is still possible to find a microorganism that will think of this chemical as a food source, and treatment would then consist of a biological contact tank in which the pure culture is maintained. Alternatively, it is now increasingly likely that specific microorganisms can be designed by gene manipulation to attack certain especially difficult organic wastes.

Finding the specific organism can be a difficult and arduous task. Since there are millions to choose from, how would we know that *Corynebacterium pyrogens* just happens to like Toxaphene, a particularly refractory organic pesticide? Second, often the pathway is convoluted and a single microorganism can only break down the chemical to another perhaps refractory compound, which would then be attacked by a different organism. For example, DDT is metabolized by *Hydrogenomonas* to *p*-chlorophenylacetic acid, which is then attacked by various *Arthrobacter* species. Tests in which only a single culture is used to study metabolism would therefore fail to note the need for a sequence of species.

One interesting development has been the use of co-metabolism to treat organic chemicals that were once considered biologically nonbiodegradable. With this technique, the hazardous material is mixed with a nonhazardous and at least partially biodegradable material and the mixture treated in a suspended or fixed film bioreactor. The microorganisms apparently are so busy making the necessary enzymes for the degradation of the biodegradable material (the metabolite) that they forget that they cannot treat the toxic material (the co-metabolite). The enzymes produced, however, will biodegrade both of the chemicals. The addition of phenol to the soil, for example, will trick some types of microorganisms such as fungus into decomposing chlorophenol, even though chlorophenol is usually toxic to the fungus.

An alternative treatment strategy, especially for inorganic wastes, is chemical treatment. In some cases a simple neutralization of the hazardous material will render the chemical harmless. In other cases, oxidation is used, such as for the destruction of cyanide. Ozone is often used as the oxidizing agent. In a case where heavy metals must be removed, precipitation is the method of choice. Most metals become extremely insoluble at high pH ranges and the treatment therefore consists of the addition of a base such as lime or caustic and the settling of the precipitate.

In addition to these techniques numerous physical-chemical methods are employed in the industry, including reverse osmosis, electrodialysis, solvent extraction, and ion exchange.

One of the most widely used treatment techniques for organic wastes is incineration. Hazardous waste incinerators ideally produce carbon dioxide, water vapor, and an inert ash. In actuality, no incinerator will achieve complete combustion, will discharge some chemicals in the emissions, and concentrate others in the bottom ash, and will additionally produce various compounds called products of incomplete combustion (PIC). For example, polychlorinated biphenyls (PCBs) are thought to decompose within the incinerator to highly toxic chlorinated dibenzo furans (CDBF) which, although organic, do not oxidize at normal incinerator temperatures.

Hazardous-waste incinerators must achieve high levels of removal efficiencies, often 99.99% or higher, commonly referred to as "four nines." In some cases, removal efficiencies require 5 or even 6 nines.

e • x • a • m • p • l • e 14.2

Problem   A hazardous waste incinerator is to burn 100 kg/hr of a PCB waste that is 22% PCB, among other organic solvents. In a trial run, the concentration of PCB in the emission is measured as 0.02 g/m^3 and the stack gas flow rate is 40 m^3/min. No PCB is detected in the ash. What removal efficiency (destruction of PCB) does the incinerator achieve?

Solution   A black box is once again useful. What comes in must go out. The rate in is

$$100 \text{ kg/hr} \times 0.22 \times 1000 \text{ g/kg} = 22,000 \text{ g/hr}$$

The rate out of the stack is

$$40 \text{ m}^3/\text{min} \times 0.02 \text{ g/m}^3 \times 60 \text{ min/hr} = 48 \text{ g/hr}$$

Since there are no PCBs in the ash, the amount of PCBs combusted must be $22,000 - 48 = 21,952$ g/hr, and the combustion efficiency is

$$\text{efficiency} = 21,952/22,000 = 0.9978$$

There are only "two nines" and this test does not, therefore, meet the "four nines" criterion.

♦

14.2.3   Disposal of Hazardous Materials

The disposal of hazardous materials is similar in many ways to the disposal of nonhazardous solid waste. Since disposal in the oceans is prohibited, and space disposal is still far too expensive, the final resting place has to be on land.

Deep well injection has been used in the past and is still the method of choice in the petrochemical industry. The idea is to inject the waste so deep into the earth so that it could not in any conceivable time reappear and cause damage. This is, of course, the problem. Once deep in the ground it is impossible to tell what the final destination of the waste will be and what groundwater it will eventually contaminate.

A second method of land disposal is to spread the waste on land and allow the soil microorganisms the opportunity to metabolize the organics. This technique was widely used in oil refineries and seemed to work exceptionally well. The EPA, however, has restricted the practice since there is little control over the chemical once it is on the ground.

The method most widely used for the disposal of hazardous material is the secure landfill. Such landfills essentially look like the RCRA Subtitle D landfills discussed in the previous chapter, only more so. Instead of one impervious liner, secure landfills require multiple liners, and all waste must be stabilized or in containers. Leachate is collected, and a cap is placed on the landfill once it is complete. Continued care is also required, although the EPA at present requires only 30-year monitoring.

Most hazardous waste engineers agree that there is no such thing as a "secure" landfill and that eventually all the material will once again find its way into the water or air. What we are betting on, therefore, is that these storage basins will eventually be dug up and the material treated. We find that it is inexpedient to do so today and leave this as a legacy to future generations.

14.3 ◆ Radioactive Waste Management

A special type of hazardous material emits ionizing radiation, and this radiation can be highly detrimental to human health. Environmental engineers do not usually get involved in radiation safety, which is a specialized field, but they nevertheless ought to be knowledgeable about both the risk and the disposal technology of radioactive materials.

14.3.1 Ionizing Radiation

Isotopes of various chemicals have different ratios of neutrons and protons in the nucleus, and some of these combinations are unstable. To regain equilibrium, these *isotopes* decay by emitting protons, neutrons, or electromagnetic radiation to carry off energy. The decay or nuclear disintegration of isotopes is called *radioactivity*. The isotopes that decay in this manner are called *radioisotopes*, and the energy emitted by this decay is called *ionizing radiation*.

The three kinds of ionizing radiation are alpha particles, beta particles, and gamma rays (and X-rays). Alpha particles consist of two protons and two neutrons and are the equivalent of the nucleus of a helium atom. Such a decay produces a new element, or a daughter, from a parent isotope. Alpha particles are quite large and do not penetrate tissue material readily; therefore they are not of major health concern.

Beta radiation results because of an instability in the nucleus between the proton and neutrons. Having too many neutrons, some of the neutrons decay into a proton and an electron. The proton stays in the nucleus to reestablish the neutron–proton balance, while the electron is ejected as the beta particle. The beta particle is much smaller than the alpha particle and can penetrate living tissue. Beta decay, as with alpha decay, changes the parent into a new element.

Both alpha and beta decay is accompanied by gamma radiation, which is a release of energy from the change of a nucleus in an excited state to a more stable state. The element thus remains the same, and energy is emitted as gamma radiation. Related to gamma radiation are X-rays, which result from an energy release when electrons transfer from a higher to a lower energy state.

All radioactive isotopes decay and will eventually reach stable energy levels. As introduced in Chapter 4, the decay of radioactive material is first order, in that the change in the number of nuclei during the decay process is directly proportional to the nuclei present, or

$$\frac{dN}{dt} = -kN$$

where N = number of neutrons
t = time, sec
k = radioactive decay constant, sec^{-1}

As before, integration yields

$$N = N_0 e^{-kt}$$

where N_0 = the number of neutrons at time zero.

Of particular interest is the half-life of the isotope, meaning that half of the nuclei have decayed in this time. Inserting $N = N_0/2$ into the above equation and solving, the half-life is calculated as

$$t_{1/2} = \frac{0.693}{k}$$

Half-lives of radioisotopes can range from the almost instantaneous (e.g., polonium-212 with a half-life of 3.03×10^{-17} seconds!) to very long (e.g., carbon-14 with a half-life of 5570 years).

14.3.2 Risks Associated with Ionizing Radiation

Soon after Wilhelm Roentgen (1845–1923) discovered X-rays, the detrimental effect of ionizing radiation on humans became known and the emergent field of health physics set about creating safety standards, with the assumption of identifying the level of exposure that would be "safe," or one that did not cause adverse health effects. They figured that there had to be a threshold, or an exposure below which no effects would be found.

Over about 30 years of studying both acute and chronic cases, they found that their threshold kept getting lower and lower, and they finally concluded that there is no threshold in radiation. Any amount of radiation can be damaging to human health.

The exposure of human tissue to ionizing radiation is complicated by the fact that different tissues absorb radiation differently. To get a grip on this, however, we must first define some unit of energy that expresses ionizing radiation. Historically this was established as the *roentgen*, defined as the exposure due of gamma or X-ray radiation equal to a unit quantity of electrical charge produced in air. The roentgen is a purely physical quantity, having nothing to do with the absorption or effect of the radiation.

Then it was necessary to look at the effect of that radiation, recognizing that different types of radioactivity can create different effects and that not all tissues react in the same way to radiation. Thus was invented a *rem*, or the "roentgen equivalent man," which takes into account the biological effect of the nuclear radiation absorbed. The rem was used to measure the extent of the biological injury that would result from the absorption of nuclear radiation. The rem, in contrast to the roentgen, is a biological dose. When different sources of radiation are compared for possible damage to human health, rems are used as the units of measurement.

Modern radiological hygiene has replaced the roentgen with a new unit, the *gray* (Gy), defined as the quantity of ionizing radiation that results in absorption of one joule of energy per kilogram of absorbing material. But the same problem exists, in that the absorption may be the same, but the damage might be different. Hence was invented the *sievert* (Sv), an absorbed radiation dose that does the same amount of biological damage to tissue as one Gy of gamma radiation or X-ray. One Sv is numerically equal to 100 rem.

The damage from radiation is chronic as well as acute. Radiation poisoning, as witnessed in the Hiroshima and Nagasaki bombings at the end of World War II, showed that radiation can kill within a few hours or days. The long-range effects of lower levels of radiation exposure can lead to cancer and mutagenic effects, and these are the exposures that are of most concern to the public.

The sources of exposure of humans to radioactivity can be classified as involuntary background radiation; voluntary radiation; involuntary incidental radiation; and involuntary radiation exposure due to accidents.

Background radiation is due mostly to cosmic radiation from space, the natural decay of radioactive materials in rocks (terrestrial), and radiation from living inside buildings (internal). A very special kind of background or natural radiation is the problem with radon. Radon-222 is a natural isotope with a half-life of 3.8 days and is the product of uranium decay in the earth's surface. Radon is a gas, and because uranium is so ubiquitous in soil and rock, there is a lot of radon present. With its relatively long half-life, it stays around long enough to build up high concentrations in basement air, and its decay products are known to be dangerous isotopes that are breathed into lungs and can produce cancer. The best techniques for reducing the risk to radon is to first monitor the basement to see if a problem exists, and if necessary to ventilate the radon to the outside.

Voluntary radiation can occur from such sources as diagnostic X-rays. One dental X-ray can produce hundreds of times the background radiation, and these should be avoided unless critically necessary. Following the discovery of X-rays (called Roentgen rays everywhere in the world except the United States), the damaging effects of X-rays began to be discovered because of the early deaths of many of Roentgen's friends.

Another source of voluntary exposure is from high-altitude flights in commercial airlines. The earth's atmosphere is a good filter for cosmic radiation, but there is little filtering at high altitudes.

Involuntary incidental radiation would be from such sources as nuclear power plants, weapons facilities, and industrial sources. While the Nuclear

Regulatory Commission allows fairly high exposure limits for workers within these facilities, it is fastidious about allowing radiation off premises. The amount of radioactivity produced in such facilities is well within the background.

The fourth type of radiation exposure is from accidents, and this is a different matter. The most publicized accidental exposure to radiation has been from accidents or near-accidents at nuclear power plants. The 1979 accident near Harrisburg, Pennsylvania, on the Three Mile Island nuclear power facility was serious enough to bring to a halt the already staggering nuclear power industry in the United States. Following a series of operating errors, large amounts of fission products were released into the containment structure. This material is still sufficiently "hot" today to prevent its cleanup. The amount of fission products released to the atmosphere was minor, so that at no time were the acceptable radiation levels around the plant exceeded. The release of radiation during this accident was so small, in fact, that it is estimated that it would result in one additional cancer death over the 50-mile radius surrounding the Three Mile Island facility. This would be impossible to detect, since within that radius there can be expected to develop more than 500,000 cancers during the lifetime of the people living in this area.

A much more serious accident occurred in Chernobyl, in the Ukraine. This actually wasn't an accident in the strict sense, since the disaster was caused by several engineers who wanted to conduct some unauthorized tests on the reactor. The warning signals went off as they proceeded with their experiment, and they systematically turned off all the alarms until the reactor experienced a meltdown. The core of the reactor was destroyed, the reactor caught fire, and the flimsy containment structure was blown off. Massive amounts of radiation escaped into the atmosphere, and while much of it landed within the first few kilometers of the facility, some reached as far as Estonia, Sweden, and Finland. In fact, the accident was first discovered when radiation safety personnel in Sweden detected unusually high levels of radioactivity. Only then did the officials of the former Soviet Union acknowledge that yes, there had been an accident. In all, 31 people died of acute radiation poisoning, many by heroically going into the extremely highly radioactive facility to extinguish the fires. The best estimates are that the accident will produce at least 2000 excess cancer deaths in Ukraine, Byelorussia, Lithuania, and the Scandinavian countries. But once again, this number is dwarfed by the expected total cancer deaths of 10,000,000 during the lifetime of the people in these areas.

The total radiation we receive is of course related to how we live our lives, where we choose to work, if we choose to smoke cigarettes, whether we have radon in our basements, and so on. It is possible, nevertheless, to calculate some averages of annual radiation dose for the average citizen of the United States. These figures, compiled by the National Academy of Sciences, are shown in Table 14-2.

It is worth noting again that this tabulation shows the *average* exposure to ionizing radiation for a person living in the United States. Obviously, the total radiation dose can be much higher for some persons.

The other point worth noting is the overwhelming importance of radon. This is especially true when it is understood that only a fraction of people

TABLE 14–2 Average Annual Dose of Ionizing Radiation to an Average Person in the United States	
Source of Radiation	*Dose (msv)*
Involuntary background	
Radon	24
Cosmic radiation	0.27
Terrestrial	0.28
Internal	0.39
Voluntary	
Medical: diagnostic X-ray	0.39
Medical: nuclear medicine	0.14
Medical: consumer products	0.10
Occupational	0.009
Involuntary incidental	
Nuclear power production	<0.01
Fallout from weapons testing	<0.01
Miscellaneous	<0.01
Total	25.64

Source: National Academy of Sciences Health Effects of Exposure to Low Levels of Ionizing Radiation (BEIR V), A. Upton, ed., Washington, D.C., National Academy Press (1990).

in the United States live in homes with basements, where radon problems generally occur.

14.3.3 Treatment and Disposal of Radioactive Waste

Radioactive wastes are classified as high-level, intermediate-level, and low-level waste. High-level wastes occur mostly from the production of electric power, and these are identified by the activities in the range of curies per liter. Intermediate-level wastes are produced by weapons manufacture, and although their activities are in the range of millicuries, the particular isotopes are long-lived, and these wastes require long-term storage. Low-level wastes, characterized as those with activities in the range of microcuries per liter, are produced in hospitals and research laboratories.

In nuclear power plants, nuclear fission occurs when fissionable material such as uranium-237 is bombarded with neutrons and a chain reaction is set up. The fissionable material then splits (hence the term *fission*) to release huge quantities of heat, used in turn to produce steam and then electricity. As the uranium-237 decays, it produces a series of daughter isotopes that themselves decay until the rate and hence heat output are reduced. What is left over, commonly known as the fission fragments, represents the high-level radioactive material that requires "cooling down," and long-term storage. The long-term storage of high-level radioactive waste has been debated for decades, with no apparent resolution, although a site in Nevada at Yucca Flats has been selected.

Low-level radioactive waste should not represent a disposal problem. Since the activity levels of these wastes is so low that they can be handled by direct contact, it would seem that with judicious volume reduction such as incineration, any secure landfill would be adequate. The U.S. Congress passed the Low-Level Waste Policy Act in 1980 that stipulated that states form a compact so that each state will take its turn in providing a disposal facility for a certain time. Unfortunately, the very mention of the word "radioactivity" is enough to arouse public opinion and often prevent the siting of these disposal sites.

14.4 ◆ Pollution Prevention

The present methods of disposing of hazardous wastes are woefully inadequate. All we are doing is simply storing them until a better idea (or more funds, or stricter laws) comes along. Would it be better to never create the waste in the first place?

The EPA defines "pollution prevention" as follows:

> The use of materials, processes, or practices that reduce or eliminate the creation of pollutants or wastes at the source. It includes practices that reduce the use of hazardous materials, energy, water or other resources and practices that protect natural resources through conservation or more efficient use.[1]

In the widest sense, pollution prevention is the idea of eliminating waste, regardless of how this might be done.

Originally, pollution prevention was applied to industrial operations with the idea of reducing either the amount of the wastes being produced or changing their characteristics to make them more readily disposable. Many industries changed to water-soluble paints, for example, thereby eliminating organic solvents, cleanup time, and so on, and often ended up saving considerable money. In fact, the concept was first introduced as "pollution prevention pays," emphasizing that many of the changes would actually save the companies money. In addition, the elimination or reduction of hazardous and otherwise difficult wastes also has a long-term effect—it reduces the liability the company carries as a consequence of its disposal operations.

With the passage of the Pollution Prevention Act of 1990, the EPA was directed to encourage pollution prevention by setting appropriate standards for pollution prevention activities, assist federal agencies in reducing wastes generated, work with industry to promote elimination of wastes by creating waste exchanges and other programs, seek out and eliminate barriers to the efficient transfer of potential wastes, and do this with the cooperation of the individual states.

In general, the procedure for the implementation of pollution prevention activities is to

- Recognize a need
- Assess the problem
- Evaluate the alterative
- Implement the solutions

Contrary to the attitude toward most pollution control activities, industries generally have welcomed this governmental action, recognizing that pollution prevention can and often does result in the reduction of costs to the industry. Thus recognition of the need quite often is internal, and the company seeks to initiate the pollution-prevention procedure.

During the assessment phase, a common procedure is to preform a "waste audit," which is nothing but the black box mass balance, using the company as the black box. Consider the following example.

e • x • a • m • p • l • e **14.3**[2]

Problem   A manufacturing company is concerned about the air emissions of volatile organic carbons. These chemicals can volatilize during the manufacturing process, but there is no way of estimating just how much, or which chemicals. The company conducts an audit of three of their most widely used volatile organic chemicals (VOCs), with the following results:

Purchasing Department Records

Material	Purchase Quantity (Barrels)
Carbon tetrachloride (CCl_4)	48
Methylene chloride (CH_2Cl_3)	228
Trichloroethylene (C_2HCl_3)	505

Wastewater Treatment Plant Influent

Material	Average Concentration (mg/L)
Carbon tetrachloride	0.343
Methylene chloride	4.04
Trichloroethylene	3.23

The average influent flow rate to the treatment plant is 0.076 m^3/sec.

Hazardous Waste Manifests
(what leaves the company by truck, headed to a hazardous waste treatment facility)

Material	Barrels	Concentration (%)
Carbon tetrachloride	48	80
Methylene chloride	228	25
Trichloroethylene	505	80

Unused Barrels at the End of the Year

Material	Barrels
Carbon tetrachloride	1
Methylene chloride	8
Trichloroethylene	13

How much VOC is escaping?

Solution   Conduct a black box mass balance, as

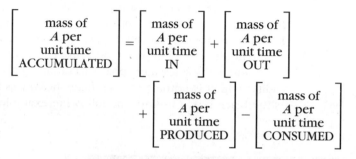

The materials A are of course the three VOCs.

We also need to know the conversion from barrels to cubic meters and the density of each chemical. Each barrel is 0.12 m³, and the density of the three chemicals is 1548, 1326, and 1476 kg/m³. The mass per year of carbon tetrachloride accumulated is

$$[\text{mass ACCUMULATED}] = 1 \text{ barrel/yr} \times 0.12 \text{ m}^3/\text{barrel} \times 1548 \text{ kg/m}^3$$

$$= 186 \text{ kg/yr}$$

Similarly,

$$[\text{mass IN}] = 48 \times 0.12 \times 1548 = 8916 \text{ kg/yr}$$

The mass out is in three parts: mass discharge to the wastewater treatment plant, mass leaving on the trucks to the hazardous waste disposal facility, and the mass volatilizing. So the [Mass OUT] is

$$[\text{mass OUT}] = [0.343 \text{ g/m}^3 \times 0.076 \text{ m}^3/\text{sec} \times 86{,}400 \text{ sec/day} \times 365 \text{ day/yr}$$

$$\times 10^{-3} \text{ kg/g}] + [48 \times 0.12 \times 1548 \times 0.80] + A_{air}$$

$$= 822.1 + 7133 + A_{air}$$

Since there is no carbon tetrachloride consumed or produced,

$$186 = 8916 - [822.1 + 7133 + A_{air}] + 0 - 0$$

and

$$A_{air} = 775 \text{ kg/yr}$$

> If a similar balance is done on the other chemicals, it appears that the loss to air of methylene chloride is about 16,000 kg/yr and of the trichloroethylene is about 7800 kg/yr. If the intent is to cut total VOC emissions, it is clear that the first target should be the methylene chloride.
>
> ◆

Once we know what and where the problems are, the next step is to figure out useful options. These options fall generally into three categories:

- Operational changes
- Materials changes
- Process modifications

Operational changes might consist simply of better housekeeping: plugging up leaks, eliminating spills, and the like. A better schedule for cleaning, as well as segregating the water might similarly yield a large return on a minor investment.

Materials changes often involve the substitution of one chemical for another that is less toxic or requires less hazardous materials for cleanup. The use of trivalent chrome for chrome plating instead of the much more toxic hexavalent chrome has found favor, as has the use of water-soluble dyes and paints. In some instances, ultraviolet radiation has been substituted for biocides in cooling water, resulting in better water quality and no waste cooling water disposal problems. In one North Carolina textile plant, biocides were used in air washes to control algal growth. Periodic "blow down" and cleaning fluids were discharged to the stream, but this discharge proved toxic to the stream and the state of North Carolina revoked the plant's discharge permit. The town would not accept the waste into its sewers, rightly arguing that this could have serious adverse effects on its biological wastewater treatment operations. The industry was about to shut down when it decided to try ultraviolet radiation as a disinfectant in its air wash system. Fortunately, they found that the ultraviolet radiation effectively disinfected the cooling water and that the biocide was no longer needed. This not only eliminated the discharge but it eliminated the use of biocides all together, thus saving the company money. The payback was 1.77 years.[3]

Process modifications usually involve the greatest investments, but can result in the most rewards. For example, countercurrent wash water use instead of once-through batch operation can significantly reduce the amount of wash water needing treatment, but such a change requires pipes, valves, and a new process protocol. In industries where materials are dipped into solutions, such as in metal plating, the use of drag out recovery tanks, an intermediate step, has resulted in the savings of the plating solution and reduction in waste generated.

In any case, the most marvelous thing about pollution prevention is that most of the time the company not only eliminates or greatly reduces the discharge of hazardous materials but it also saves money. Such savings are in several forms, including of course the direct savings in processing costs as with the ultraviolet disinfection example. But there are other savings, including the savings in not having to spend time on submitting compliance permits and suffering potential fines for noncompliance. Future liabilities weigh heavily where hazardous wastes have to be buried or injected. In addition, there are the intangible benefits of employee relations and safety. Of course, finally, there is the benefit that comes from doing the right thing, something not to be sneezed at.

In hazardous waste management, doing the right thing often involves the elimination of a potential hazard for future generations. Much has been written about this commitment. Do we, indeed, owe anything to these yet-to-be-born people?

14.5 ◆ Hazardous Waste Management and Future Generations

Why should we even worry about all that? Is it not true that we will certainly never be personally affected by hazardous wastes? And it is unlikely even that any of our children will suffer from inadequate disposal of these materials. Why then are we so concerned with the future and with future generations?

The Resource Conservation and Recovery Act 1976 requires waste facility operators to provide "perpetual care" for depositories; however, this is defined as a period of only 30 years. "Perpetual care" means only one and one half generations! The primary hazardous waste law in the United States apparently does not even care about future generations, so why should we?

After all, these people do not exist, nor might they ever exist. They are not real persons, and ethics can only apply to interactions among real people. Also, we don't know what the future will bring. We cannot assume that future people will have the same preferences or needs and, therefore, we cannot predict the future. We may be well intentioned, but utterly wrong. Also, history teaches us that at least to this point in the history of civilization, each successive generation has been better off than the previous one. We have better health, better communication, better food, more time, and greater opportunity for personal growth and the enjoyment of the quality of life than ever before. Because we can expect future generations to continue this trend and to have an ever-improving life, we should do nothing for them. Besides, as the old saying goes "What has posterity ever done for me?" So it may be very noble to take account of future generations, but there appears to be no basis for the claim that we ought to do so.

For all these reasons, then, it may seem that there is nothing wrong with our deciding to generate nuclear wastes in order to enjoy the benefits of electricity, to use up the nonrenewable resources such as oil, or to interpret our obligations of perpetual care as extending for only 30 years. Planning for hazardous waste management (or for anything else), therefore, needs to take into account only the interests of present generations.

But are there not counterarguments? Surely we cannot just blunder along hedonistically without concern for future humanity. What reasons can we give for a concern for future generations?

Alastair Gunn[4] argues that there are indeed strong reasons for our obligations to future generations. It is true that (by definition) future generations do not exist, but that does not mean we do not have obligations to them. Certainly we do not have the kind of one-on-one obligations that identifiable individuals have to each other. For instance, debts and promises can be owed only by one person to another. But we also have obligations to anyone who might be harmed by our actions even though we do not know their identity.

Gunn uses an analogy of a terrorist who plants a bomb in a primary school.

> Plainly the act is wrong, and in breach of a general obligation not to cause (or recklessly risk) harm to our fellow citizens. This is so even though the terrorist does not know who the children are, or even whether any children will be affected (the bomb may not explode, or may be discovered in time, or the school may be closed down). It is equally wrong to place a time bomb, set to go off in the school in 20 years time, even though any children that may be in the school when it goes off are not yet born. An unsafe hazardous waste dump may be seen as analogous to a time bomb.

The analogy is not perfect because the owners of the chemical company do not set out to kill children. But intent is not necessary if there is good reason to believe that an action would cause harm. A person throwing a cherry bomb into a crowd does not intend to harm anyone, but if harm does occur, he or she is as guilty as if the original intent was to hurt people. Whether the people harmed are here today, or will be here in the future, is irrelevant. It is the harm that counts and causing the harm is unethical.

The counterarguments for the second claim, that we will never know what future generations will value, are that yes we do, in fact, have some pretty good ideas of what they will *not* value. For example, it is highly unlikely that they will value cancer, AIDS, or polluted groundwater. We can expect future generations to be much more similar to ourselves than different. To suggest that we should contaminate the earth because we don't know what future people will value is not a valid argument.

The third objection to caring about future generations is that, because they will be better off than we are, there is no need to worry about their well-being. This makes no sense because the economic condition of a person is irrelevant to the harm that might be done. It is not legal or ethical to rob a person just because he or she might be wealthy. Besides, there is no guarantee that enhanced technology will bring continued greater well-being. There are more starving people in the world today than ever before and the continued growth of human population and the destruction of natural habitats may well bring the world into an economic and ecological collapse. We have stretched the carrying capacity of the earth to its limits, and we have no guarantee that technology (or social and political progress) will enable us to even *have* a future generation much less be able to care for it. As Gunn concludes:

> We have an obligation to avoid creating detrimental environmental conditions for future generations that we do not want to create for ourselves. This means that in terms of waste management, it is

necessary to develop methods for very long term acceptably safe storage and, more importantly, steadily reducing the quantity, toxicity and persistence of the wastes we produce so that acceptably safe management becomes possible. Further, there seems little doubt that future generations will appreciate clean air and water, adequate wilderness areas, and the preservation of natural monuments, and that this is an obligation of our present generation.

◆ Abbreviations

CERCLA	= Comprehensive Environmental Response, Compensation, and Liability Act	N	= number of neutrons in radioactive decay
Gy	= Gray, ionizing radiation measure	RCRA	= Resource Conservation and Recovery Act
HRS	= Hazards Ranking System	rem	= roentgen equivalent man
k	= radioactive decay constant, sec^{-1}	Sv	= Sievert, measure of radioactive damage
LD$_{50}$	= lethal dose where 50% of animals die	$t_{1/2}$	= half life, y
LDC$_{50}$	= lethal concentration dose where 50% of animals die	TCLP	= Toxic Characteristics Leaching Procedure
		TOSCA	= Toxic Substances Control Act
		VOC	= volatile organic carbon

◆ Problems

14-1   The following paragraph appeared as part of a full-page ad in a professional journal:

> Today's Laws
> A reason to act now
> The Resource Conservation and Recovery Act provides for corporate fines up to $1,000,000 and jail sentences up to five years for officers and managers, for the improper handling of hazardous wastes. That's why identifying waste problems, and developing economical solutions is an absolute necessity. O'Brien & Gere can help you meet today's strict regulations, and today's economic realities, with practical, cost-effective solutions.

What do you think of this ad? Do you find anything unprofessional about it? It is, after all, factual. Suppose you are the president of O'Brien & Gere (one of the most reputable environmental engineering firms in the business) and you saw this in the journal. Write a one-page memo to your marketing director commenting on it.

14-2 Some old electrical transformers were stored in the basement of a university maintenance building and were "forgotten." One day a worker entered the basement and saw that some sticky, oily substance was oozing out of one of the transformers and into a floor drain. He notified the director of grounds, who immediately realized the severity of the problem. They called in hazardous waste consulting engineers, who first took out the transformers and eliminated the source of the polychlorinated biphenyls (PCBs) that were leaking into the storm drain. Then they traced the drain to a little stream and started taking water samples and soil samples in the stream. They discovered that the water was at 0.12 mg/L PCB and the soil ranged from 32 mg PCB/kg (dry soil) to 0.5 mg/kg. The state environmental management required that streams contaminated with PCBs be cleaned so that the stream is "free" of PCBs. Recall that PCBs are very toxic and extremely stable in the environment, and will biodegrade very slowly. If nothing was done, the contaminant in the soil would remain for perhaps hundreds of years.

a. If you were the engineer, how would you approach this problem? What would you do? Describe how you would "clean" this stream. Be as detailed as possible. Will your plan disturb the stream ecosystem? Is this of concern to you?

b. Will it ever be possible for the stream to be "free of PCBs"? If not, what do you think the state means by this? How would you know that you have done all you can?

c. There is no doubt that if we wait long enough, someone will eventually come up with a really slick way of cleaning PCB-contaminated streams, and that this method will be substantially cheaper than anything we have available today. Why not just wait, then, and let some future generation take care of the problem? The stream does not threaten us directly, and the worst part of it can be cordoned off so people will not be able to get close to the stream. Then someday, when technology has improved, we can clean it up. What do you think of this approach? (It's of course illegal, but this is not the question. Is it ethical? Would you be prepared to promote this solution?)

14-3 In Problem 14-2, if the cleanup resulted in the PCB concentration in the water being at 0.000073 mg/L, what is the percent reduction in the contaminant? How many "nines" have been achieved?

14-4 A dry cleaning establishment buys 500 gallons of carbon tetrachloride every month. As a result of the dry cleaning operation, most of this is lost to the atmosphere, and only 50 gallons remains to be disposed of. What is the emission rate of carbon tetrachloride from this dry cleaner, in lb/day? (Density of carbon tetrachloride is about 1.6 g/mL.)

14-5 A 200-ft deep free aquifer is contaminated with a hazardous material. The aquifer is composed of sand with a permeability (K) of 6×10^{-4} m/sec. The contaminated groundwater is being pumped up for treatment at a steady-state pumping rate of 0.02 m³/sec, and a drawdown of 21 ft occurs in a monitoring well 30 ft away from the extraction well. If there is a second monitoring well, how far away is it if it's drawdown is 6 ft?

14-6   A hazardous waste landfill is to be constructed, and some of the clay on site is being considered for a clay liner. The requirement is that the permeability of the clay be less than 1×10^{-7} cm/sec. A permeameter is set up and a test is conducted. If the depth of the sample in the permeameter is 10 cm, a head of 1 m water is placed on the sample, the diameter of the permeameter is 4 cm, and it takes 240 hr to collect 100 mL of water draining out of the permeameter, what is the sample permeability?

14-7   A tank truck loaded with liquid ammonia (NH_3) is hit by a train, resulting in a steady and uncontrollable leak of ammonia gas at an estimated rate of 1 kg/ sec. You estimate the wind speed at 5 km/hr, and it's a sunny afternoon. There is a small community about 1 mile directly downwind. Realizing that the health threshold limit of ammonia is about 100 ppm, should you recommend evacuation of the town?

14-8   In the 1970s, we discovered many hazardous waste dumping sites in the United States. Some of these had been in operation for decades, and some were still active. Most of them, however, were abandoned, such as the notorious "Valley of Drums" in Tennessee— a rural valley where an almost unimaginable concoction of all manner of hazardous materials had been dumped, and the owner had simply abandoned the site after collecting hefty fees for "getting rid" of the hazardous materials for large chemical companies.

Why did we wait until the 1970s to discover the problem? Why did we not discover the problem of hazardous waste years ago? Why is the concern with hazardous waste so recent? Discuss your thoughts in a one-page paper.

14-9   Peter has been working with the Bigness Oil Company's local affiliate for several years and has established a strong, trusting relationship with Jesse, the manager of the local facility. The facility, on Peter's recommendation, has followed all the environmental regulations to the letter and has a solid reputation with the state regulatory agency. The local company receives various petrochemical products via pipeline and tank truck and blends them for resale to the private sector.

Jesse has been so pleased with Peter's work that he has recommended that Peter be retained as the corporate consulting engineer. This would be a significant advancement for Peter and his consulting firm, cementing Peter's steady and impressive rise in his firm. There is talk of a vice presidency in a few years.

One day, over a cup of coffee, Jesse starts telling Peter a wild story about a mysterious loss in one of the raw petrochemicals he receives by pipeline. Sometime during the 1950s, when operations were pretty lax, a loss of one of the process chemicals was discovered when the books were audited. There were apparently 10,000 gallons of the chemical missing. After running pressure tests on the pipelines, the plant manager found that one of the pipes had corroded and had been leaking the chemical into the ground. After stopping the leak, the company sank observation and sampling wells and found that the product was producing a vertical plume, slowly diffusing into a deep aquifer. Because there was no surface or groundwater pollution off the plant property, the plant manager decided to do nothing. Jesse thought that somewhere under the plant there still sits this plume, slowly diffusing into the aquifer, although the last tests from the sampling wells showed that the concentration of the chemical in the groundwater within 400 ft of the surface was essentially zero. The wells were capped, and the story never appeared in the press.

Peter is taken aback by this apparently innocent revelation. He recognizes that the state law requires him to report all spills, but what about spills that occurred years ago, and where the effects of the spill seem to have dissipated? He frowns and says to Jesse, "We have to report this spill to the state, you know."

Jesse is incredulous.

"But there *is* no spill. If the state made us look for it, we probably could not find it, and even if we did, it makes no sense whatever to pump it out or contain it in any way."

"But the law says that we have to report," argues Peter.

"Hey look. I told you this in confidence. Your own engineering code of ethics requires client confidentiality. And what would be the good of going to the state? There is nothing to be done. The only thing that would happen is that the Company would get into trouble and have to spend useless dollars to correct a situation that cannot be corrected and does not need remediation."

"But . . ."

"Peter. Let me be frank. If you go to the state with this, you will not be doing anyone any good—not the Company, not the environment, and certainly not your own career. I cannot have a consulting engineer who does not value client loyalty."

What should be Peter's response to the situation? What should he do? Who are the parties concerned? (Don't forget future generations!) What are all his alternatives? List these in order of detrimental effect on Peter's career; then list them in order of detrimental effect on the environment. All things considered, what ought Peter do?

◆ Endnotes

1. Quoted in H. Freeman et al., "Industrial Pollution Prevention: A Critical Review," presented at the Air and Waste Management Association Meeting, Kansas City, MO (1992); Environmental Protection Agency Pollution Prevention Directive, U.S. EPA (13 May 1990).
2. This example is based in part on a similar example in M. L. Davis and D. A. Cornwell, Introduction to Environmental Engineering, McGraw-Hill, New York (1991).
3. S. Richardson, "Pollution Prevention in Textile Wet Processing: An Approach and Case Studies," *Proceedings, Environmental Challenges of the 1990's,* U.S. EPA 66/9-90/039 (September 1990).
4. P. Aarne Vesilind and Alastair Gunn, *Making a Difference: Engineering, Ethics and the Environment,* Cambridge University Press, New York (1997).

NOISE
POLLUTION

If a tree falls in the forest, does it make a sound?

You have heard this old question many times. Now you will have an answer for it. In fact, if a tree falls in the forest, it does indeed make a sound, since sound is pressure waves in the air. But it might not make a *noise*, which is defined as unwanted sound—unwanted by humans. Thus if there are no humans around, the tree falling in the forest does not make noise.

In this chapter we discuss first the basics of sound, defining some of the terms that acoustic engineers use to describe and control unwanted sound (noise). We then discuss some of the effects of noise, ending with a discussion on noise control.

15.1 ◆ Sound

Pure sound is described by pressure waves traveling through a medium (air, in almost all cases). These pressure waves are described by their amplitude and their frequency. With reference to Figure 15-1, note that the pure sound wave can be described as a sinusoidal curve, having positive and negative pressures within one cycle. The number of these cycles per unit time is called the sound *frequency*, often expressed as cycles per second, or hertz (Hz).[1] Typical sounds that healthy human ears hear range from about 15 Hz to about 20,000 Hz, a huge range. Low-frequency noise is a deep sound, while high-frequency noise represents high-pitched sounds. For example, a middle A on the piano is at a frequency of 440 Hz. Speech is usually in the range of 1000 to 4000 Hz.

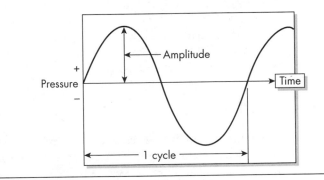

FIGURE 15-1 Pure sound travels as a perfect sinusoidal wave.

The wide frequency spectrum is significantly reduced by age and environmental exposure to loud noise. The most significant sources of such damaging noise come from occupational sources and loud music, particularly rock concerts and earphones turned too high. High-quality sound-reproduction equipment is designed to reproduce a full spectrum of frequencies, although advertising that the equipment is capable of less than 10 Hz reproduction is somewhat silly, since most people can only hear down to 20 Hz.

Young, healthy persons (particularly young women) can hear very high frequencies, often including such signals as automatic door openers. With age, and unhappily with the damage so many young people do to their ears, the ability to detect a wide frequency range drops. Older people especially tend to lose the high end of the hearing spectrum and might start to complain that "everyone is mumbling."

The loudness of a noise is expressed by its *amplitude*. With reference again to Figure 15-1, the energy in a pressure wave is the total area under the curve. Because the first half is positive pressure and the second is negative, adding these would produce zero net pressure. The trick is to use a root mean square analysis of a pressure wave, first multiplying the pressure by itself and then taking its square root. Since the product of two negative numbers is a positive number, the result is a positive pressure number. In the case of sound waves, the pressure is expressed as Newtons per square meter (N/m^2), although sometimes sound pressure is also expressed in bars or atmospheres. In this text we use the modern designation of N/m^2.

The human ear is a phenomenal instrument, being able to hear sound at both a huge frequency range and an even more spectacular range of pressures. In fact, the audible range of hearing in humans ranges from 1 to 10^{18} — a range that makes it difficult to express sound pressure in some meaningful and useful way, and as a result ratios are used to express sound amplitude.

Psychologists have known for many years that human responses to stimuli are not linear. In fact, the ability to detect an incremental change in any response such as heat, cold, odor, taste, or sound depends entirely on the level of that stimulus. For example, suppose you are blindfolded and are holding a 20-pound weight in your hand, and 0.1 pound is added to it. You will probably not detect the change. However, if you are holding 0.1 pound in your hand, and another 0.1 is added (a 100% increase), you would be able to detect the change. Thus the ability to detect stimuli can be described on a logarithmic scale.

With respect to sound, it makes sense to consider not pressure but the energy level of the sound, so that the ratio of two sound energies are

$$\log_{10}\left(\frac{W}{W_{ref}}\right)$$

where the W and W_{ref} represent the energy in watts of the sound wave, and some reference energy. Since the energy is proportional to the pressure squared, this ratio can also be expressed as

$$\log_{10}\left(\frac{P^2}{P_{ref}^2}\right)$$

where P = sound pressure, N/m^2
P_{ref} = some reference pressure, N/m^2

This expression can be simplified to

$$2\log_{10}\left(\frac{P}{P_{ref}}\right)$$

By convention, this expression defines *bel*.[2] Dividing this unit into ten makes it easier to use and avoids fractions; it is then known as the *decibel*.

But what reference pressure to use? It appears that the least pressure a human ear is able to detect is about 2×10^{-5} N/m^2 and this would make a convenient datum. Using 2×10^{-5} N/m^2 as a reference value, the *sound pressure level* (SPL) is then defined in terms of *decibels*, designated by the symbol *dB*, as

$$\text{SPL} = 20 \log_{10}\left(\frac{P}{P_{\text{ref}}}\right)$$

where SPL = sound pressure level, dB
        P = the sound pressure as measured in N/m^2
      P_{ref} = the reference sound pressure, 2×10^{-5} N/m^2

The highest possible SPL, at which point the air molecules can no longer carry pressure waves, is 194 dB, while 0 dB is the threshold of hearing. A typical classroom might be at about 50 dB, whereas mowing your lawn could subject you to about 90 dB. Conversational speech is at about 60 dB. A ringing alarm clock (next to your head) is about 80 dB, whereas a passenger jet on takeoff can produce over 115 dB. Rock concerts often average 110 dB, far above the threshold permitting conversation. The loudest sound recorded is the Saturn rocket on take off, at 134 dB.

Remember that the decibel scale is logarithmic. One dB difference from 40 to 41 is considerably less energy than a one dB difference from 80 to 81. Every 10 dB increase produces a doubling of the energy level, and a doubling of the danger of damage from excessive noise.

Because sound levels are logarithmic ratios, they cannot be numerically added without some calculation. If two sources of sound are combined, the pressures have to be calculated from the SPL equation, these added, and a new SPL calculated. As a rule of thumb, two equal sounds added result in a 3-dB increase in overall sound level. If the difference between two sounds is greater than 10 dB, the lesser of the two does not contribute to the overall level of sound.

e • x • a • m • p • l • e **15.1**

Problem A machine shop has two machines, one producing a sound pressure level of 70 dB and one producing 58 dB. A new machine producing 70 dB is brought into the room. What is the new sound pressure level in the room?

Solution Add 70 and 58 dB. Since the difference is greater than 10, the effect of the 58 dB sound is negligible, and the room would have a sound pressure level of 70 dB. If another 70 dB machine is brought in, the two sounds are equal, producing an increase of 3 dB. Thus, the sound pressure level in the room would be 73 dB.

◆

Sound in the atmosphere travels uniformly in all directions, radiating from its source. The sound intensity is reduced as the square of the distance away from the source of the sound according to the *inverse square law*. That is, the sound pressure level is proportional to $1/r^2$, where r is the radial distance from the source.

An approximate relationship can be developed if the sound power is expressed as a logarithmic ratio based on some standard reference power, such that

$$SPL_r \approx SPL_0 - 10 \log r^2$$

where SPL_r = sound pressure level at some distance r from the sound source, dB

SPL_0 = sound pressure level at the source, dB

r = distance away from the sound, m

e • x • a • m • p • l • e   **15.2**

Problem   A sound source generates 80 dB. What would the SPL be 100 m from the source?

Solution

$$SPL_r = 80 - 10 \log (100)^2 = 40 \text{ dB}$$

◆

This is of course an approximation. We assume that the sound propagates in all directions evenly, but in the real world this is never true. If, for example, the sound occurs on a flat surface so that the area through which it propagates is a hemisphere instead of a sphere, the approximate addition is 3 dB to the SPL calculated in Example 15.2. In enclosed spaces, reverberation can also greatly increase the sound pressure level since the energy is not dissipated. The most important point to remember is that the SPL is reduced approximately as the log of the square of the distance away from it.

The frequency in cycles per second (Hz) and the amplitude in decibels describe a pure sound at a specific frequency. All environmental sounds, however, are quite "dirty," with many frequencies. A true picture of such sounds is obtained by a *frequency analysis*, in which the sound level at a number of different frequencies is measured and the results plotted as shown in Figure 15-2. Often a frequency analysis is useful in noise control because the frequency of the highest sound pressure level can be identified and corrective measures taken.

By convention, frequencies are designated as *octave bands* that represent a narrow range of frequencies, as shown in Table 15-1. The average sound pressure can be calculated from a frequency diagram by adding the sound pressure levels at the individual frequencies. But since sound pressure level is not an arithmetic quantity, but rather a logarithmic ratio, such addition must be done by the following equation:

$$SPL_{ave} = 20 \log \left(\frac{1}{N}\right) \sum_{j=1}^{N} 10^{(L_j/20)}$$

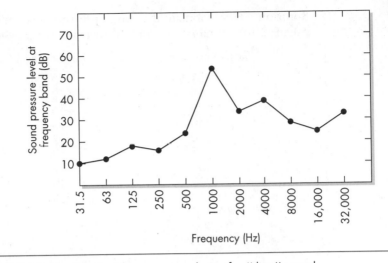

FIGURE 15-2 A typical frequency analysis of a "dirty" sound.

where SPL$_{ave}$ = average sound pressure level, dB
 N = number of measurements taken
 L_j = the jth sound pressure level, dB
 j = 1, 2, 3, . . . , N

TABLE 15-1 Octave Bands	
Octave Frequency Range (Hz)	*Geometric Mean Frequency (Hz)*
22–44	31.5
44–88	63
88–175	125
175–350	250
350–700	500
700–1400	1000
1400–2800	2000
2800–5600	4000
5600–11,200	8000
11,200–22,400	16,000
22,400–44,800	32,000

e • x • a • m • p • l • e **15.3**

Problem What is the average sound pressure level of the sound pictured in Figure 15-2?

Solution   First, tabulate the sound pressure level at each octave band, then calculate $10^{(L_i/20)}$, sum these, and calculate the SPL_{ave} from the previous equation.

Octave Band (Hz)	Sound Pressure Level (dB)	$10^{(L_i/20)}$
31.5	10	3
63	12	4
125	16	6
250	15	6
500	22	13
1000	52	398
2000	32	40
4000	40	100
8000	28	25
16,000	27	22
32,000	34	50
		667

$$SPL_{ave} = 20 \log (1/11)(667) = 35.7 \text{ dB}$$

15.2 ◆ Measurement of Sound

Sound is measured with an instrument that converts the energy in the air pressure waves to an electrical signal. A microphone picks up the pressure waves and a meter reads the sound pressure level, directly calibrated into decibels. Data thus obtained with a *sound pressure level meter* represent an accurate measurement of energy level in the air.

But this sound pressure level is not necessarily what the human ear hears. Although we can detect frequencies over a wide range, this detection is not equally effective at all frequencies (our ear does not have a *flat response* in audio terms). If the meter is to simulate the efficiency of the human ear in detecting sound, the signal has to be filtered.

Using thousands of experiments, researchers discovered that on average the human ear has an efficiency over the audible frequency range that resembles Figure 15–3. At very low frequencies, our ears are less efficient than at middle frequencies, say between 1000 and 2000 Hz. At higher frequencies, the ear becomes increasingly inefficient, finally petering out at some high frequency where no sound can be detected. This curve is called the *A-weighed* filtering curve, since there are other filtering curves for different purposes. Interestingly, the band of greatest efficiency for the human ear is very close to human speech range, which may have been due to Darwinian selection.

Using the curve shown in Figure 15–3, instrument designers have constructed a meter that filters out some of the very low-frequency sound and much of the very high-frequency sound, so that the sound measured somewhat represents the hearing of a human ear. Such a measurement is called a *sound level* and is designated dB(A), since it represents modified dB value with the A filter. The meter is often called a *sound level meter* to distinguish it from a

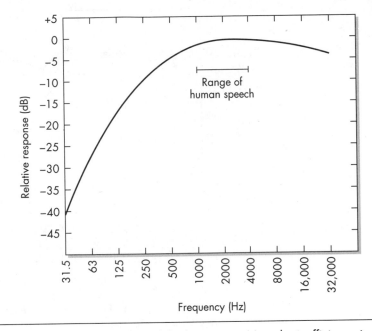

FIGURE 15-3 Typical response of the human ear. Note the inefficiency in hearing low frequencies and high efficiencies in the range of human speech.

sound *pressure* level meter, which measures sound as a flat response. Almost all sound measurements related to human hearing use the dB(A) scale for measurement because this approximates the efficiency of the human ear.

Frequency analyses are useful in measuring hearing ability. Using the hearing ability of an average young ear as the standard, an *audiometer* measures the hearing ability at various frequencies producing an *audiogram*. Such audiograms are then used to identify the frequencies where the hearing aids must be able to boost the signal.

Figure 15-4, for example, shows three audiograms. Person *A* has excellent hearing with a basically flat response. Person *B* has lost hearing at a specific frequency range, in this case around 4000 Hz. Such hearing losses are often due to industrial noise that often destroys the ear's ability to hear sounds at a specific frequency. Since speech is near that range, this person already has difficulty hearing. The third curve on Figure 15-4, Person *C*, shows a typical audiogram either for an older person who has lost much of the higher-frequency range and probably is a candidate for a hearing aid or a young person who has sustained severe damage to his or her ears due to exposure to loud music.

15.3 ◆ Effect of Noise on Human Health

Figure 15-5A shows a schematic of a human ear. The air pressure waves first hit the ear drum (*tympanic membrane*), causing it to vibrate. The cavity leading to the tympanic membrane and the membrane itself are often called the *outer*

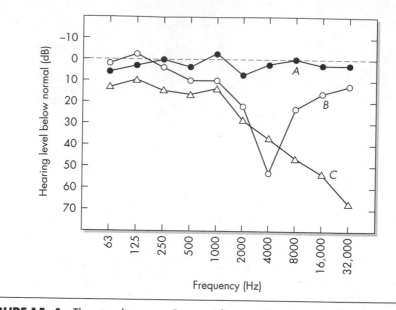

FIGURE 15–4   Three audiograms. Person *A* has excellent hearing, being able to hear at normal levels at all frequencies. Person *B* has lost considerable hearing in the 2000 to 8000 Hz range. Since this is the approximate speech range of frequencies, this person would have great difficulty hearing speech. Person *C* has a typical audiogram either for an older person who also has lost hearing in the higher frequencies or a younger person with damage to the ears from loud music or other insults.

ear. The tympanic membrane is attached physically to three small bones in the middle ear, which start to move when the membrane vibrates. The purpose of these bones, called colloquially the *hammer, anvil*, and *stirrup* because of their shapes, is to amplify the physical signal (to achieve some *gain* in audio terms). This air-filled cavity is called the *middle ear*. The amplified signal is then sent to the *inner ear* by first vibrating another membrane called the *round window membrane*, which is attached to a snail-shaped cavity called the *cochlea*.

Within this fluid-filled cochlea, a cross section of which is shown in Figure 15–5B, is another membrane, the *basilar membrane*, which is attached to the round window membrane. Attached to the basilar membrane are two sets of tiny *hair cells*, pointing in opposite directions. As the round window membrane vibrates, the fluid in the inner ear is set in motion and the thousands of hair cells in the cochlea shear past each other, setting up electrical impulses that are then sent to the brain through the *auditory nerves*. The frequency of the sound determines which of the hair cells will move. The hair cells close to the round window membrane are sensitive to high frequencies and those in the far end of the cochlea respond to low frequencies.

Damage to the human ear can occur in several ways. For example, very loud impulse noise can burst the ear drum, causing mostly temporary loss of hearing, although frequently a torn eardrum heals poorly and can result in

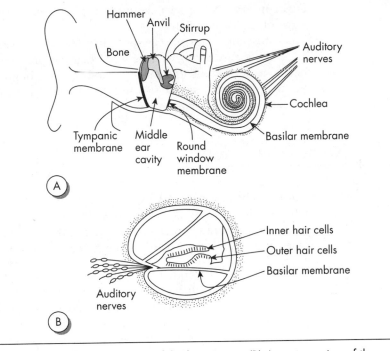

FIGURE 15-5 (A) A schematic of the human ear. (B) A cross section of the cochlea.

permanent damage. The bones in the middle ear are not usually damaged by loud sounds, although they can be hurt by infections. Because our sense of balance depends very much on the middle ear, a middle ear infection can be debilitating. Finally, the most important and most permanent damage can occur to the hair cells in the inner ear. Very loud sounds will stun these hair cells and cause them to cease functioning. Most of the time this is a temporary condition, and time will heal the damage. Unfortunately, if the insult to the inner ear is prolonged, the damage can be permanent. This damage cannot be repaired by an operation or corrected by hearing aids. It is this permanent damage to young people, inflicted by loud music, that is the most frequent, insidious, and sad. Is it really worth it to spend your time in front of huge loudspeakers in concerts, or turn up the volume of your "Walkman," when the result will be that you will not be able to hear *any* music by the time you are 40 years old?

But loud noise does more than cause permanent hearing damage. Noise is, in the Darwinian sense, synonymous with danger. Thus the human body reacts to loud noise so as to protect itself from imminent danger. The bodily reactions are amazing—the eyes dilate, the adrenaline flows, the blood vessels dilate, the senses are alerted, the heartbeat is altered, blood thickens—all to get the person "up." Apparently, such an "up" state, if prolonged, is quite unhealthy. People who live and work in noisy environments have measurably greater general health problems, are grouchy and ill-tempered, and have trouble concentrating. Noise that deprives one of sleep carries with it an additional array of health problems.

Most important, we cannot "adapt" to high noise levels. Industrial workers are often seen walking around with their ear protectors hanging around their necks, as if this is some macho thing to do. They think they can "take it." Perhaps they can "take it" because they already are deaf at the frequency ranges prevalent in that environment.

15.4 ◆ Noise Abatement

Noise has been a part of urban life since the first cities. The noise of chariot wheels in Rome, for example, caused Julius Caesar to forbid chariot riding after sunset so he could sleep. In England, the birthplace of the Industrial Revolution, cities were incredibly noisy. As one American visitor described it:

> The noise surged like a mighty heart-beat in the central districts of London's life. It was a thing beyond all imaginings. The streets of workaday London were uniformly paved in "granite" sets . . . and the hammering of a multitude of iron-shod hairy heels, the deafening side-drum tattoo of tyred wheels jarring from the apex of one set [of cobblestones] to the next; the sticks dragging along a fence; the creaking and groaning and chirping and rattling of vehicles, light and heavy, thus maltreated, the jangling of chain harness; augmented by the shrieking and bellowings called for from those of God's creatures who desired to impart information or proffer a request vocally—raised a din that is beyond conception. It was not any such paltry thing as noise.[3]

In the United States, cities grew increasingly noisy, and all manner of ordinances were contrived to hold down the racket. In Dayton, Ohio, the city proclaimed that "It is illegal for hawkers or peddlers to disturb the peace and quiet by shouting or crying their wares." Most of these ordinances were, of course, of little use and soon disappeared into oblivion, as the noise increased.

The most effective single legislation to pass Congress in the noise control area is the Occupational Health and Safety Act of 1970 in which OSHA was given authority to control industrial noise levels. One of the most important regulations promulgated by OSHA is the limits of industrial noise. Recognizing that both the intensity and the duration are important in preventing damage to hearing, the OSHA regulations limit noise as shown in Table 15-2. Sound levels greater than 115 dB(A) are clearly detrimental and should not be permitted.

Industrial noise is mostly constant over an eight-hour working day. Community noise, however, is intermittent noise. If an airplane goes overhead only a few times a day, on average the sound level is quite low, but the irritation factor is high. For this reason, there have been a plethora of noise indexes developed, all of which are intended to estimate the psychological effect of

TABLE 15–2 OSHA Industrial Noise Limits

Duration (hr)	Sound Levels [dB(A)]
8	90
6	92
4	95
3	97
2	100
1	105
0.5	110
0.25	115

the noise. Most of these begin by using the cumulative distribution technique, similar to the analysis of natural events such as floods, as discussed in Chapter 2.

The U.S. Department of Transportation has been concerned with traffic noise and has established maximum sound levels for different vehicles. Modern trucks and automobiles are considerably quieter than they were only a few years ago, but much of this has been forced not by governmental fiat but by public demand. The exception is renegade trucks that continue to be the major noisemakers on the highways and have little chance of being caught by police who do not consider noise as a serious problem compared to other highway safety concerns.

Excellent work has also been done by acoustic engineers working to quiet commercial aircraft. The modern airliners are amazingly quiet compared to the much louder models such as the Boeing 727, one of the loudest large planes still in service.

The control of noise in the urban environment or noise produced by machines and other devices has not received much governmental or public support. Although Congress passed the Noise Control Act of 1972, little progress has been made in controlling noise. In the 1980s, the noise office of the U.S. EPA was shut down in the near-scuttling of the agency and has not been reestablished.

15.5 ◆ Noise Control

The level of noise can be reduced by using one of three strategies: protect the recipient, reduce the noise at the source, or control the path of sound.

Protecting the recipient usually involves the use of earplugs or other ear protectors. The small earplugs, although easy and cheap, are not very effective for many frequencies of noise. The ear can detect sound not only coming through the ear canal, but also from the vibration of the bones surrounding the ear. Thus the small earplugs are only partially effective. Far better are the ear muffs that cover the entire ear and protect the wearer from most of the surrounding noise. The trouble with the large ear protectors, however, is psychological. Too often people will shun what they consider clumsy and uncomfortable ear protectors and decide to take their chances, thus negating the effectiveness of the protection.

Reducing the source of the noise is often the most effective means of noise control. The redesign of commercial airplanes has already been mentioned as an example of the effectiveness of this control strategy. Changing the type of motors used in and around the home also often effectively reduces noise. For example, changing from a two-stroke gasoline motor for a lawn mower to an electric mower effectively eliminates a common and insidious source of neighborhood noise. The most offensive new noisemakers now are the leaf blowers, and often the noise level of leaf blowers is far above what OSHA would allow for industrial noise.

Changing the path of the noise is a third alternative. The most visible evidence of this tactic is the growth of the noisewalls around our highways. Noise produces a shadow, and the loudest noise comes when the recipient is in a direct line with the source. Placing the walls between the trucks and the people in the homes is intended to take advantage of the noise shadow effect. Unfortunately, noise is not like light. A noise shadow is not perfect and noise can bend and bounce off air, depending on the atmospheric conditions. Sometimes noises miles away can be heard as the pressure waves bounce off inversion layers. The noise fences along highways are for the most part psychological and do not provide much protection except for the homes immediately in the lee of the fence. Whether this is sufficient reason to spend public funds is questionable.

Highway sound can also be abated by heavy natural growth. Trees by themselves are not very effective, but a dense growth will reduce the sound pressure level by several decibels per 100 feet of dense forest. Cutting down natural growth to widen a highway invariably causes increased noise problems. Indirectly, therefore, a tree falling in the forest *will* make noise!

◆ Abbreviations

dB	= decibels		P	= reference sound pressure, usually assumed as 0.00002 N/m^2
dB(A)	= decibels on the A scale approximating human hearing			
Hz	= Hertz, cycles per second		r	= distance away from a sound source, m
L_j	= the jth sound pressure level, dB		SPL	= sound pressure level, dB
N	= number of measurements taken in a frequency analysis		SPL_r	= sound pressure level at some distance r, dB
P	= sound pressure, N/m^2		W	= sound energy, watts

Problems

15–1   A dewatering building in a wastewater treatment plant has a centrifuge that operates at 89 dB(A). Another similar machine is being installed in the same room. Will the operator be able to work in this room?

15–2   A lawn mower emits 80 dB(A). Suppose two other lawn mowers that produce 80 and 61 dB(A), respectively, join the mowing operation.
   a.   What will be the cacophony, in dB(A)?
   b.   How many 80 dB(A) lawnmowers are needed to reach 90 dB(A)?

15–3   A sludge pump emits the following sound as measured by a sound level meter at different octave bands:

Octave Band (Hz)	Sound Level [dB(A)]
31.5	60
63	85
125	82
250	105
500	95
1000	120
2000	80
4000	21
8000	100

   a.   What is the overall sound level, in dB(A)?
   b.   Does this meet OSHA criteria for an eight-hour work day?
   c.   What approaches could be taken to reduce the sound pressure level?
   d.   If the least expensive approach seems to be to use noise protectors (ear plugs), should this be used? What ethical considerations might there be involved in choosing this approach to noise control?

15–4   Using the frequency analysis in Problem 15–3, what would the sound *pressure* level be, in dB? (Note: You first have to correct the decibel readings to dB instead of dB(A), and then add them.)

15–5   A 20-year-old rock star has an audiogram done. The results are shown in Figure 15–6.
   a.   What percent normal hearing has he lost?
   b.   If the curve for person *C* shown in Figure 15–4 represents a normal loss of hearing due to aging for a 65-year-old person, draw the audiogram for the rock star when he reaches 65 years old.
   c.   Will he be able to hear anything at all?

15–6   A frequency analysis was conducted at an industrial site with the following results:

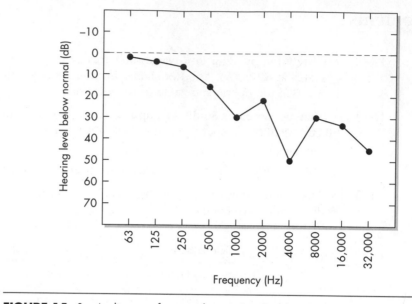

FIGURE 15–6   Audiogram for a rock star. See Problem 15–5.

Octave Band (Hz)	Sound Pressure Level (dB)
31.5	30
63	40
125	10
250	30
500	128
1000	80
2000	10
4000	2
8000	0

a.   What is the average sound pressure level?

b.   What would the average sound pressure level be 100 m from this source?

c.   If you had to reduce the sound pressure level, what would be your strategy for noise reduction?

15–7   A frequency analysis is conducted in an industrial plant with the following results:

Octave Band (Hz)	Sound Pressure Level (dB)
31.5	5
63	10
125	40
250	30
500	82
1000	48
2000	70

4000	125
8000	20
16,000	40
32,000	40

a. Suppose Person *B*, whose audiogram is shown in Figure 15–4, works at this plant. What can you say about the relationship between her audiogram and the frequency analysis conducted at the plant?

b. What is the average sound pressure level at the plant?

c. What is the average sound level [dB(A)] at the plant? (Note: Approximate by correcting the sound pressure levels using Figure 15–3.)

15–8   *Paternalism* is, as the name implies, a process of sheltering or caring for someone who cannot or will not care for himself or herself, as in the case of a parent and child. The name takes on a negative meaning, however, if the "child" is actually an adult and the "parent" is the government. The setting of OSHA industrial noise standards is a clear case of governmental paternalism.

Suppose an engineering friend of yours is working as an environmental control engineer for a corporation and as part of her duties she is required to monitor the OSHA noise levels in her plant. She discovers that in one shop the noise level exceeds the allowable eight-hour OSHA level. There seems to be no means of reducing the source of the noise, nor can the path of the noise be altered. She suggests that the workers be required to wear ear protectors.

You are an OSHA compliance monitor and happen to be at the plant when your friend tells you about the shop that will require ear protection. But your friend tells you that the employees at this shop have asked her to please speak to you to see if you could allow them to not wear what they consider cumbersome and useless protectors. They insist that they know what is best for themselves, and that they do not want to be told how to protect themselves.

You meet with them and tell them and your friend, the plant engineer, that this is what the law says and that your hands are tied. But the workers insist that there must be reasonableness in the law, that it was passed so that workers who *want* protection can have the weight of the government on their side. But in this case the workers want the government to butt out, and leave them alone.

You have the power to overlook this small infraction in noise level in an otherwise safe industrial environment if you write a memo justifying the waiver of the regulations. Do you allow the waiver, or do you insist that the workers be protected from what appears to be excessive noise?

Write a one-page memorandum, addressed to your boss, recommending action.

◆ Endnotes

1. In honor of German physicist Heinrich Hertz.
2. In honor of Alexander Graham Bell.
3. From H. Still, *In Quest of Quiet*, Stackpole Books, Harrisburg, PA (1970).

EPILOG

When we were kids we all made up games to play. The rules of a game often evolved, but there was always an agreement as to how the game was played, and there was no compulsion to cheat because cheating destroyed the game. If we were playing hide-and-seek and agreed that some area was out of bounds, then to hide there would have stopped the game. We enjoyed the game more than we enjoyed winning by under-handed means and therefore agreed to abide by our rules.

The rules for more complex and organized games are of two kinds—formal rules and those rules that are simply understood to be part of the game. In bridge, for example, following suit is a written rule, while peeking at your neighbors cards, even if they are visible to you, is simply not done.

The violation of either a written or unwritten rule may result in alienation of others and possible exclusion from the game. In baseball, for example, trying to score by running over the catcher is within the formal rules, but this action is not considered appropriate in an all-star game where unnecessary injury could cost a career. When Pete Rose did exactly that he was roundly criticized for unnecessary roughness. A cricket player would have said that what Rose did was "not cricket," meaning that even though it was not forbidden under the rules of the game, a player who truly valued the game would not have done it.

In many ways both science and engineering are games where cooperation and competition are part of the process, and as with games, there are many unwritten rules that determine how the game is played. To violate these rules "just isn't cricket."

But science and engineering differ in how the rules of the game are enforced. Although both science and engineering have built-in enforcement mechanisms, these mechanisms are different, and it is this difference that places a special burden on engineers.

When we need information on any technical topic, we go to the library and "look it up." In so doing, we assume that the technical material in print is true; that it is not made up or manipulated in some fashion. If we are interested in a chemical procedure to synthesize an organic compound, for example, and we find a procedure to do this in the literature, we can be reasonably certain that this procedure will work and that the published material is based on actual laboratory experience. We assume that one of the principal rules of scientific and technical literature is that only truth reaches print.

This is not a bad assumption, of course. The scientific and technical establishment has created a filter through which all published material must pass before it sees print. First, most laboratories and other research establishments have an internal check and would not allow a paper that is not based on truthful science to leave the laboratory. The organization of the laboratories, including the managerial control by the laboratory director, should prevent the preparation of papers that don't meet the truthfulness criterion.

More important, however, the papers are peer reviewed before publication, and the more prestigious the journal, the harsher the review. Misrepresentations can and often are caught at this stage and never appear in the journal. Finally, once the papers are published, interested researchers in other laboratories may duplicate the procedures and confirm the results. In looking for technical material in the journals, therefore, you have some genuine assurance that the papers you are reading can be believed and that the authors themselves actually did obtain all the data reported.

The need for truthfulness in the reporting and publishing of technical information is easy to understand if we consider what would happen if this rule were flagrantly broken. Nothing in the literature could then be believed, and all research projects would have to begin from basic principles and primary sources. The publication of truthful research and scholarship would eventually cease, and the game would be destroyed.

Bronowski summed it up like this:

> All this knowledge, all our knowledge, has been built up communally; there would be no astrophysics, there would be no history, there would not even be language, if man were a solitary animal. What follows? It follows that we must be able to rely on other people; we must be able to trust their work. That is, it follows that there is a principle which binds society together, because without it the individual would be helpless to tell the truth from the false. This principle is truthfulness.[1]

But this does not apply to engineering.

In every engineering office or department there is a designated "engineer in responsible charge" whose job it is to make sure that every project is completed successfully and within budget. This responsibility is often indicated by the fact that the engineer in responsible charge places his or her professional engineering seal on the design drawings or the final report. By this action, the engineer is telling the world that the drawings or plans or programs or whatever are correct and accurate and that they will work. (In some countries, not too many years ago, the engineer in responsible charge of building a bridge was actually required to stand under the bridge while it was tested for bearing capacity.) In sealing drawings or otherwise accepting responsibility, the engineer in charge places his or her professional integrity and professional honor on the line. There is nowhere the engineer in charge can hide if something goes wrong. If something *does* go wrong, "One of my younger engineers screwed up" is not a defense, since it is the engineer in charge who is supposed to have overseen the calculations or the design.

For very large projects, where the responsible engineer may not even know all the engineers working on the project, much less be able to oversee their calculations, personally supervising all engineers is clearly impossible. But this does not absolve the engineer of his or her responsibility.

In a typical engineering office, the responsible engineer depends on a team of senior engineers who oversee other engineers, who oversee others, and so on down the line. At every level, all engineers must be playing the game by the rules. Only then can the responsible engineer at the top of the pyramid be confident that the product of collective engineering skills meets the requirements set out by the client.

Fortunately, the rules governing this activity are fairly simple. Central to the rules is the concept of truthfulness in engineering communication. Such technical communication up and down the organization requires an uncompromising commitment to tell the truth no matter what the next level of engineering wants to hear.

Unfortunately, organizations such as engineering firms do not have the same built-in truth detectors as the scientific establishment. There is seldom a peer review, and the client depends entirely on the engineer for the right answer to his or her engineering needs. Thus it is theoretically possible for an engineer in the lower ranks to develop spurious data, lie about test results, or generally manipulate the basic design components. The reason that this is not immediately detected, as it most certainly would be in science, is that if the bogus information is beneficial to the company, few would question its validity. If it is not beneficial, however, everyone along the chain of engineering responsibility will give it a hard critical look. Therefore, the inaccurate information, if it is the desired information, steadily moves up the engineering ladder because at every level the tendency is not to question good news. The superiors at the next level also want good news and to know that everything is going well with the project. They do not want to know that things may have gone wrong somewhere at the basic level. Knowing this, and fearing being shot as the messenger, engineers tend to accept good news and question bad news. In short, the axiom that "good news travels up the organization—bad news travels down" holds for engineering as well. And bad news will travel down very quickly (for example, "You're fired!").

The only correcting mechanism in engineering, unlike that of science, comes at the very end of the project where a failure occurs: for example, the software crashes, the bridge falls down, the project is grossly overbid, or the refinery explodes. Then the search begins for what went wrong. Eventually the truth emerges, and often the problems can be traced to the initial level of engineering design, the development of data, and the interpretation of test results.

It is for this reason that engineers, especially young engineers, must be extremely careful of their work. It is one thing to make a mistake (we all do), but it is another thing totally to have misinformation on which the design or calculation is based. Fabricated or spurious test results can lead to catastrophic failures because there is an *absence* of a failure-detection mechanism in engineering until the project is completed.

It will be some time until you the student engineer will find yourself in "responsible charge," and you may own a professional engineering seal for years before you first use it to seal drawings. But when you do, that first time you place your name, reputation, and career on the line, you want to know that all the engineers who worked on this project did their very best engineering, and at the absolute minimum, were truthful. Without such trust in engineering, the system will fail. It is your responsibility now as a young engineering student to do your part in the engineering enterprise by being totally truthful in all you do. If I may paraphrase Bronowski:

> All engineering projects are communal; there would be no computers, there would be no airplanes, there would not even be civilization, if engineering was a solitary activity. What follows? It follows that we must be able to rely on other engineers; we must be able to trust their work. That is, it follows that there is a principle which binds engineering together, because without it the individual engineer would be helpless. This principle is truthfulness.

◆ Endnote

1. Quoted by Ian Jackson, "Honor in Science," *Sigma Xi* (1994), p. 7, from J. Bronowski, *Science and Human Values,* Messner, New York (1956), p. 73.

INDEX

A-weighed scale, 448
Acid formers, 302
Acid producing bacteria, 302
Acid rain, 108, 341
Activate sludge system, 100, 223
Activated carbon adsorption, 295
Adding sounds, 445
Adiabatic lapse rate, 321
Adsorption, 295, 364
Aeration
  diffused, 285
  extended, 275
  mechanical, 285
Aeration tank, 273
Aerobic cycle, 174
Aerobic decompositions, 174
Air, clean, 323
Air change (in rooms), 348
Air classifier, 93, 108, 398
Air/fuel ratio, 377
Air pollutants
  gases, 331
  particulates, 328
  primary, 341
  secondary, 432
Air pollution, effect of, 337
Air quality standards, 351, 352
Air toxics, 352
Air treatment
  bag filter, 363
  cyclone, 362
  electrostatic precipitator, 107, 364
  scrubber, 364, 368
Albedo, 345
Algae blooms, 184
Alpha particles, 427
Alum, 249, 297
Aluminum sulfate, 249, 297
Alveoli, 338
Ambient air quality standards, 352
American Society for Testing and Materials (ASTM), 399
Amoebic dysentery, 221
Amplitude (of sound), 443
Anaerobic digestion, 102, 301
Animistic religions, 37
Annual cost, 6
Anoxic environment, 297
Anthropocentric environmental ethics, 31
Anvil (bone in middle ear), 450
Approximations in engineering calculations, 53
Aquiclude, 239

Aquifer, 236
  confined, 236
  unconfined, 236
Arbitrary flow reactors, 138
Arithmetic probability plot, 60
Artesian well, 236
Asbestos, 347, 353
Assessment, in an EIS, 21
Asthma, 339
Atmospheric stability, 325, 371, 373
Audiogram, 449
Audiometer, 449
Auditory nerves, 450
Automobile emissions, 347, 375

Bacillary dysentery, 221
Backwashing, 256
Bacteriological measurement, 220
Bag filter, 363
Bar screen (water), 267
Basilar membrane, 450
Batch reactors, 127
Beer's Law, 220
Bel, 444
Bell-shaped curve, 57
Belt filter, 306
Benefit/cost analysis, 9
Benzene, 353
Beta particle, 427
Binary separator, 90
Biocentric environmental ethics, 35
Biocentric equality in Deep Ecology, 119
Biochemical oxygen demand (BOD), 206
  carbonaceous, 215
  nitrogenous, 215
  ultimate, 216
Bioconcentration, 417
Biological process dynamics, 276
Biological (secondary) sludge, 275, 298
Biological treatment of hazardous waste, 425
Biomagnification, 181
Biomass, 158
Bismuth, 350
Black box, 71
Black lung disease, 338
Blank, in a BOD test, 211
Blender, as a black box, 73
BOD bottle, 208
Bomb calorimeter, 154
Box model, 87
Bridging, in flocculation, 247
Broad Street pump, 222

Btu, 152
Bubbler, 334
Bulking sludge, 285

C-distribution curve, 128, 129, 139
Cake (from a centrifuge), 307
Calories, 152
Calorimeter, 154
Capital recovery factor, 6
Carbon adsorption, 295
Carbon dioxide, 172, 174, 182, 332, 345, 346, 352, 405
Carbon monoxide, 332, 336, 338, 339, 347, 352, 353, 375, 376
Carbonaceous BOD, 215
Carboxyhemoglobin, 338, 354, 379
Carburetor, emissions from, 375
Carnot engine, 162
Catalytic converter, 377
Catalytic reactor, 377
Centrate, 307
Centrifuge, 95, 307
Charge neutralization (in coagulation), 247
Chernobyl, Ukraine, 430
Chlorination, 294
Chlorofluorocarbons (CFCs), 114, 333, 344
Cholera, 222
Cigarette smoking, 350, 430
Cilia, 338
Clarifier
  final, 273
  primary, 267
Clean Air Act, 352
Cleaning up old hazardous waste sites, 421
Clear well, 257
Coagulation, 246
Coal-fired power plant, 161
Cochlea, 450
Code (in materials separation), 89, 395
Coefficient of variation, 57
Coffee cup controversy, 408
Coliforms, 222
Collecting sewers, 264
Collection vehicle
  commercial, 388
  recycling, 389
  residential, 386
Colorimetric measurement of nitrogen, 219
Combustion of solid waste, 397, 409
Comminutor, 266

Common parameter weighed checklist, 24
Comparison of reactors, 145
Competition in ecosystems, 179
Completely mixed flow reactors, 130
Completely mixed flow reactors in series, 132
Completely stirred tank reactors (CSTR), 130
Compound interest factors, 7
Comprehensive Environmental Response, Compensation and Liability Act (CERCLA), 421
Concentration, 48
Cone of depression, 240
Confidence interval, 60
Confined aquifer, 236
Consecutive reactions, 127
Conservative assumption in materials flow, 76
Conservative signals, 127
Conservative tracers, 127
Consumers (in ecosystems), 172, 182
Containment of hazardous wastes, 423
Continuous signals, 137
Cooling tower, 163
Corrosivity, 417
Cost effectiveness analysis, 5
Council on Environmental Quality, 20
Criteria pollutants, 417
Critical particle settling velocity, 252
Cryptosporidiosis, 221
Cryptosporidium, 221
Cumulative function, 58
Cyclone, 108, 110, 362

Dalton's Law, 288
DDT, 180
Deadheading, 387
Decibel (dB), 444
Decomposers, 172, 182
  aerobic, 172
  anaerobic, 174
Deep Ecology, 36, 119
Deep well injection, 426
Deficit in dissolved oxygen, 187
Delisting a hazardous waste, 421
Density, 47
Density of solid waste, 404
Deontological ethical theories, 30
Deoxygenation constant, 185, 213

Dependent variables, 59
Desalinization, 110
Detention time, 52
Detritus, 172
Dewatering of sludge
  belt filter, 306
  sand beds, 305
  solid bowl centrifuge, 307
Diapers, 167
Dibenzodioxins, 401, 425
Dibenzofurans, 401
Diesel engine, 378
Diffused aeration, 285
Digester
  aerobic, 301
  anaerobic, 102, 301
  egg-shaped, 304
  primary, 102
  secondary, 102
Dilution in the BOD test, 210
Dimensions
  derivative, 47
  fundamental, 47
Dioxins, 401, 425
Disinfection, 257, 294
Dispersion coefficients, 373
Dispersion of air pollutants, 369
Disposal of hazardous waste, 426
Dissolved oxygen measurement, 204
Dissolved oxygen sag curve, 187
Dissolved solids, 217
Donora, PA, 88, 337
Dose-response curve, 16
Drawdown, 240
Drinking water standards, 226
Dump, 402
Dysentery, 221

Ear, human
  damage to by noise, 449
  functioning of, 451
Ecocentric environmental ethics, 36
Ecofeminism, 180
Ecology, 171
Ecosystems, 50, 171
  forest, 50
  lake, 182
  materials flow in, 172
  pesticides in, 180
  stream, 184
  terrestrial, 176
Effect of air pollution, 377
Effective stack height, 372
Effectiveness of materials separation, 93

Effluent, 71
Effluent water quality standards, 227
Egg-shaped digesters, 304
Electric cars, 377
Electric power production, 161
Electrostatic precipitator, 107, 364
Emission (air) standards, 352
Emissions, automobile, 347, 375
Emphysema, 339
Energy
  balances, 153
  conversions, 153
  equivalence, 159
  flow in ecosystems, 172
  global flows, 167, 356
  sources and availability, 159
  units of measure, 152
Environmental ethics, 31
Environmental impact analysis, 20
Environmental Impact Statements, 20
Environmental risk analysis, 15
Environmental risk management, 19
Epilimnion, 183
Equivalence in energy, 159
Ethical analysis, 28
Ethics (defined), 29
Euler's tour, 387
Eutrophication, 184
Evaluation, in environmental impact statements, 27
Evaporation, 231
Evaporation from gas tanks, 375
Evapotranspiration, 231
Extended aeration, 275
Extensionist ethics, 32
Extraction and treatment of hazardous waste, 423
Extinction coefficient, 336

Facultative microorganisms, 175
Fall turnover, 183
Filamentous microorganisms, 293
Filter
  backwashing (water), 256
  bag, 363
  fabric, 363
  rapid sand, 255, 295
  tricking, 272
  water, 255, 295
Final clarifier, 273
First order reactions, 114, 117
Fish, effect of temperature on, 164
Fission, 237
Fixed film reactors, 272

Fixed solids, 217
Fixing hazardous waste, 423
Flammability, 471
Flare, 366
Flat response, 448
Flocculation, 246
Flocculator, 249
Flocs, 247
Flow rate, 50
Flue gas desulfurization, 368
Fluid-bed incinerator, 311
Food chain, 176
Food/microorganism ratio, 275,
    285, 294, 299
Food web, 176
Force mains, 264
Forest ecosystem, 110
Formaldehyde, 347
Frequency analysis, 61, 244
    of sound, 446
Frequency (of sound waves), 443
Fume, 328
Furbish lousewort, 166
Future generations, 158, 436

Gaia hypothesis, 345
Gamma radiation, 427
Garbage, 386
Gas transfer, 286
Gas transfer coefficient, 291
Gaseous air pollutants
    control of, 364
    measurement of, 331
Gases from landfills, 405
Gastrointestinal disease, 221
Gaussian distribution, 57, 371
Giardia lamblia, 221
Giardiasis, 221
Global energy flow, 167, 356
Global warming, 345
Gooch crucible, 216
Gravity settling tank, 249
Gray, 429
Green cans, 386, 390
Greenhouse effect, 345, 405
Greenhouse gases, 345, 405
Grit chamber, 266
Groundwater, 236
    supplies, 236
    table, 236
Grouped data analysis, 63

Hair cells (in the ear), 450
Half life, 122, 428
Hammer (in the ear), 450
Hazard Ranking System, 422

Hazardous substances, 417
Hazardous waste, 386, 416, 417
    cleanup of old sites, 421
    definition of, 417
    disposal of, 426
    fixing of, 423
    treatment of, 424
Heat engine, 161
Hemoglobin, 338, 339, 354
Henry's Law, 288
Hepatitis, 221
Hertz (Hz), 443
Heuristic routing, 391
High density polyethylene (HDPE),
    98, 396
High-level radioactive waste, 431
High rate activated sludge system,
    275
High volume sampler, 329
Highway noise, 454
Hi-vol, 329
Homeostasis, 176
Hydrocarbons, 333, 336, 343, 353,
    375, 376
Hydrogen acceptor, 175
Hydrogen sulfide, 174, 332, 336,
    368
Hydrologic cycle, 235
Hydropower, 158, 169
Hypolimnion, 186

Ideal reactor performance, 147
Ideal tank theory, 251
Incineration
    air quality, 366
    hazardous waste, 425
    materials balance, 107
    multiple hearth, 311
    sludge, 311
    waste-to-energy, 397
Independent variables, 59
Indicator organisms, 222
Indoor air, 347
Industrial noise, 452
Industrial scrap, 393
Infiltration (to sewers), 264
Inflow (to sewers), 264
Influent, 71
Information analysis, 56
Inner ear, 450
In-situ treatment of hazardous
    waste, 423
Instantaneous signal, 127
Instantaneous tracer, 127
Instrumental value, 31

Integrated Risk Information System,
    17
Integrated Solid Waste Manage-
    ment, 408
Interaction matrix, 23
Intrinsic value, 32
Inventory, of an environmental im-
    pact statement, 21
Inverse square law, 445
Inversion, 109, 325
    radiation, 327
    subsidence, 327
Involuntary risk, 14, 402, 429
Ionizing radiation, 427
Isotopes, 427

Joule, 152

Kilowatts, 152
Kjeldahl nitrogen, 216, 219

Lake
    ecosystems, 182
    stratification, 183
Land spreading of sludge, 311
Land treatment (of hazardous
    waste), 426
Landfill, 402, 409
Lapse rate
    adiabatic, 321
    inversion, 109, 325
    prevailing, 325, 370
    subadiabatic, 325
    superadiabatic, 325
LD_{50}, 417
LDC_{50}, 417
Leachate, 405
Lead, 350, 353, 376
Lethal dose, 417
Life cycle analysis, 407
Lifeboat ethics, 81
Lime, 368
Lime stabilization of sludge, 301
Limestone, 368
Liners for landfills, 405
Links, 387
Listed pollutants, 352
Loading, of activated sludge system,
    275
London air pollution, 340
Love Canal, Niagara Falls, NY, 422
Low density polyethylene, 98, 396
Low-level radioactive waste, 432
Low Level Waste Policy Act, 432
Low sulfur coal, 367
Lung cancer, 339, 350

Magnet, 398
Mass curve, 243
Materials balance, 70
  multiple materials, 83, 95
  with reactors, 100
  rules for solving, 82
  single material, 71
Materials flow in ecosystems, 172
Materials recovery facility (MRF), 394
Maximum growth rate constant, 276
Mean, 57
Mean cell residence time, 277
Mechanical aeration, 285
Mesosphere, 324
Metalimnion, 183
Meteorology, 323
Methane, 102, 301, 333, 336, 345, 405
Methane formers, 301
Methane producing bacteria, 301
Middle ear, 450
Mixed batch reactors, 127
Mixed liquor, 275
Mixed liquor suspended solids (MLSS), 275
Mixing depth, 87
Mixing materials, 83
Mixing models of reactors, 127
Monod model, 276
Moral agent, 31
Moral community, 31
Most probable number (MPN) test, 224
Moving sources, control of, 375
Mud valve, 250
Multiple hearth incinerator, 311
Municipal solid waste (MSW), 386

National Academy of Science, 430
National Environmental Policy Act, 20
National Pollution Discharge Elimination System (NPDES), 227
National Priority List, 322
Native American religions, 37
Nephthelometer, 331
Nessler reagent, 219
Newsprint, 111
Nitric oxide, 332, 344, 368
Nitrification, 296
Nitrobacter, 296
Nitrogen
  balance in ecosystems, 176
  Kjeldahl, 216, 219

measurement, 219
  removal in wastewater treatment, 295
  in streams, 193
Nitrogen dioxide, 332, 341, 343
Nitrogen oxides, 353, 375, 376
Nitrogenous BOD, 215
Nitrosomonas, 296
Nitrous oxide, 332, 345, 346
Nodes, 387
Noise abatement, 452
Noise control, 453
Noise Control Act, 453
Noise pollution, 442
  effect on human health, 449
Non-attainment areas, 353
Non-integer order reactions, 121
Normal distribution, 57
Normalizing, 61
Nuclear power, 158, 430
Nuclear Regularity Commission, 430
Nutrients, 172
  in ecosystems, 182

Obligate anaerobes, 175
Observation well, 241
Occupational Health and Safety Act, 452, 457
Octave bands, 446
Ogallala aquifer, 81
Operating curve (centrifuge), 308
Organisms, number in a stream, 193
Orthophosphate, 297
OSHA, 452, 457
Outer ear, 450
Overflow rate, 253
Oxidation pond, 295
Oxygen solubility, 186, 288
Ozone, 333, 339, 343, 344, 345, 348, 353

Packers (refuse collection), 386
Paper, price of, 397
Partial pressure of a gas, 288
Particulates, 328, 336, 348, 352, 353
  control, 364
  dust, 328
  fume, 328
  measurement, 329
  respirable, 331
  smoke, 329
  spray, 329
  treatment of, 362

Paternalism, 457
Pathogenic microorganisms, 220
PCB, 425, 439
PCDD, 401
PCDF, 401
Permeability, coefficient of, 238
Pesticides, 180
PETE, 396
Phosphorus, 182
Phosphorus removal, 297
Photochemical smog, 341, 351
Photometric analysis, 219
Photosynthesis, 172, 182
Phytotoxicity, 417
Pickers (in solid waste processing), 396
Plastics separation, 98, 395
Plug flow reactor, 128
Plume rise, 327, 369
Plume types
  coning, 369
  fanning, 369
  fumigation, 369
  looping, 369
Poison gas, 112
Pollution prevention, 432
Pollution Prevention Act, 432
Polonium, 350
Polychlorinated biphenyl, 425
Polychlorinated dibenzodioxins, 401
Polychlorinated dibenzofurans, 401
Polyethylene, high density, 98, 396
Polyethylene, low density, 98, 396
Polyethylene terephthalate, 98, 396
Polymer, 306
Polynary separator, 94
Polyphosphate, 297
Polypropylene, 98, 396
Polystyrene, 98, 396
Polyvinyl chloride, 98, 396
Ponding, in trickling filter, 315
Population dynamics, 177
Porosity, 236
Positive crankcase ventilation valve, 375
Potency factor, 17
Pounds of BOD, 212
Power production (electric), 161
Precipitation, 235
Predators, 177
Preliminary wastewater treatment, 264
Present worth, 6
Present worth factor, 6
Preservation principle, 36
Prevailing lapse rate, 325, 370

Primary air pollutants, 341
Primary clarifier, 268
Primary wastewater treatment, 266
Prisoner dilemma, 245
Probability, 56
Probability plot, 59
Process loading factor, 281
Producers, 172
Products of incomplete combustion, 425
Pseudomonas, 296
Purity, 90
PVC valve, 375

Quantified checklist, 20

Radiation inversion, 327
Radiation poisoning, 429
Radioactive waste, 427
   high level, 431
   low level, 432
Radioactivity, 427
Radioisotopes, 427
Radon, 348, 350, 429
Rapid sand filter, 255, 295
Raw primary sludge, 265, 298
Reactions
   consecutive, 122
   first order, 117
   non-integer order, 121
   second order, 121
   zero order, 115
Reactive waste, 417
Reactor model of reactors
   comparison of, 145
   completely mixed flow, 143
   completely mixed flow in series, 145
   ideal performance, 147
   mixed batch, 139
   plug flow, 141
Reactors
   arbitrary flow, 138
   comparison of, 145
   completely mixed flow, 130
   fixed film, 272
   mixed batch, 127
   plug flow, 128
   in series, 132
   suspended growth, 273
Reciprocity, in ethics, 31
Recovery, 90, 360, 392
Recycling, 393, 409
Recycling symbol, 413
Refractory hazardous waste, treatment of, 425

Refuse derived fuel, 397, 399
Refuse processing, 398
Rem, 429
Remedial action (hazardous waste), 422
Removal action (hazardous waste), 422
Reoxygenation, 185
Reoxygenation constant, 187
Residence time, 52, 128, 277
Resource Conservation and Recovery Act (RCRA), 421, 426, 436
Respirable particulates, 331
Respiration, 172
Respiratory system, 338
Retention time, 52
Return activated sludge, 274, 285
Return period, 61, 244
Reuse, 393
Reverence for life, 35
Rietema effectiveness of separation, 93
Ringlemann scale, 331
Risk
   acceptable, 17
   analysis, 12
   assessment, 12
   involuntary, 14
   ionizing radiation, 428
   management, 12
   voluntary, 14
Roentgen, 428
Roof drains, 315
Rotating biological contactors, 273
Round window membrane, 450
Route optimization, 387
Routing, heuristic, 391
Rubbish, 386

Sag curve, dissolved oxygen, 187
Salmonella, 221
Salmonella typhosa, 221
Salmonellosis, 221
Sanitary landfill, 402, 409
Sanitary sewers, 264
Saturation constant, 276
Saturation of oxygen in water, 186, 288
Saturation, zone of, 236
Screens
   in refuse processing, 398
   in wastewater treatment, 266
Scrubber, 364, 368
Second order reactions, 115, 121
Secondary air pollutants, 342

Secondary ambient air quality standards, 352
Secondary digesters, 102
Secondary sludge, 275, 298
Secondary treatment of wastewater, 271
Sedimentation tank, 268
Seeding, in BOD bottles, 211
Self-purification of streams, 192
Self-realization, in Deep Ecology, 119
Separator, black box, 71, 89, 90
   binary, 90
   polynary, 94
Settling chamber (air pollution control), 362
Settling rate, 251
Settling tank (water treatment), 249, 267
Sewers
   collecting, 264
   sanitary, 264
   storm, 264
   trunk, 264
Shigellosis, 221
Sievert, 429
Signals, in reactors, 127
Significant figures, 54
Sleepy driver syndrome, 350
Sludge
   bulking, 285
   dewatering, 304
   digestion, 102
   production, 298
   raw primary, 268, 298
   secondary, 275, 298
   stabilization, 301
   toxicity, 311
   ultimate disposal, 311
   waste activated, 101, 275
Sludge age, 276
Sludge volume index (SVI), 293
Smoke, 335
Solar power, 158
Solid bowl centrifuge, 307
Solid waste disposal, air quality, 347
Solids
   dissolved, 217
   fixed, 217
   suspended, 217
   total, 216
   volatile, 102, 217
   volatile suspended, 218
Solids retention time, 277
Solubility constant (in Henry's Law), 288

Solubility of oxygen, 186, 288
Sound
  amplitude, 443
  frequency, 443
  measurement of, 448
Sound level meter, 448
Sound pressure level, 444
Sound pressure level meter, 448
Source reduction, 409
Source separation, 394
Source strength, 349
Sources of air pollution, 337
Species in steams, 193
Specific growth rate, 276
Specific yield, 236
Spray tower, 364
Stability, of atmosphere, 325, 371, 373
Stack height, 372
Stacks, tall, 364
Standard deviation, 57, 60
Standard Methods, 225, 231
Standards
  air quality, 351
  ambient (air), 352
  drinking water, 226
  effluent (water), 227
  emission (air), 352
  stream (water), 227
Steady state, 71
Steady state assumption in materials flow, 76
Stirrup (in middle ear), 450
Stratification of lakes, 183
Stratosphere, 324
Stream
  classification, 228
  ecosystem, 184
  standards, 227
Strong lapse rate, 325
Styrene, 353
Styrofoam, 408
Subadiabatic lapse rate, 325
Subsidence inversion, 327
Substrate, 276
Substrate removal velocity, 280
Sulfur dioxide, 332, 339, 341, 348, 353, 366, 368
Sulfur oxide, control of, 366
Sulfur trioxide, 332, 366
Sunk cost, 11
Superadiabatic lapse rate, 325
Superfund, 421
Surface water supplies, 243
Suspended growth reactors in wastewater treatment, 273

Suspended solids, 217
Switch, in separation, 90, 395
Synergism, 338
Systéme International (SI) á Unités, 47

Table, groundwater, 236
Technical analysis, 4
Temperature
  absolute, 157
  aquatic life, 164
  of the earth, 345, 356
  effect on fish, 164
  effect on reoxygenation, 187
  effect on solubility, 191
  profile in earth's atmosphere, 324
  reactions, 144
  in sludge digestion, 302
Thermal pollution, 163
Thermal radiation inversion, 328
Thermal stratification of lakes, 183
Thermocline, 183
Thickening, 92
Three Mile Island, 430
Threshold, 337
Tip, 402
Total body burden, 337
Total solids, 216
Total suspended particulates, 331
Toxic Substances Control Act (TOSCA), 421, 422
Toxicity (hazardous waste), 417
Toxicity of sludge, 311
Tracers, in reactors, 127
Trachea, 338
Transpiration, 235
Transportation, Department of, 452
Trash, 386
Trichloroethane, 357
Trichloroethylene, 357
Trickling filter, 272
Trophic level, 176
Troposphere, 323
Trunk sewer, 264
Tuning (automobile engine), 377
Turbidity, 226
Turnover, 183
Tympanic membrane (in the ear), 449

Ultimate biochemical oxygen demand, 210, 213, 216
Ultimate disposal of refuse, 402
Ultimate disposal of sludge, 311
Ultimate oxygen demand, 189
Unconfined aquifer, 236

Units, 47
Universalization, 30
Unsteady state assumption, 81
Utilitarianism, 29

V-notch weir, 268, 273
Variables
  dependent, 59
  independent, 59
Visibility, 335
Volatile solids, 102, 217
Volatile suspended solids, 218
Voluntary risk, 14, 402, 429

Waste activated sludge, 101, 275
Waste-to-energy facility (refuse), 397
Wastewater transport, 264
Wastewater treatment
  activated sludge, 273
  chlorination, 294
  comminutor, 266
  disinfection, 294
  grit chamber, 266
  final clarifier, 273
  land treatment, 296
  nutrient removal, 296
  phosphorus removal, 297
  preliminary, 266
  primary treatment, 295
  rotating biological contactor, 273
  secondary, 271
  sludge production, 298
  tertiary, 295
  trickling filter, 272
Water quality
  effluent, 227
  measurement of, 204
  standards for, 226
  stream, 227
  surface water, 227
Water treatment
  coagulation and flocculation, 246
  disinfection, 257
  filtration, 255
  settling, 249
Weir, 268, 273
Well, artesian, 236
Wellpoint, 239
Wind, 324
Worrell-Stessel effectiveness, 93

X-rays, 427

Yield, 276
Yucca Flats, NV, 432

Zero order reactions, 114, 115